Modern
Analytic
Geometry

Modern
Analytic
Geometry

**Gerald C. Preston
Anthony R. Lovaglia**

SAN JOSE STATE COLLEGE

Harper & Row, Publishers
New York/Evanston/San Francisco/London

MODERN
ANALYTIC
GEOMETRY

*Copyright © 1971 by Gerald C. Preston and
Anthony R. Lovaglia*
Printed in the United States of America. All rights reserved. No part of this book may be used or reproduced in any manner whatsoever without written permission except in the case of brief quotations embodied in critical articles and reviews. For information address Harper & Row, Publishers, Inc., 49 East 33rd Street, New York, N.Y. 10016.

Standard Book Number: 06-045256-0

Library of Congress Catalog Card Number: 74-127342

Contents

Preface, ix

1 Fundamentals

    1. Logic: Propositions and Connectives   1
    2. Sets   2
    3. Inequalities and Absolute Value   5
    4. Review of Trigonometry   10

2 Functions, Relations, and Their Graphs

    1. Functions and Relations   16
    2. The Algebra of Functions; Graphing by Addition of Ordinates   22
    3. Inverse Functions   24
    4. Techniques for Sketching Graphs   27
    5. Curves; Parametric Equations   34

3 Vectors and Line Segments in the Plane

    1. Arrows   40
    2. Vectors   44
    3. The Angle Between Two Vectors; Dot Product   49
    4. Geometric Problems   55

4 Lines, Circles, and Convex Sets in the Plane

    1. Lines   63
    2. The General Linear Equation   68
    3. Angle of Inclination; Angle Between Two Lines; Parallel and Perpendicular Lines   72
    4. Distance from a Point to a Line; the Normal Form; Families of Lines   77
    5. Circles and Tangent Lines   85
    6. Families of Circles   90
  ★ 7. Convex Sets   96

5 Polar Coordinates

    1. Definition of Polar Coordinates   101
    2. Polar Curves   105
    3. Transforming Equations from Rectangular to Polar Coordinates and Vice Versa; Equations of Lines and Circles in Polar Coordinates   110
    4. Intersections of Polar Curves   114

6 Translation and Rotation of Axes; Isometries of the Plane

    1. Translation and Rotation of Axes   117
  ★ 2. Translations and Rotations in Matrix Form   123
  ★ 3. Isometries   127

7 The Conic Sections

    1. Definition and Equations of the Conic Sections   131
    2. The Parabola ($e = 1$)   136
    3. The Ellipse ($e < 1$)   142
    4. The Hyperbola ($e > 1$)   147
    5. The General Quadratic Equation in Two Variables—First Method   152
  ★ 6. The General Quadratic Equation in Two Variables—Second Method   157

8 Three-Dimensional Vectors

    1. Three-Dimensional Space R^3 167
    2. Vector Operations   172
    3. Length; Dot Product   177
    4. The Angle Between Two Vectors; Direction of a Vector   180
    5. The Cross Product   189

9 Lines, Planes, Spheres, and Convex Sets

    1. Line Segments, Rays, and Lines   194
    2. Symmetric Equations of a Line; Angle Between Two Lines   199
    3. Planes   204
    4. The Angle Between Two Planes; Intersections of Planes   208
    5. The Normal Form of the Equation of a Plane; Families of Planes   214
★ 6. The Relationship Between a Plane and R^2 219
    7. Spheres and Tangent Planes   222
★ 8. Convex Sets   225

10 Surfaces and Curves

    1. Surfaces   229
    2. Cylinders   237
    3. Curves   241
    4. Surfaces of Revolution   246
    5. Cylindrical and Spherical Coordinates   248

★ 11 Isometries in Space and Quadric Surfaces

    1. Quadric Surfaces   255
    2. Isometries of R^3 261
    3. The General Quadratic Equation   270
    4. Enumeration of the Quadric Surfaces   279

Appendix

    Matrices   282
    Determinants   286

Answers to Selected Exercises, 291

Index, 315

Preface

In addition to presenting the traditional topics of analytic geometry essential to the study of the Calculus, the authors have made an attempt to reflect the growing trend to introduce linear algebra in the first two years of college mathematics.

To this end we have developed all the concepts in a purely algebraic or set-theoretic manner, divorcing ourselves (except for purposes of motivation, terminology, and pictorial representation) from synthetic geometry. We do not *assume* such objects as point, line, plane, and so on, but rather, we *define* them and develop their properties using sets and the properties of the real-number system.

Departures from a traditional course in analytic geometry include the treatment of *convexity* (in connection with the conic sections, lines, and planes), the discussion of *isometries* (in connection with translation and rotation of axes), and the use of matrices, *eigenvalues*, and *eigenvectors* in an alternative method for analyzing second-degree equations in two variables. (The traditional method is presented first.) The matrix method is extended to the case of second-degree equations in three variables in the last chapter. The material is so arranged that matrix methods may be omitted entirely without loss of continuity and without omitting any of the traditional topics.

Chapter 1, "Fundamentals," contains the prerequisite material on logic, sets, the real-number system, and trigonometry. This can be omitted or dealt with briefly. The appendix, "Matrices and Determinants," need be taken up only if (and when) one chooses to use the matrix methods mentioned above. Those problems which are of particular importance or which may be needed later are marked with a dagger (†). The more difficult problems are marked with an asterisk (*). In a minimal course, one may omit all sections marked with stars (★).

GERALD C. PRESTON
ANTHONY R. LOVAGLIA

Modern
Analytic
Geometry

chapter 1
Fundamentals

In this introductory chapter, we present a brief outline of those topics that are requisite to a study of *analytic geometry*. We shall give the rudiments of *logic* and *sets*, introducing notation and terminology that will be used in other parts of the book. This will be followed by a short treatment of *inequalities* and *absolute value*.

1. LOGIC: PROPOSITIONS AND CONNECTIVES

A *proposition* is a statement that is either *true* (T) or *false* (F), but not both. Propositions will be denoted by $p, q, r, \ldots$. *Connectives* are operations by means of which propositions are combined (or modified) to form new propositions. The five connectives used in this book are defined as follows:

Conjunction. The *conjunction* of p with q, written "p and q," is true only when *both* p, q are true; otherwise it is false.

Disjunction. The *disjunction* of p with q, written "p or q," is true whenever *at least one* of p, q is true; otherwise it is false.

Negation. The *negation* of p is true when p is false, and false when p is true.

Implication. The *implication* $p \Rightarrow q$ (read "p implies q") is defined to be true in all cases *except that in which p is true and q is false*. Accordingly, when we say "$p \Rightarrow q$ is true," we merely assert that "it is not the case that p is true and q is false." We may also read "$p \Rightarrow q$" as "If p, then q." We call p the *antecedent* or *hypothesis* and q the *consequent* or *conclusion* of the implication.

Equivalence. The *equivalence* $p \Leftrightarrow q$ (read "p is equivalent to q") is defined to be true whenever p, q are *both true* or *both false*; that is, p, q have the same truth value. We may also read $p \Leftrightarrow q$ as "p if and only if q," which is often abbreviated "p iff q."

REMARK. To prove a statement of the form $p \Leftrightarrow q$, it is sufficient to prove the two implications: $p \Rightarrow q$ and $q \Rightarrow p$. ($q \Rightarrow p$ is called the *converse* of $p \Rightarrow q$.)

Example. Determine whether the following propositions are *true* or *false*.

(a) $2 + 3 = 5$ and $7 - 4 = 2$.
(b) $6 = \sqrt{30} \Rightarrow 2 \cdot 7 = 14$.
(c) It is not the case that "5 is greater than 2 or 7 is less than 3."

DISCUSSION
(a) This is a conjunction with the first part *true* and the second *false*. Hence, the proposition is *false*.
(b) This is an implication with antecedent *false* and consequent *true*. Hence, the proposition is *true*.
(c) The proposition "5 is greater than 2 or 7 is less than 3" is a disjunction with the first part *true* and second part *false*. Hence, it is *true*. Thus the given proposition, which is the *negation* of the above, is *false*.

EXERCISE 1

Determine whether the following propositions are true or false.

1. $6 \cdot 3 = 9$ and $4 + 3 = 6$.
2. $6 \cdot 3 = 9$ or $4 + 3 = 6$.
3. $6 \cdot 3 = 18$ and $4 + 3 = 7$.
4. 2 is greater than 5 or $6 = 4 + 2$.
5. 3 is less than $6 \Rightarrow 4 + 3 = 12$.
6. 5 is a prime number $\Rightarrow 3 + 0 = 0$.
7. $6 = \sqrt{36} \Rightarrow 0 \cdot 5 = 0$.
8. It is not the case that "There is a real number whose square equals -1."
9. It is not the case that "$4 + 2 = 6 \Rightarrow 4 \cdot 2 = 6$.."
10. $-(2 + 3) = (-2) + (-3) \Leftrightarrow 1/(2 + 3) = \frac{1}{2} + \frac{1}{3}$.

2. SETS

The word *set* is synonymous with the terms *collection*, *class*, or *family* of objects. The objects that comprise a set are called *elements* of the set. Sets will

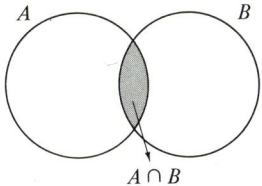
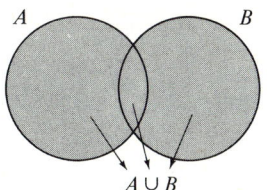

Figure 1-1   Figure 1-2

be denoted by upper-case letters, $A, B, C, \ldots, X, Y, Z$, whereas elements of a set will usually be denoted by lower-case letters, $a, b, c, \ldots, x, y, z$. To denote the fact that an object a is an element (or member) of a set A, we write $a \in A$ (read "a is an element of A" or "a belongs to A"). If a does not belong to A, we write $a \notin A$ (read "a is not an element of A"). For example, $A = \{1, 2, 3\}$ is the set whose elements are the numbers 1, 2, 3. Thus, $2 \in A$, but $5 \notin A$. A set is often characterized by giving a property shared by those, and only those, objects that belong to the set. We write $A = \{x|p(x)\}$ (read "A equals the set of all objects x such that $p(x)$ is true") where $p(x)$ is a statement about x. If R is the set of all real numbers, then $A = \{x|x \in R \text{ and } 1 \leq x \leq 2\}$ is the set of all real numbers between 1 and 2 inclusive. The *void* set (*empty* set) is the set $\emptyset$ that contains no elements. For example, we may characterize $\emptyset$ by $\emptyset = \{x|x \in R \text{ and } x^2 < 0\}$.

We may combine sets to form new sets. The manner in which we do this is called a *set operation*. There are three basic set operations.

Intersection. $A \cap B = \{x|x \in A \text{ and } x \in B\}$: The *intersection* of A with B is *the set of all objects that belong to both A and B*. We picture the intersection as in Figure 1-1.

Union. $A \cup B = \{x|x \in A \text{ or } x \in B\}$: The *union* of A with B is *the set of all objects that belong to A or B or both* (Figure 1-2).

Difference. $A - B = \{x|x \in A \text{ and } x \notin B\}$: $A - B$ consists of *those elements that belong to A but do not belong to B* (Figure 1-3). We also call $A - B$ the *complement* of B in A.

In addition to the above set operations, we have certain set *relations*.

Inclusion. $A \subseteq B$ iff $x \in A \Rightarrow x \in B$. Thus, $A \subseteq B$ iff every element of A is also an element of B (Figure 1-4). In this case, A is said to be a *subset* of B.

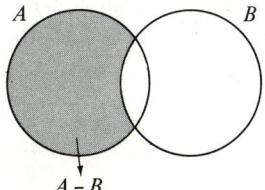

Figure 1-3

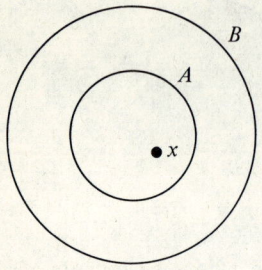 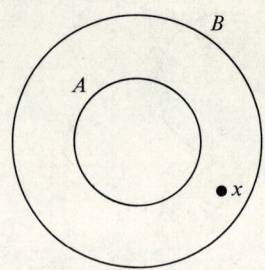

Figure 1-4 **Figure 1-5**

Equality. $A = B$ iff $A \subseteq B$ and $B \subseteq A$. Thus, two sets are equal iff they consist of precisely the same objects.

Proper Inclusion. $A \subset B$ iff $A \subseteq B$ and $A \neq B$. Thus, A is a *proper subset* of B iff A is a subset of B, and there is at least one element of B which is not in A (Figure 1-5).

The sets A, B are said to be *disjoint* iff $A \cap B = \emptyset$; disjoint sets have no elements in common (Figure 1-6).

Example 1. Let $A = \{1, 2, 3, 5, 6, 9\}$, and $B = \{0, 2, 4, 6, 8\}$. Then $A \cap B = \{2, 6\}$, $A \cup B = \{0, 1, 2, 3, 4, 5, 6, 8, 9\}$, $A - B = \{1, 3, 5, 9\}$ and $B - A = \{0, 4, 8\}$.

Ordered Pairs. The *ordered pair* with *first element* x and *second element* y is denoted by (x, y). We shall assume that

$(x, y) = (u, v)$ iff $x = u$ and $y = v$

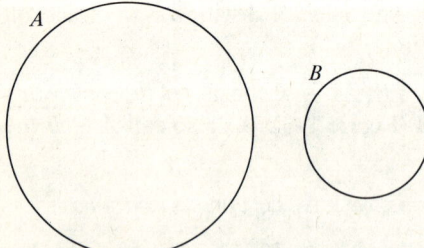

Figure 1-6

Although it is possible to give a definition of the ordered pair (x, y), it is not necessary for our purposes. The important aspect, of course, is the order in which the elements are written. For example, $(2, 3) \neq (3, 2)$, whereas $(2, 3) = (7 - 5, 4 - 1)$.

Ordered Triples. The *ordered triple* (x, y, z) is defined in terms of ordered pairs:

$(x, y, z) = ((x, y), z)$

From this it follows that

$(x, y, z) = (u, v, w)$ iff $x = u$, $y = v$, and $z = w$

Example 2. Find all real numbers x, y such that

$(2x + y, y - x) = (4y - 1, x + 2)$

We have $2x + y = 4y - 1$ and $y - x = x + 2$. Solving simultaneously, we obtain

$x = -\frac{5}{4}, \quad y = -\frac{1}{2}$

Cartesian Product. If A, B are sets, the *Cartesian product* of A with B *(in that order)* is defined by

$A \times B = \{(x, y) | x \in A \text{ and } y \in B\}$

The Cartesian product $A \times B \times C$ is defined by

$A \times B \times C = \{(x, y, z) | x \in A, y \in B, z \in C\}$

EXERCISE 2

1. Let $A = \{1, 4, 3, 8, 7\}$, $B = \{0, 2, 3, 8, 5, 9, 10\}$, $C = \{3, 0, 8\}$, and $D = \{2, 5, 9\}$.
   (a) Find $A \cap B$, $A \cup C$, $B - C$, and $C - B$.
   (b) Determine whether the following are true or false.
      (i) $C \subseteq B$
      (ii) $C \subset B$
      (iii) $A \cap D = \emptyset$
      (iv) $A \subseteq B$
      (v) $A \cap B = \emptyset$
      (vi) $A \cap (C \cup D) = (A \cap C) \cup (A \cap D)$
   (c) Find $C \times D$ and $D \times C$.
2. (a) Find numbers x, y such that $(x - 3, y + 2) = (y, -x)$.
   (b) Find numbers x, y, z such that $(2x - 3, z - 4, 3y - 1) = (4, 3, 2)$.
   (c) Find numbers x, y, z such that $(2x - y, y - z, x + z) = (1, 2, 3)$.

3. INEQUALITIES AND ABSOLUTE VALUE

The *real-number system* is a set R (whose elements are called *real numbers*) together with two binary operations: $+$ (addition) and $\cdot$ (multiplication), and a relation $<$ ("less than") satisfying certain axioms. We assume that the reader is familiar with the basic properties of the real numbers and list here only those laws pertaining to inequalities and absolute value. We use ab as an abbreviation for $a \cdot b$; we read $a \leq b$ as "a is less than or equal to b."

(1) For each pair (a, b) of real numbers exactly one of the following holds:
$a < b$; $a = b$; $b < a$ (trichotomy law)

(2) $a < b$ and $b < c \Rightarrow a < c$ (transitive law)
(3) $a < b \Rightarrow a + c < b + c$ (addition law)
(4) $a < b$ and $0 < c \Rightarrow ac < bc$ (multiplication law)

If $a < b$, we also write $b > a$ (b is greater than a). The set
$$R^+ = \{x \in R | x > 0\}$$
is called the set of *positive* reals, and $R^- = \{x \in R | x < 0\}$ is the set of *negative* reals. Thus, $R = R^+ \cup R^- \cup \{0\}$. Following are several important subsets of the reals:

The Integers
$$Z = \{0, \pm 1, \pm 2, \pm 3, \ldots\}$$

The Positive Integers
$$N = \{1, 2, 3, \ldots\}$$

The Rational Numbers
$$Q = \{x \in R | x = \frac{a}{b}, a, b \in Z, b \neq 0\}$$
= the set of all *quotients* of integers with nonzero denominators.

Observe that $N \subset Z \subset Q \subset R$.

A real number is said to be *irrational* iff it is not *rational*; for example, $\sqrt{2}, \pi, \sqrt[3]{4}$ are irrational. Every real number is rational or irrational, but not both.

Theorem 1. *For all real numbers a, b, c,*

(1) (a) $a > b \Leftrightarrow a - b > 0$
    (b) $a > 0 \Leftrightarrow -a < 0$
(2) $a \neq 0 \Rightarrow a^2 > 0$
(3) $a < b$ and $c < 0 \Rightarrow ac > bc$
(4) $0 < a < b \Rightarrow 0 < \frac{1}{b} < \frac{1}{a}$
(5) $a < b$ and $c < d \Rightarrow a + c < b + d$
(6) $0 < a < b$ and $0 < c < d \Rightarrow ac < bd$
(7) (a) $ab > 0 \Leftrightarrow (a > 0$ and $b > 0)$ or $(a < 0$ and $b < 0)$
    (b) $ab < 0 \Leftrightarrow (a < 0$ and $b > 0)$ or $(a > 0$ and $b < 0)$

Theorem 2. *For $a, b, \in R$ and $n \in Z$,*

(1) (a) If $a, b, n > 0$, then $a^n < b^n \Leftrightarrow a < b$
    (b) If $n < 0, a, b > 0$, then $a^n > b^n \Leftrightarrow a < b$
(2) (a) If $a > 1$, then $a^n < a^m \Leftrightarrow n < m$
    (b) If $0 < a < 1$, then $a^n > a^m \Leftrightarrow n < m$

Inequalities and Absolute Value

If $a > 0$ and $n \in N$, then $\sqrt[n]{a}$ (nth root of a) is the *unique positive number* x such that $x^n = a$. Thus, $x = \sqrt[n]{a} \Leftrightarrow x^n = a$, by definition. If $a < 0$ and n is odd, then $\sqrt[n]{a}$ is the *unique negative number* x such that $x^n = a$. Thus, $\sqrt[3]{-8} = -2$, since $(-2)^3 = -8$. $\sqrt[n]{a}$ is not defined for $a < 0$ and n even, since an even power of a real number cannot be negative. If $a > 0$ and $r = m/n$ is rational, then a^r is defined by $a^r = a^{m/n} = \sqrt[n]{a^m}$. This is also equal to $(\sqrt[n]{a})^m$. We often write $\sqrt[n]{a} = a^{1/n}$. If $a < 0$, $a^{m/n}$ is defined as above, provided n is odd.

Theorem 3. For $a, b > 0$ and $n, m \in N$,

(1) $\sqrt[n]{ab} = \sqrt[n]{a}\sqrt[n]{b}$

(2) $\sqrt[n]{\dfrac{1}{a}} = \dfrac{1}{\sqrt[n]{a}}$

(3) $\sqrt[m]{\sqrt[n]{a}} = \sqrt[mn]{a}$

(4) $a < b \Leftrightarrow \sqrt[n]{a} < \sqrt[n]{b}$

(5) If $a > 1$, then $n < m \Leftrightarrow \sqrt[n]{a} > \sqrt[m]{a}$

The *absolute (numerical) value* of a real number x is defined as follows:

$$|x| = \begin{cases} x & \text{if } x \geq 0 \\ -x & \text{if } x < 0 \end{cases}$$

Note that for all x, $|x| \geq 0$. For example, $|0| = 0$, $|2| = 2$, and $|-3| = -(-3) = 3$

Theorem 4. For all numbers x, y,

(1) $-|x| \leq x \leq |x|$

(2) (a) $|x|^2 = |x^2| = x^2$
    (b) $|x| = |y| \Leftrightarrow x^2 = y^2$
    (c) $|x| < |y| \Leftrightarrow x^2 < y^2$

(3) (a) $|x| = \sqrt{x^2}$
    (b) $|-x| = |x|$

(4) (a) $|xy| = |x| \cdot |y|$
    (b) $\left|\dfrac{x}{y}\right| = \dfrac{|x|}{|y|}, \quad y \neq 0$

(5) (a) $|x + y| \leq |x| + |y|$ (triangle inequality)
    (b) $|x - y| \leq |x| + |y|$

(6) $|x - y| \geq \big||x| - |y|\big|$

(7) If $d > 0$, then
  (a) $|x| = d \Leftrightarrow x = d \text{ or } x = -d$
  (b) $|x| < d \Leftrightarrow -d < x < d$
  (c) $|x| > d \Leftrightarrow x < -d \text{ or } x > d$

Intervals on the Real Line. Intervals are subsets of the reals defined and represented geometrically as follows: Let a, b be real numbers with $a < b$.

(1) *Closed interval* $[a, b]$ (Figure 1-7(a))
  $[a, b] = \{x | a \leq x \leq b\}$
(2) *Open interval* (a, b) (Figure 1-7(b))
  $(a, b) = \{x | a < x < b\}$
(Observe that the symbol (a, b) denotes both an open interval and an ordered pair. The meaning will be clear from the context.)
(3) *Left closed–right open interval* $[a, b)$ (Figure 1-7(c))
  $[a, b) = \{x | a \leq x < b\}$
  Similarly we can define $(a, b]$.

(a)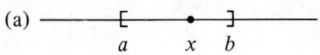

(b)

(c)

Figure 1-7

(4) *Infinite intervals* (∞ is read "infinity.")
  (a) $(a, \infty) = \{x | x > a\}$ (Figure 1-8(a))
  (b) $[a, \infty) = \{x | x \geq a\}$ (Figure 1-8(b))
  $(-\infty, a)$ and $(-\infty, a]$ are defined analogously.
  $(-\infty, \infty)$ denotes R.

Linear Inequalities. A *linear inequality* in one unknown, x, is an inequality which may be written in the form $ax + b < 0$ or $ax + b > 0$. The *solution set* is the set of all real numbers which satisfy the inequality.

(a)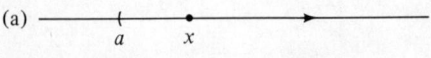

(b)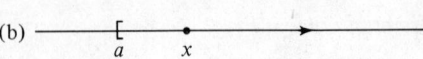

Figure 1-8

Example 1. Solve $-3x + 2 > 0$.

SOLUTION

$-3x + 2 > 0$
$\Leftrightarrow 3x < 2$ (adding $3x$ to both sides)
$\Leftrightarrow x < \frac{2}{3}$ (dividing by 3)

The solution set is $S = \{x | x < \frac{2}{3}\} = (-\infty, \frac{2}{3})$.

Quadratic Inequalities. A *quadratic inequality* in one unknown, x, is an inequality which may be written in the form

$$ax^2 + bx + c < 0 \quad \text{or} \quad ax^2 + bx + c > 0, \quad a \neq 0$$

Example 2. Solve $x^2 - x - 6 > 0$.

SOLUTION. By factoring,

$x^2 - x - 6 > 0$
$\Leftrightarrow (x - 3)(x + 2) > 0$
$\Leftrightarrow (x - 3 > 0 \text{ and } x + 2 > 0) \text{ or } (x - 3 < 0 \text{ and } x + 2 < 0)$
$\Leftrightarrow (x > 3 \text{ and } x > -2) \text{ or } (x < 3 \text{ and } x < -2)$
$\Leftrightarrow (x > 3) \text{ or } (x < -2)$

The solution set is

$S = \{x | x > 3\} \cup \{x | x < -2\}$
$= (3, \infty) \cup (-\infty, -2)$

Example 3. Solve $1 - 2x - x^2 > 0$.

SOLUTION. By completing the square,

$1 - 2x - x^2 > 0$
$\Leftrightarrow x^2 + 2x < 1$ (transposing terms)
$\Leftrightarrow x^2 + 2x + 1 < 1 + 1$
$\Leftrightarrow (x + 1)^2 < 2$
$\Leftrightarrow |x + 1| < \sqrt{2}$ (Theorem 4(2))
$\Leftrightarrow -\sqrt{2} < x + 1 < \sqrt{2}$ (Theorem 4(7))
$\Leftrightarrow -\sqrt{2} - 1 < x < \sqrt{2} - 1$

The solution set is

$S = \{x | -\sqrt{2} - 1 < x < \sqrt{2} - 1\}$
$= (-\sqrt{2} - 1, \sqrt{2} - 1)$

We now solve an equation involving absolute values.

Example 4. Solve $|2x - 3| = |1 - 3x|$.

SOLUTION

$|2x - 3| = |1 - 3x|$

⇔ $(2x - 3)^2 = (1 - 3x)^2$ (Theorem 4(2))
⇔ $(2x - 3)^2 - (1 - 3x)^2 = 0$
⇔ $[(2x - 3) - (1 - 3x)] \cdot [(2x - 3) + (1 - 3x)] = 0$
⇔ $[5x - 4] \cdot [-x - 2] = 0$
⇔ $(5x - 4 = 0)$ or $(-x - 2 = 0)$
⇔ $x = \frac{4}{5}$ or $x = -2$

EXERCISE 3

Find the solution set for each of the following and represent them on the real line.

1. (a) $13 > 5 - 2x$
   (b) $\frac{1}{4} - 2x < 5x - 1$
   (c) $3x - 2 < \sqrt{2}x + \frac{2}{3}$
2. (a) $x^2 - 6x + 8 < 0$
   (b) $3x^2 + 2x - 8 \leq 0$
   (c) $-x^2 + 4x - 4 < 0$
   (d) $x^2 + 6x - 7 \leq 0$
   (e) $x^2 - ax \geq a^2$, $a > 0$
3. (a) $\frac{1}{x} + \frac{x}{3} < 2x$
   (b) $\frac{1}{x^2} + \frac{2}{3} \leq \frac{1}{x}$
   (c) $\frac{x + 3}{2x - 4} < 4$
4. (a) $|x - 3| = |2x + 4|$
   (b) $|2x - 3| = 5$
   (c) $|x^2 - 4| = 3$
   (d) $\left|\frac{2x - 1}{x + 4}\right| = 2$

4. REVIEW OF TRIGONOMETRY

The following facts concerning trigonometry are listed for reference.

(1) We regard a real number θ as the *radian measure* of an angle. *Degree* and radian measure are related by the equation $2\pi = 360°$. (When no units are specified, radian measure is assumed.)
(2) In a circle of radius r, a central angle of θ radians subtends an arc of length $s = r\theta$.
(3) The trigonometric functions of θ are defined as follows. (See Figure 1-9.)

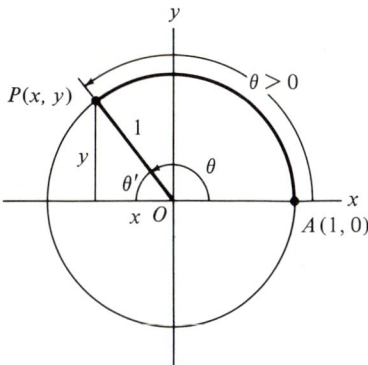

Figure 1-9

$\sin \theta = y$

$\cos \theta = x$

$\tan \theta = \dfrac{y}{x}, \quad x \neq 0$

$\csc \theta = \dfrac{1}{y}, \quad y \neq 0$

$\sec \theta = \dfrac{1}{x}, \quad x \neq 0$

$\cot \theta = \dfrac{x}{y}, \quad y \neq 0$

(4) The *related angle* θ' is the nonnegative acute angle determined by the terminal side $\overline{OP}$ of θ and the x-axis (Figure 1–9). The trigonometric functions of θ and θ' have the same numerical value. The sign of a function of θ is determined by the quadrant in which P lies. The following table is useful in determining the function values of certain "common" angles.

θ	0	$\dfrac{\pi}{6}$	$\dfrac{\pi}{4}$	$\dfrac{\pi}{3}$	$\dfrac{\pi}{2}$
$\sin \theta$	$\dfrac{\sqrt{0}}{2}$	$\dfrac{\sqrt{1}}{2}$	$\dfrac{\sqrt{2}}{2}$	$\dfrac{\sqrt{3}}{2}$	$\dfrac{\sqrt{4}}{2}$
$\cos \theta$	$\dfrac{\sqrt{4}}{2}$	$\dfrac{\sqrt{3}}{2}$	$\dfrac{\sqrt{2}}{2}$	$\dfrac{\sqrt{1}}{2}$	$\dfrac{\sqrt{0}}{2}$
$\tan \theta$	0	$\dfrac{1}{\sqrt{3}}$	1	$\dfrac{\sqrt{3}}{1}$	∞

For example,

$$\cos \frac{7\pi}{6} = -\cos \frac{\pi}{6} = -\frac{\sqrt{3}}{2}$$

since $\frac{7}{6}\pi$ is a third quadrant angle.

(5) *Trigonometric identities* are as follows.
  (a) *Fundamental relations:*

$$\left. \begin{array}{l} \sin^2 \theta + \cos^2 \theta = 1 \\ \tan^2 \theta + 1 = \sec^2 \theta \\ \cot^2 \theta + 1 = \csc^2 \theta \end{array} \right\} \text{(Pythagorean relations)}$$

$$\left. \begin{array}{l} \tan \theta = \dfrac{\sin \theta}{\cos \theta} \\ \cot \theta = \dfrac{\cos \theta}{\sin \theta} \end{array} \right\} \text{(ratio relations)}$$

$$\left. \begin{array}{l} \csc \theta = \dfrac{1}{\sin \theta} \\ \sec \theta = \dfrac{1}{\cos \theta} \\ \cot \theta = \dfrac{1}{\tan \theta} \end{array} \right\} \text{(reciprocal relations)}$$

  (b) *Reduction formulas:*

$$\sin(-\theta) = -\sin \theta$$
$$\cos(-\theta) = \cos \theta$$
$$\sin(\tfrac{1}{2}\pi - \theta) = \cos \theta$$
$$\cos(\tfrac{1}{2}\pi - \theta) = \sin \theta$$

  (c) *Addition formulas:*

$$\cos(\theta + \varphi) = \cos \theta \cos \varphi - \sin \theta \sin \varphi$$
$$\cos(\theta - \varphi) = \cos \theta \cos \varphi + \sin \theta \sin \varphi$$
$$\sin(\theta + \varphi) = \sin \theta \cos \varphi + \cos \theta \sin \varphi$$
$$\sin(\theta - \varphi) = \sin \theta \cos \varphi - \cos \theta \sin \varphi$$

  (d) *Double-angle formulas:*

$$\sin 2\theta = 2 \sin \theta \cos \theta$$
$$\cos 2\theta = \cos^2 \theta - \sin^2 \theta$$
$$= 1 - 2 \sin^2 \theta$$
$$= 2 \cos^2 \theta - 1$$

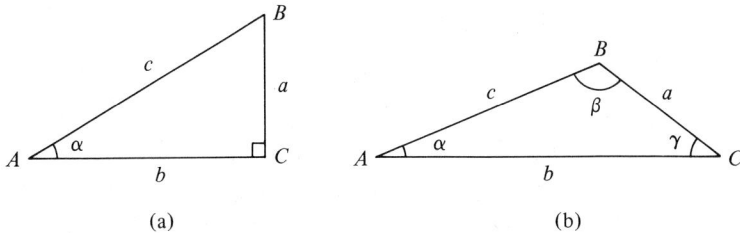

Figure 1-10

(e) *Half-angle formulas:*

$$\sin^2 \frac{\theta}{2} = \frac{1 - \cos \theta}{2}$$

$$\cos^2 \frac{\theta}{2} = \frac{1 + \cos \theta}{2}$$

(6) *The trigonometry of triangles* is given below.
  (a) *Right triangle* (Figure 1–10(a)):

$$\sin \alpha = \frac{a}{c} = \frac{\text{side opposite}}{\text{hypotenuse}}$$

$$\cos \alpha = \frac{b}{c} = \frac{\text{side adjacent}}{\text{hypotenuse}}$$

$$\tan \alpha = \frac{a}{b} = \frac{\text{side opposite}}{\text{side adjacent}}$$

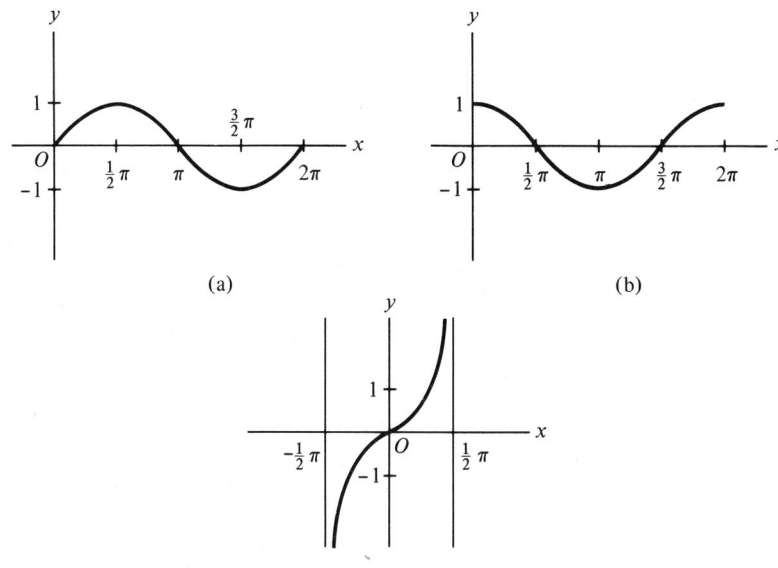

Figure 1-11 (a) $y = \sin x$. (b) $y = \cos x$. (c) $y = \tan x$.

(b) *General triangle* (Figure 1–10(b)):
$$c^2 = a^2 + b^2 - 2ab\cos\gamma \quad \text{(law of cosines)}$$
$$\frac{a}{\sin\alpha} = \frac{b}{\sin\beta} = \frac{c}{\sin\gamma} \quad \text{(law of sines)}$$

(7) The graphs of the trigonometric functions (for a period interval) are shown in Figure 1–11. Here x denotes the radian measure of the angle, and y the corresponding function value.

EXERCISE 4

1. Evaluate the following, if they exist.
   (a) $\sec\frac{11}{4}\pi$
   (b) $\tan\frac{2}{3}\pi$
   (c) $\csc(-\frac{5}{4}\pi)$
   (d) $\cos\frac{28}{3}\pi$
   (e) $\sec 17\pi$

2. Express the following in degrees.
   (a) $\frac{7}{6}\pi$
   (b) $\frac{7}{4}\pi$
   (c) $-\frac{2}{3}\pi$
   (d) 2
   (e) $\frac{7}{8}\pi$

3. Express the following in radians.
   (a) $120°$
   (b) $330°$
   (c) $75°$
   (d) $\pi°$
   (e) $-150°$

4. Given the following information, find all possible values of all trigonometric functions of θ.
   (a) $\sin\theta = \frac{3}{4}$ (θ in the second quadrant)
   (b) $\csc\theta = -2$ ($\tan\theta < 0$)
   (c) $\cos\theta = \frac{2}{\sqrt{5}}$
   (d) $\csc\theta = \frac{1}{\sqrt{2}}$
   (e) $\tan\theta = 3$ ($\cos\theta < 0$)

5. Prove the following identities.
   (a) $\dfrac{\sin^2\theta}{(1-\cos\theta)^2} = \dfrac{(1+\cos\theta)^2}{\sin^2\theta}$
   (b) $\dfrac{\sec\theta + 1}{\tan\theta} = \dfrac{\tan\theta}{\sec\theta - 1}$

(c) $\dfrac{1}{1+\cos\theta} + \dfrac{1}{1-\cos\theta} = 2(\cot^2\theta + 1)$

(d) $\dfrac{\sin^3\theta + \cos^3\theta}{1 - \sin\theta\cos\theta} = \sin\theta + \cos\theta$

(e) $\sin(\tfrac{1}{2}\pi - \theta) + \sin(\pi - \theta) + \sin(\tfrac{3}{2}\pi - \theta) + \sin(2\pi - \theta) = 0$

(f) $\cos 4\theta = 8\cos^4\theta - 8\cos^2\theta + 1$

(g) $\sin(\theta + \varphi)\sin(\theta - \varphi) = \cos^2\varphi - \cos^2\theta$

chapter 2
Functions, Relations, and Their Graphs

1. FUNCTIONS AND RELATIONS

Much of mathematics is concerned with the interdependence of two "variable quantities," say x and y. For example, if x denotes the side length of a square and y its area, then $y = x^2$. Thus, for a given length x, this equation determines one (and only one) value of the area y. We say "y is a function of x." Observe further that the equation $y = x^2$ determines a set of ordered pairs (x, y), wherein $x > 0$, and the second element y is the square of the first. Denoting this set of ordered pairs by f, we have

$$f = \{(x, y) \mid x > 0 \text{ and } y = x^2\}$$

Note that there are two sets of numbers involved in this example: the set of values of x, called the *domain* of f, and the set of corresponding values of y, called the *range* of f. This is an example of a *function*, according to the following definition:

Definition. A function f, with *domain* D_f and range R_f is a set of ordered pairs (x, y) in $D_f \times R_f$ with the property that for each x in D_f there is *exactly one* y in R_f such that (x, y) is in f. Moreover, for each y in R_f there is *at least one* x in D_f such that the pair (x, y) is in f. y is called the *value* of f at x, and we write

$y = f(x)$ (read "y equals f of x"). x is called the "independent variable" and y the "dependent variable."

The *domain* of f consists of all first elements, and the *range* consists of all second elements of pairs in f. Thus, in the above example, D_f is the set of all positive real numbers, as is R_f. Moreover, $f(x) = x^2$. For future reference, a function f is often called a *mapping* and any set B containing the range is called a *codomain* of f. In this terminology, we say "f maps D_f into B." If $B = R_f$, we say "f maps D_f onto B."

From the above definition we see that a set of ordered pairs may be construed as a function, *provided no two pairs have the same first element*. We often wish to consider relationships between two variables x and y in which certain values of x are associated with one *or more* values of y. For example, $y^2 = x$. Thus, if $x = 4$, then $y = 2$ or $y = -2$. If $x = 0$, then $y = 0$. This defines a set of ordered pairs (x, y) which we shall call a *relation*.

Definition. A relation F with domain A and codomain B is a set of ordered pairs (x, y) in $A \times B$ with the property that for each x in A there is *at least one* "corresponding" y in B such that (x, y) is in F.

Thus, every function is a relation but not vice versa. The distinction between a function and a relation lies in the following: A relation F is not a function provided there is at least one x in the domain to which there corresponds more than one y in the range.

In analytic geometry, the domains and codomains of relations are subsets of R, so that the relations themselves are subsets of $R \times R$. We denote $R \times R$ by R^2 (read "R two"), and call R^2 the *plane*. Elements of R^2 are called *points* and are denoted by capital letters, $A, B, C, \ldots, P, Q, \ldots$. We write $P = (x, y)$ or $P(x, y)$. For $P(x, y)$ in R^2, x is called the *x-coordinate* or *abscissa* of P and y the *y-coordinate* or *ordinate*. We call the set of points $(x, 0)$ the *x-axis*, and the set of points $(0, y)$ the *y-axis*. The point $O = (0, 0)$ is called the *origin*. We represent R^2 and its four *quadrants* as in Figure 2–1.

When the emphasis is on the relationship between the variables x and y, we often refer to the set of ordered pairs as the *graph* of the relation. The term "graph" also refers to a pictorial representation or drawing of the set of

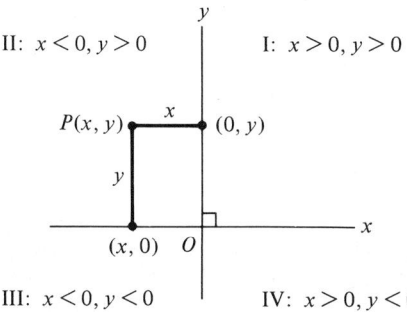

Figure 2–1

ordered pairs. We shall be concerned with rules for drawing graphs of relations. In this chapter, we list a few elementary rules. The list will be extended in subsequent chapters. At first we simply plot points and join them in some "reasonable" fashion.

Example 1. Let $f = \{(x, y) | y = x^2\}$. The graph of f is shown in Figure 2–2. Observe that f is a function.

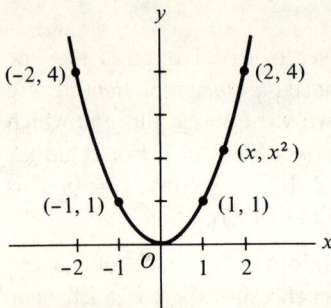

Figure 2–2

Example 2. Let $D_f = \{x | -2 \leq x \leq 2\}$, and

$$y = f(x) = \begin{cases} -2 & \text{if } -2 \leq x \leq 0 \\ x & \text{if } 0 < x \leq 2 \end{cases}$$

A table of pairs (x, y) and a sketch of the graph are shown in Figure 2–3. Again, f is a function.

If the domain of a function φ is itself a set of ordered pairs (x, y), φ is called a *function of two variables*. The function values are written $\varphi(x, y)$, and x and y are both called *independent variables*. Every relation may be written in the form $\{(x, y) | \varphi(x, y) = 0\}$ for some function φ of two variables. Although we shall not show how to determine φ in general, in cases where the relation is given by an equation, $\varphi(x, y)$ can be obtained by simply transposing all terms to one side. Thus, the equation $y^2 = x$ may be written $y^2 - x = 0$, so that $\varphi(x, y) = y^2 - x$. For brevity, we call the equation $\varphi(x, y) = 0$ the relation.

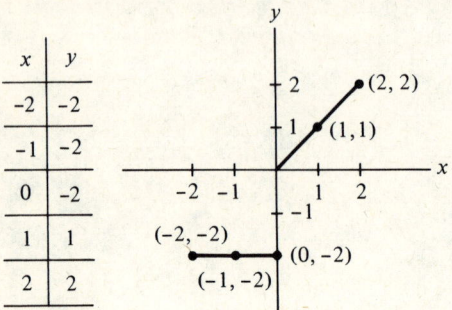

Figure 2–3

We adopt the convention that, unless otherwise specified, the domain of the relation $\varphi(x, y) = 0$ is the set of *all* x such that for some y, $\varphi(x, y) = 0$. A sketch of the graph of a relation F must have the following properties:

(1) Every vertical line through a point $(x, 0)$, $x \in D_F$, must intersect the graph of F in at least one point (*exactly* one point if F is a function).
(2) Every horizontal line through a point $(0, y)$, $y \in R_F$, must intersect the graph in at least one point.

Example 3. Sketch the graph of $\varphi(x, y) = |x| + |y| - 1 = 0$. We first compute the domain and range of the relation. Since $|x| + |y| = 1$, we must have $|x| \le 1$. For any such x, the corresponding y must satisfy $|y| = 1 - |x|$; that is, $y = \pm(1 - |x|)$. Since such y exists for each x with $|x| \le 1$, we have shown that $D_F = \{x \mid -1 \le x \le 1\}$. Similarly, $R_F = \{y \mid -1 \le y \le 1\}$. The table and graph are shown in Figure 2–4.

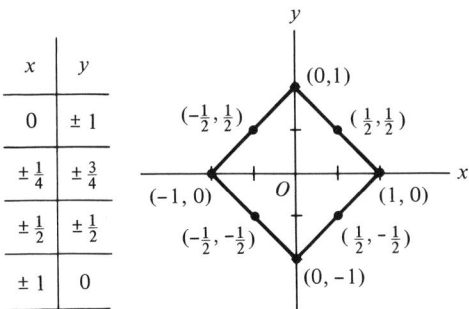

Figure 2-4

Example 4. Sketch the graph of the relation $y < x$. This relation is not given in the form $\varphi(x, y) = 0$. Although such a function φ exists, it is easier to find the graph from the given form. Any point (x, y) with $y < x$ is on the graph. Since the point (x, y_1) is plotted below (x, y_2) iff $y_1 < y_2$, the graph consists of all points below a point of the form (x, y) with $y = x$, that is, all points *below* the graph of the function $y = x$. The graph is the shaded region in Figure 2–5.

If $\varphi(x, y) = 0$ is a relation and if f is a function with the property that for

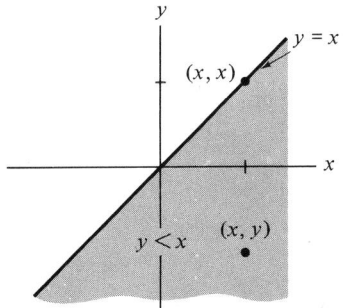

Figure 2-5

all $x \in D_f$, $\varphi(x, f(x)) = 0$, the function f is said to be *defined implicitly by the relation* $\varphi(x, y) = 0$. This simply means that the graph of the function $y = f(x)$ is contained in the graph of the relation $\varphi(x, y) = 0$. In some cases, several functions may be defined by the same relation. For example, the relation $\varphi(x, y) = x^2 - y^2 - 1 = 0$ defines implicitly the function $y = f(x) = \sqrt{x^2 - 1}$, as well as the function $y = g(x) = -\sqrt{x^2 - 1}$, since $\varphi(x, y) = 0$ for all x in the domain of each function. Note that these functions were obtained by solving the equation for y in terms of x. The graph of a relation $\varphi(x, y) = 0$ may be the union of the graphs of several functions. The determination of these functions may expedite the sketching of the graph of the relation.

Example 5. Sketch the graph of $y^2 - x^2 = 0$. This equation is satisfied iff $y = x$ or $y = -x$. Hence the graphs of these two functions comprise that of the given relation. The domain and range of each function are clearly R (Figure 2-6).

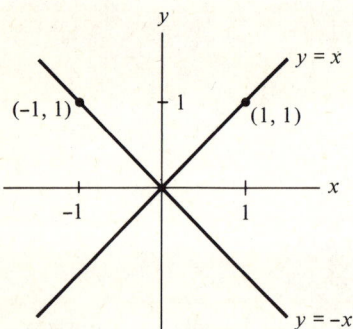

Figure 2-6

EXERCISE 1

In the following exercises, x is the independent variable.

1. Let $y = f(x) = 3x^2$. Find
   (a) $f(0)$
   (b) $f(1)$
   (c) $f(-2)$
   (d) $f(\sqrt{2})$
   (e) $f(\pi)$
   (f) $f(2 - \sqrt{3})$
   (g) $f(a + h)$, for any $a, h \in R$
   (h) $\dfrac{f(a + h) - f(a)}{h}$, for any $a, h \in R, h \neq 0$

2. Let $y = f(x) = 2$ for all $x \in R$. Sketch the graph. A function of this type (that is, $y = f(x) = c$ for some fixed number c) is called a *constant function*.

Functions and Relations

3. Find the domain of each of the following relations and state whether they are functions. Sketch the graphs.
   (a) $x^2 + y^2 = 16$
   (b) $x^2 - 2y = 0$
   (c) $y^2 - 2x = 0$
   (d) $y^2 = 4x^2$
   (e) $y = 2x$
   (f) $y = 2x - 4$
   (g) $y = 2x^2 + 1$
   (h) $2x - 4y = 8$
   (i) $y^2 = x + 2$

4. Sketch the graphs of the following functions and find their domains.
   (a) $y = 2x^2$
   (b) $y = \dfrac{1}{x}$
   (c) $y = \sqrt{x}$
   (d) $y = \sqrt{x^2 + 4}$
   (e) $y = -\sqrt{x^2 + 4}$
   (f) $y = \dfrac{1}{x - 1}$

5. Sketch the graphs of the following functions.
   (a) $f(x) = \begin{cases} x^2, & \text{if } 0 < x \leq 2 \\ x - 4, & \text{if } -1 \leq x \leq 0 \end{cases}$ (here $D_f = \{x \mid -1 \leq x \leq 2\}$)
   (b) $f(x) = |2x|$
   (c) $f(x) = \begin{cases} |2x|, & \text{if } x < 0 \\ 1, & \text{if } x \geq 0 \end{cases}$

6. Sketch the graphs of the following relations.
   (a) $y \leq -1$
   (b) $x = 3y^2$
   (c) $x < 3y^2$
   (d) $|x| \leq 1$
   *(e) $|x| - |y| = 1$
   *(f) $x^{2/3} + y^{2/3} = 1$
   (g) $2x + y \leq 4$

7. Find functions defined by the following relations. Sketch the graphs of the functions and of the relations.
   (a) $x^2 - 2y^2 = 1$
   (b) $x = 3y^2$
   (c) $|y| = x + 1$
   (d) $x^2 - y^2 = 0$
   (e) $x^3 - y^3 = 0$

*8. For each $x \in R$, let n be the unique integer such that $n \le x < n + 1$, and let $f(x) = n$. f is called "the greatest integer function" and is denoted by $f(x) = [x]$. For example, $[-0.1] = -1$, $[\frac{5}{2}] = 2$, and $[-\frac{5}{2}] = -3$. Sketch the graph of $y = [x]$ for $-5 \le x \le 5$.

*9. For each $x \in R$, let $g(x) = x - [x]$. (See Problem 8.) This function is called "the fractional part of x" and is denoted by $\{x\}$. Thus, $\{x\} = x - [x]$. For example, $\{\frac{1}{2}\} = \frac{1}{2} - [\frac{1}{2}] = \frac{1}{2}, \{-\frac{1}{4}\} = -\frac{1}{4} - [-\frac{1}{4}] = -\frac{1}{4} - (-1) = \frac{3}{4}$. Note that $0 \le \{x\} < 1$ for all x. Sketch the graph of $y = \{x\}$ for $-5 \le x \le 5$.

2. THE ALGEBRA OF FUNCTIONS; GRAPHING BY ADDITION OF ORDINATES

Let f, g be functions whose domains and ranges are subsets of R. We define the *sum* $f + g$, *difference* $f - g$, and *product* $f \cdot g$ of f and g as follows: For all $x \in D_f \cap D_g$,

$(f + g)(x) = f(x) + g(x)$
$(f - g)(x) = f(x) - g(x)$
$(f \cdot g)(x) = f(x) \cdot g(x)$

For each $x \in D_f \cap \{x \in D_g | g(x) \neq 0\}$, we define the *quotient* f/g by

$$\left(\frac{f}{g}\right)(x) = \frac{f(x)}{g(x)}$$

Example 1. Let $f(x) = x^3, g(x) = \sqrt{x}$. Then $D_f = R$ and $D_g = \{x | x \ge 0\}$. Hence for $x \ge 0$,

$(f + g)(x) = x^3 + \sqrt{x}$
$(f - g)(x) = x^3 - \sqrt{x}$
$(f \cdot g)(x) = x^3 \cdot \sqrt{x} = x^{7/2}$

For $x > 0, (f/g)(x) = x^3/x^{1/2} = x^{5/2}$. Note that the formula $(f/g)(x) = x^{5/2}$ does not exhibit the fact that $x \neq 0$, so this must be explicitly stated.

An important operation on functions that has no analogue in the real numbers is *composition of functions*. We define the *composite function of g by f* (in that order), written $g(f)$, as follows:

$g(f)(x) = g(f(x))$ (for all $x \in D_f$ such that $f(x) \in D_g$)

Thus, to compute $g(f)(x)$, substitute $f(x)$ for the independent variable in the formula for g. The domain of $g(f)$ is $\{x | x \in D_f \text{ and } f(x) \in D_g\}$.

The Algebra of Functions; Graphing by Addition of Ordinates

Example 2. Let $f(x) = \sqrt{x}$ and $g(x) = 1/(x+1)$. Then $D_f = \{x|x \geq 0\}$ and $D_g = \{x|x \neq -1\}$. We have

$$g(f)(x) = g(f(x)) = \frac{1}{f(x)+1} = \frac{1}{\sqrt{x}+1} \qquad (x \geq 0)$$

Also

$$f(g(x)) = \sqrt{g(x)} = \sqrt{\frac{1}{x+1}} = \frac{1}{\sqrt{x+1}} \qquad (x > -1)$$

Note that $f(g(x)) \neq g(f(x))$ in general.

When a function is the sum (or difference) of two functions, its graph may be sketched by the method of *addition of ordinates* as illustrated in the next example.

Example 3. Sketch the graph of $y = f(x) = x + \sqrt{x}$. Letting

$$y_1 = f_1(x) = x \quad \text{and} \quad y_2 = f_2(x) = \sqrt{x}$$

we have $y = y_1 + y_2 = f_1(x) + f_2(x)$. We first plot the graphs of $y_1 = x$ and $y_2 = \sqrt{x}$ on the same set of axes. We add (geometrically) the corresponding values of $y_1 = f_1(x)$ and $y_2 = f_2(x)$ to obtain the value of $y = f(x)$ (Figure 2–7).

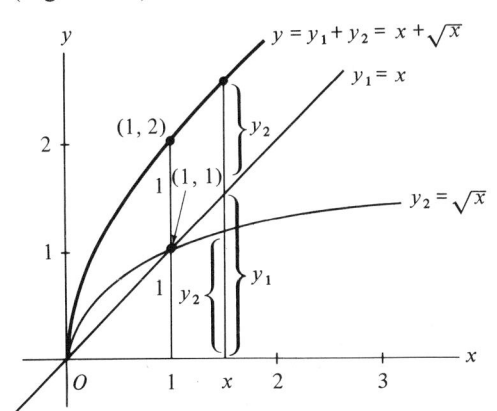

Figure 2-7

EXERCISE 2

1. Let $f(x) = |x|$ and $g(x) = x$. Find the domains and sketch the graphs of $(f+g)(x)$, $(f-g)(x)$, $(f \cdot g)(x)$, and $(f/g)(x)$.
2. Give the domains and find formulas for $(f \cdot g)(x)$ and $(f/g)(x)$, where $f(x) = \sqrt{x}$ and $g(x) = |x|$.

3. Compute $g(f)(x)$ and $f(g)(x)$, and give their domains.
   (a) $f(x) = \sqrt{x}$, $g(x) = |x|$
   (b) $f(x) = \sin x$, $g(x) = \sqrt{x}$
   (c) $f(x) = x^2 - 2$, $g(x) = \dfrac{1}{x+1}$
   (d) $f(x) = x^2 - 5x + 6$, $g(x) = x - 3$
   (e) $f(x) = \dfrac{1}{x}$, $g(x) = x^2 + 2$
   (f) $f(x) = \dfrac{1}{x}$, $g(x) = \dfrac{1}{x}$

4. Sketch the graphs by addition of ordinates.
   (a) $y = x + \sin x$
   (b) $y = x - x^2$
   (c) $y = x^2 - x$
   (d) $y = x^2 - \cos x$
   (e) $y = |x| + x^2 - \cos x$
   (f) $y = 3x + 1 + \sqrt{x}$
   (g) $y = x^2 + 2x$
   (h) $y = \sin x + \cos x$
   (i) $y = \sin x + \cos 2x$
   (j) $y = x^2 + \cos x$

*5. Sketch the graphs of $|x| - [x]$, $|x| \cdot [x]$, and $|x|/[x]$, where $[x]$ is the "greatest integer" function. (See Problem 8, Exercise 1.) Find at least five points on each graph.

3. INVERSE FUNCTIONS

Let f be a function and suppose that f has the property that no two ordered pairs have the same *second* element; that is, if (x_1, y_1) and (x_2, y_2) belong to f and $x_1 \neq x_2$, then $y_1 \neq y_2$. (In this case, we say that f is a *one-to-one* function.) Now let $g = \{(x, y) | (y, x) \in f\}$. Thus, g is that set of ordered pairs obtained by interchanging the first and second elements of all pairs in f. Since f is one to one, g is a function, namely, that function f^{-1} (read "f inverse") whose domain is R_f, whose range is D_f, and which is defined by $y = f^{-1}(x)$ iff $x = f(y)$. f^{-1} has the following properties:

(1) $f(f^{-1}(x)) = x$, for all $x \in R_f$
(2) $f^{-1}(f(x)) = x$, for all $x \in D_f$

Property 1 may be used to obtain a formula for f^{-1} if a formula for f is given.

Example 1. Find the inverse of the function $f(x) = x^2$, where $D_f = \{x | x \geq 0\}$

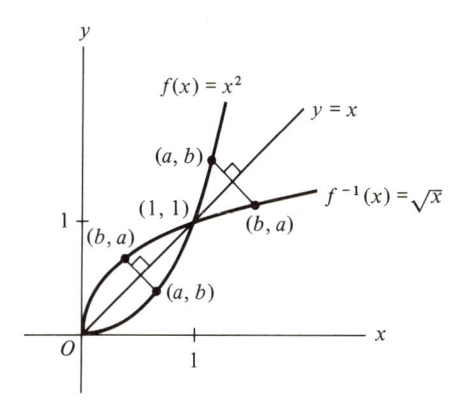

Figure 2-8

SOLUTION. Note that f is one to one ($x_1 \neq x_2$ implies $x_1^2 \neq x_2^2$, since $x_1, x_2 \geq 0$), so f^{-1} exists. Moreover, $R_f = D_f$. By Property 1 we have

$$f(f^{-1}(x)) = x$$
$$(f^{-1}(x))^2 = x$$
$$f^{-1}(x) = \sqrt{x}$$

where the nonnegative root is chosen since $f^{-1}(x) \geq 0$, due to the nature of D_f. If the function f is written $y = x^2$, its inverse f^{-1} is written $y = \sqrt{x}$. Their graphs are shown in Figure 2-8.

REMARK. Since the point (a, b) is on the graph of f iff the point (b, a) is on the graph of f^{-1}, and since these two points are symmetric to each other with respect to the line $y = x$, the graph of f^{-1} may be obtained by reflecting the graph of f about this line (Figure 2-8). The points (a, b) and (b, a) are said to be symmetric with respect to the line $y = x$ because this line is the perpendicular bisector of the line segment joining the two points. (These concepts will be defined in Chapter 4.)

Example 2. Find a formula for the inverse of $f(x) = x^3$ and sketch its graph.

SOLUTION. Note that f is one to one and that $D_f = R_f = R$. Using Property 1 we have

$$f(f^{-1}(x)) = x$$
$$(f^{-1}(x))^3 = x$$
$$f^{-1}(x) = \sqrt[3]{x}$$

The graphs of $y = x^3$ and $y = \sqrt[3]{x}$ are shown in Figure 2-9.

The procedure for finding the inverse of certain functions may be described as follows:

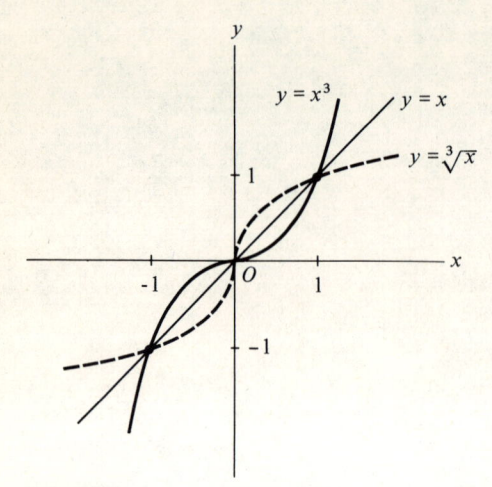

Figure 2-9

(1) Solve the equation $y = f(x)$ for x in terms of y.
(2) Interchange x and y in the equation obtained in Part 1.

Example 3. Find the inverse of $y = f(x) = (2x + 1)/(x - 1)$.

SOLUTION

(a) Solve for x in terms of y:

$$y = \frac{2x + 1}{x - 1}$$
$$xy - y = 2x + 1$$
$$xy - 2x = y + 1$$
$$x(y - 2) = y + 1$$
$$x = \frac{y + 1}{y - 2}$$

(b) Interchange x and y:

$$y = \frac{x + 1}{x - 2}$$
$$\therefore f^{-1}(x) = \frac{x + 1}{x - 2}$$

CHECK

$$f(f^{-1}(x)) = \frac{2 \cdot f^{-1}(x) + 1}{f^{-1}(x) - 1} = \frac{2\left(\dfrac{x + 1}{x - 2}\right) + 1}{\dfrac{x + 1}{x - 2} - 1} = x$$

$$f^{-1}(f(x)) = \frac{f(x)+1}{f(x)-2} = \frac{\frac{2x+1}{x-1}+1}{\frac{2x+1}{x-1}-2} = x$$

EXERCISE 3

For Problems 1 to 11 compute the inverses of the given functions. Verify Properties 1 and 2. Sketch the graphs of the functions and their inverses.

1. $f(x) = \sqrt{x} - 2$
2. $f(x) = 2x - 3$
3. $f(x) = -3x + 5$
4. $f(x) = x^{2/3} + 3 \quad (x \geq 0)$
5. $f(x) = x^3 - 5$
6. $f(x) = |x| \quad (x \leq 0)$
7. $f(x) = x^2 + 2 \quad (x \leq 0)$
8. $f(x) = x^3 + 8$
9. $f(x) = x^2 - 4x + 7 \quad (x \geq 2)$
10. $f(x) = \begin{cases} x^2 & \text{for } x \geq 0 \\ x - 1 & \text{for } x < 0 \end{cases}$
11. $f(x) = \sqrt{x+5}$

†*12. $f(x)$ is said to be an *increasing* function iff

$$x_1 < x_2 \Rightarrow f(x_1) < f(x_2)$$

and a *decreasing* function iff

$$x_1 < x_2 \Rightarrow f(x_1) > f(x_2).$$

(a) Prove that if f is increasing (or decreasing), it is one to one and hence has an inverse.
(b) Prove that if f is increasing (or decreasing), then f^{-1} is also increasing (or decreasing).

*13. Assuming that $f(x) = 2^x$ defines an increasing function with domain R and hence that the inverse exists, sketch the graph of $f(x)$ and of $f^{-1}(x)$.

4. TECHNIQUES FOR SKETCHING GRAPHS

The purpose of this section is to analyze certain aspects of relations which serve to expedite the sketching of their graphs. For example, we may wish to know where the function is positive, where it is zero, and where it is negative. Other aspects which we shall investigate are *extent, symmetry, intercepts,* and *asymptotes*.

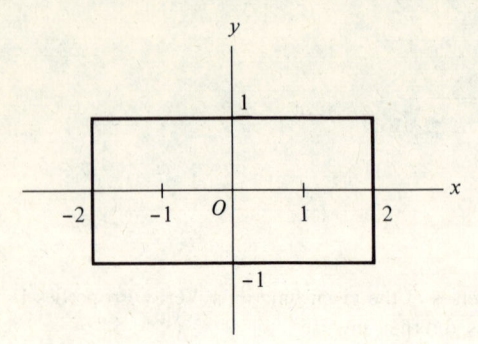

Figure 2-10

Extent. The *extent* of the graph of $\varphi(x, y) = 0$ *with respect to x* is simply the smallest interval (or union of intervals) which contains the domain. The extent with respect to y is defined similarly with "domain" replaced by "range." If the extent is $a \leq x \leq b$, $c \leq y \leq d$, then the graph must lie between the vertical lines $x = a$, $x = b$, and between the horizontal lines $y = c$, $y = d$. These lines determine a rectangle containing the graph.

Example 1. Find the extent of $x^2 + 4y^2 - 4 = 0$.

SOLUTION. Since $y^2 = (4 - x^2)/4$, the domain is $|x| \leq 2$; so the extent with respect to x is $-2 \leq x \leq 2$. Similarly, the extent with respect to y is $-1 \leq y \leq 1$. The graph is contained in the rectangle shown in Figure 2-10.

Example 2. Find the extent of $x^2 - 4y^2 - 4 = 0$.

SOLUTION. Since $y^2 = (x^2 - 4)/4$, $|x| \geq 2$. Since $x^2 = 4y^2 + 4$, the range is R. Hence the graph lies in the "half planes," $x \geq 2$, $x \leq -2$ (Figure 2-11).

Symmetry. If $\varphi(x, y) = 0$ implies $\varphi(-x, y) = 0$, we say that the graph of the relation $\varphi(x, y) = 0$ is *symmetric with respect to the y-axis.* Thus, if the point (x, y) is on the graph, the point $(-x, y)$ is also. Two such points are said to be symmetric with respect to the y-axis. Similarly, if $\varphi(x, y) = 0$ implies $\varphi(x, -y) = 0$, the graph is said to be *symmetric with respect to the x-axis.* If $\varphi(x, y) = 0$ implies $\varphi(-x, -y) = 0$, the graph is said to be

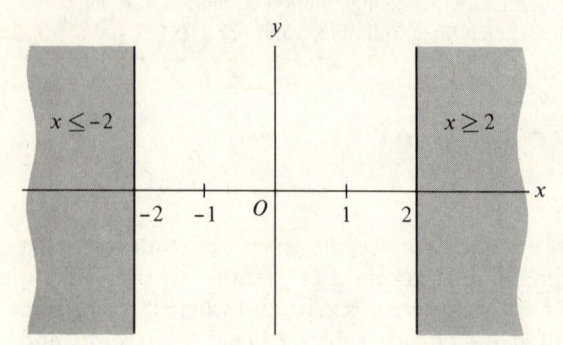

Figure 2-11

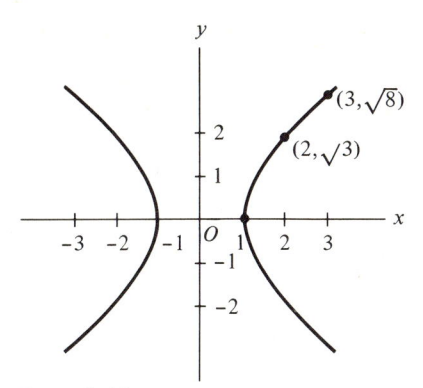

Figure 2-12

symmetric with respect to the origin. Note that if the graph of $\varphi(x, y) = 0$ is symmetric with respect to both the x- and y-axes, we have

$$\varphi(x, y) = 0 \Rightarrow \varphi(-x, y) = 0 \Rightarrow \varphi(-x, -y) = 0$$

so that the graph is symmetric with respect to the origin.

Although we could define symmetry with respect to any line, we shall restrict our discussion to symmetry about the coordinate axes.

Example 3. Plot $x^2 - y^2 = 1$. We first note that the graph is symmetric with respect to both the x- and y-axes and therefore with respect to the origin. Hence it is sufficient to sketch the graph in the first quadrant, since the portions in the other quadrants may be obtained by symmetry. For (x, y) in the first quadrant, $x^2 - y^2 = 1$ iff $y = \sqrt{x^2 - 1}$. The extent with respect to x is $x \geq 1$. Plotting the points $(1, 0)$, $(2, \sqrt{3})$, and $(3, \sqrt{8})$, we see that the graph appears as in Figure 2-12.

It is important to note that any relation, $\varphi(x, y) = 0$, in which x occurs only to even powers $(x^0, x^2, x^4, \ldots)$ is symmetric about the y-axis, whereas if y occurs only to even powers $(y^0, y^2, y^4, \ldots)$, the graph is symmetric about the x-axis. The following example shows that a graph may be symmetric about the origin without having symmetry about either axis.

Example 4. The graph of $xy - 1 = 0$ (that is, $y = 1/x$) is symmetric with respect to the origin, since $xy - 1 = 0$ implies $(-x)(-y) - 1 = 0$. However, it is symmetric with respect to neither axis, as $(1, 1)$ is on the graph, but neither $(-1, 1)$ nor $(1, -1)$ is (Figure 2-13).

Intercepts. A point where the graph intersects the x- or y-axis is called an x- or y-intercept, respectively. To find the x-intercepts, then, for the relation $\varphi(x, y) = 0$, we must solve $\varphi(x, 0) = 0$ for x; to find the y-intercepts, we solve $\varphi(0, y) = 0$ for y. If $(a, 0)$ is an x-intercept, we often refer to a as an x-intercept. We adopt a similar convention for y-intercepts. The graph of Example 4 has no intercepts.

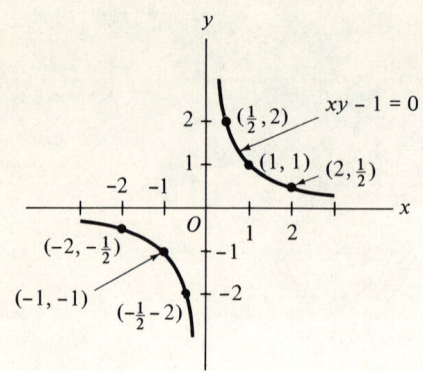

Figure 2-13

Example 5. $x^2 + 3y^2 - 1 = 0$ has intercepts $(0, 1/\sqrt{3})$, $(0, -1/\sqrt{3})$, $(1, 0)$, $(-1, 0)$.

Asymptotes. We shall not attempt a complete discussion of *asymptotes*, but shall give a few "working definitions" which are simply techniques for finding asymptotes in special cases. A complete definition requires the concept of *limit*, which is beyond the scope of this book. Roughly speaking, an *asymptote* is a straight line which the graph "approaches." For a nonvertical asymptote, this means that if we imagine a point "moving along the line," the distance between it and a point on the graph having the same abscissa becomes as small as we choose, provided the abscissa is sufficiently large in absolute value. This means that portions of the graph sufficiently far away from the origin are almost indistinguishable from the line. We shall limit our discussion of asymptotes to *polynomial relations*.

Definition. A *polynomial* is a function of the form

$$p(x) = a_0 + a_1 x + a_2 x^2 + \ldots + a_n x^n$$

where a_i ($i = 0, 1, 2, \ldots, n$) are real numbers, $a_n \neq 0$, and n is a nonnegative integer. n is called the *degree* of the polynomial. A *rational function* f is of the form

$$f(x) = \frac{p(x)}{q(x)}$$

where $p(x)$ and $q(x)$ are polynomials.

For example, $p(x) = 5 + 2x - 4x^3$ is a polynomial of degree *three* with $a_0 = 5, a_1 = 2, a_2 = 0$, and $a_3 = -4$.
$f(x) = (5 + 6x - x^3)/(2 + x^2)$ is a rational function.

Techniques for Sketching Graphs

(1) Suppose the graph of the relation $\varphi(x, y) = 0$ contains the graph of the rational function

$$y = \frac{p(x)}{q(x)}$$

where $p(x)$ and $q(x)$ are polynomials. Then if $q(a) = 0$ and $p(a) \neq 0$, the line $x = a$ is called a *vertical asymptote* of the graph of $\varphi(x, y) = 0$.

(Intuitively, as x "approaches" a, $q(x)$ approaches zero, while $p(x)$ does not. This implies that $p(x)/q(x)$ becomes arbitrarily large (numerically). We say "$p(x)/q(x)$ approaches infinity as x approaches a.")

(2) Similarly, if the graph of $\varphi(x, y) = 0$ contains the graph of the function

$$x = \frac{p(y)}{q(y)}$$

where $p(y)$ and $q(y)$ are polynomials, $q(b) = 0$, and $p(b) \neq 0$, then $y = b$ is called a *horizontal asymptote* of $\varphi(x, y) = 0$.

(3) If the graph of $\varphi(x, y) = 0$ contains the graph of a relation of the type

$$y^k = \frac{a_0 + a_1 x + \ldots + a_m x^m}{b_0 + b_1 x + \ldots + b_n x^n}$$

where $0 \leq m \leq n$ and $a_m, b_n \neq 0$, then the graph of $\varphi(x, y) = 0$ has the *horizontal asymptote* $y = 0$ if $m < n$, and

$$y = \sqrt[k]{\frac{a_m}{b_n}}$$

if $m = n$, provided this kth root exists. (If $m > n$, there is no horizontal asymptote.)

(4) By interchanging x and y in Part 3, one obtains a rule for *vertical asymptotes*.

Example 6. Find all horizontal and vertical asymptotes of $xy - y - x - 2 = 0$, and sketch.

SOLUTION. The relation may be written $y = (2 + x)/(-1 + x)$, which is of the form $p(x)/q(x)$ as in Part 1, with $q(1) = 0$ and $p(1) = 3 \neq 0$. Hence the graph has a vertical asymptote at $x = 1$. The relation is also of the form given in Part 3 with $k = 1, m = n = 1$, and $a_1 = b_1 = 1$. Therefore, the graph has a horizontal asymptote

$$y = \frac{a_1}{b_1} = 1$$

We note that $(-2, 0)$ and $(0, -2)$ are the intercepts.

Since there is an intercept at $x = -2$ and a vertical asymptote at $x = 1$, we examine the sign of y for $x \leq -2$, $-2 < x < 1$, and $x > 1$, using $y = (x + 2)/(x - 1)$.

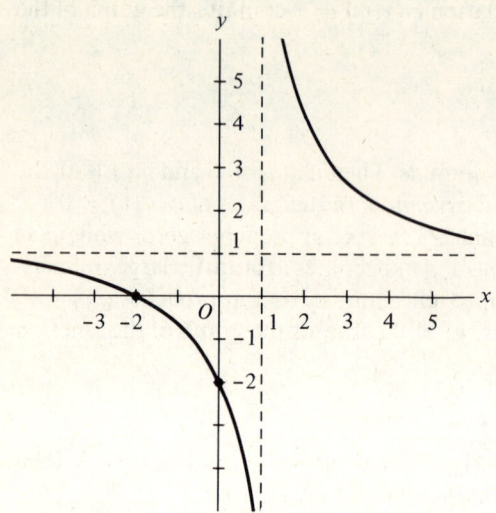

Figure 2-14

If $x < -2$, we have $x + 2 < 0$ and $x - 1 < 0$; so $(x + 2)/(x - 1) > 0$. Moreover, $x + 2 > x - 1$, so that $(x + 2)/(x - 1) < 1$, since $x - 1 < 0$.
If $-2 < x < 1$, we have $x + 2 > 0$ and $x - 1 < 0$; so $(x + 2)/(x - 1) < 0$.
If $x > 1$, we have $x + 2 > 0$ and $x - 1 > 0$; so $(x + 2)/(x - 1) > 0$. Moreover, $x + 2 > x - 1$, and so $(x + 2)/(x - 1) > 1$, since $x - 1 > 0$.

Note that if we write the relation in the form

$$y = \frac{1 + \frac{2}{x}}{1 - \frac{1}{x}}$$

(dividing numerator and denominator by the highest power of x), we see that if $|x|$ is very "large," y will be "close" to 1, since $2/x$ and $1/x$ are "small." A sketch showing the above aspects of the graph is shown in Figure 2-14.

We have listed four characteristics of graphs—extent, symmetry, intercepts, and asymptotes—which in general should be investigated. These, together with the technique of expressing a graph as the union of simpler graphs, are valuable aids in curve sketching. For some relations, however, certain of these aspects are too difficult to investigate, in which case we confine our analysis to those that can readily be determined. For example, in some instances, the extent may be the easiest to compute and the most informative, whereas it may be extremely difficult to ascertain whether any asymptotes exist. In other cases, asymptotes may be relatively easy to find, and the extent may be very difficult to determine. One must use some judgment as to the amount of effort that should be expended on any aspect of the graph.

Example 7. Sketch $-x^2y^2 + 4x^2 - y = 0$.

Techniques for Sketching Graphs

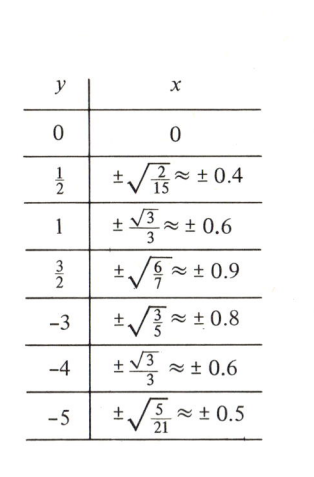

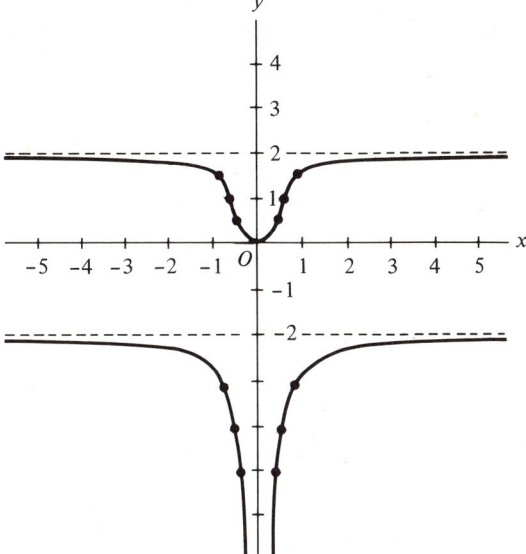

y	x
0	0
$\frac{1}{2}$	$\pm\sqrt{\frac{2}{15}} \approx \pm 0.4$
1	$\pm\frac{\sqrt{3}}{3} \approx \pm 0.6$
$\frac{3}{2}$	$\pm\sqrt{\frac{6}{7}} \approx \pm 0.9$
-3	$\pm\sqrt{\frac{3}{5}} \approx \pm 0.8$
-4	$\pm\frac{\sqrt{3}}{3} \approx \pm 0.6$
-5	$\pm\sqrt{\frac{5}{21}} \approx \pm 0.5$

Figure 2-15

SOLUTION
(a) The graph is symmetric with respect to the y-axis.
(b) $x = 0$ iff $y = 0$, so $(0, 0)$ is the only intercept.
(c) $x^2 = y/(4 - y^2)$. Since $x^2 \geq 0$, either $y = 0$, or $4 - y^2 > 0$ and $y > 0$, or $4 - y^2 < 0$ and $y < 0$. Thus, the extent with respect to y is $0 \leq y < 2$ or $y < -2$. Solving for y by the quadratic formula, we have, if $x \neq 0$,

$$y = \frac{-1 \pm \sqrt{1 + 16x^4}}{2x^2}$$

which is real for all $x \neq 0$. Since $(0, 0)$ is on the graph, there is no limitation on the x-extent.

(d) From $x^2 = y/(4 - y^2)$ we see that the lines $y = \pm 2$ are horizontal asymptotes. Since the degree of the numerator is less than that of the denominator, it follows that the line $x = 0$ is a vertical asymptote. The graph is shown in Figure 2-15.

EXERCISE 4

Sketch the graphs of the following relations, giving as much information as possible.

1. (a) $2x^2 + 3y^2 = 6$
   (b) $x^2 - y^2 = 4$
   (c) $y^2 = x - 3$
   (d) $y^2 = (x - 1)(x - 2)$
2. (a) $2(x^2 - 1)y^2 = 0$
   (b) $x^3 - xy^2 = 0$
   (c) $(x^2 + y^2 - 4)(y - x) = 0$

3. (a) $x^2y = 1$
   (b) $xy^2 = 1$
   (c) $(x-2)(x+3)y = 1$

4. (a) $y^2 = \dfrac{1}{x-3}$
   (b) $y = \dfrac{1}{x-3}$
   (c) $y = \dfrac{1}{(x-3)^2}$
   (d) $y = \dfrac{3x}{5x-7}$

5. (a) $2(x^2-1)y^2 = 1$
   (b) $(x^2-4)y = 4$

6. (a) $y^3 = \dfrac{3x^2}{x^2-5x+6}$
   (b) $y^4(x^2-1) = 1$

7. (a) $x = \dfrac{y^3}{2y^3-y^2}$
   (b) $x^2 = \dfrac{y}{(y-3)(y+4)}$
   (c) $x^3 + 3xy^2 = 1$

5. CURVES; PARAMETRIC EQUATIONS

Definition. A (planar) *curve C* is a function whose domain is some interval I (possibly infinite) of the real numbers and whose range is a subset of R^2.

If F is such a function with independent variable t (called a *parameter*), we may write $F(t) = (x, y)$. Since x, y depend on t, we write $x = f(t)$, $y = g(t)$, so that $F(t) = (f(t), g(t))$. The equations $x = f(t)$, $y = g(t)$, $t \in I$, are called the *parametric equations* of the curve. The collection of points comprising the range is called the *graph* of the curve. We write

$$C: \{x = f(t), y = g(t), t \in I\}$$

If $I = (-\infty, \infty) = R$, we sometimes omit the phrase "$t \in I$."

Note that if $y = f(x)$ is a function whose domain is an interval I, we may let $x = t$ and write $x = t$, $y = f(t)$, $t \in I$. Thus a function may be regarded as a curve.

We use the word "curve" for either the pair of parametric equations or the set of points in the range (that is, the graph) of the curve, the meaning usually being clear from the context. If we wish to emphasize the distinction, we shall use the symbol C for the collection of points and refer to $x = f(t)$, $y = g(t)$, $t \in I$ as a *parametrization of C*. C may have more than one parametrization, as shown by the following example.

Example 1. Consider the curve

$$C: \begin{cases} x = 2t + 1 \\ y = t^2, \end{cases} \quad 0 \le t \le 1$$

We may introduce another parameter s by letting $t = 2 - s$. We then obtain

$$C: \begin{cases} x = 5 - 2s \\ y = (2 - s)^2, \end{cases} \quad 1 \le s \le 2$$

(Observe that if $0 \le t \le 1$, then $0 \le 2 - s \le 1$, and so $1 \le s \le 2$.) Clearly, corresponding values of s and t yield the same point (x, y) on C.

Definition. If $C: \{x = f(t), y = g(t), t \in I\}$ is a curve and if $\varphi(x, y) = 0$ is a relation for which $\varphi(f(t), g(t)) = 0$ for all $t \in I$ (that is, the graph of $\varphi(x, y) = 0$ includes the graph of C), then $\varphi(x, y) = 0$ is called an *eliminant of the equations* $x = f(t)$, $y = g(t)$.

If in addition to the above, we have the condition that for each (x, y) for which $\varphi(x, y) = 0$, there exists $t \in I$ such that $x = f(t)$ and $y = g(t)$, the parametric equations for C are said to be a *parametrization of* $\varphi(x, y) = 0$. The word "curve" may refer to the relation $\varphi(x, y) = 0$, its graph, or the parametric equations $x = f(t)$, $y = g(t)$, $t \in I$.

Clearly, an eliminant of a pair of parametric equations is not unique. The reason for computing an eliminant is that in some cases, it is easier to sketch its graph than to plot the curve directly. We must then determine that portion of the graph of the eliminant which comprises the curve C. In other cases, however, it may be difficult to eliminate the parameter, or the eliminant may be too complicated to analyse. In such cases, one simply constructs a table of corresponding values of t, x, and y, plots the points (x, y), and then connects them by some sort of "plausible" drawing. The following examples illustrate the process of *eliminating the parameter*.

Example 2. $x = t^2 - 2$, $y = \cos t$, $0 \le t \le 1$. We may solve for t in terms of x to obtain $t = \sqrt{x + 2}$ since $t \ge 0$. Accordingly, an eliminant is $y = \cos \sqrt{x + 2}$.

Example 3. $x = 3 \cos t$, $y = 4 \sin t$. We may use the identity $\sin^2 t + \cos^2 t = 1$, noting that

$$\sin t = \frac{y}{4}, \quad \cos t = \frac{x}{3}.$$

Thus,

$$\frac{y^2}{4^2} + \frac{x^2}{3^2} = 1 \text{ is an eliminant}$$

Example 4. $x = t^2, y = 3t^2 + 2$. Substituting x for t^2 in the equation for y, we obtain the eliminant $y = 3x + 2$. Note that the domain of the eliminant is R, while the values of x for points on the given curve are all nonnegative. In this case, the graph of the eliminant contains more than the given curve.

In some cases, a geometric description of a relation may lead to a parametric representation.

Example 5. A *cycloid* is a curve usually described in the following way: A wheel "rolls along the x-axis without slipping." A cycloid is the curve traced out by a fixed point on its rim. If the fixed point P starts at the origin and the circle of radius r has "rolled" through angle φ, the angle between a vertical line through the center and the line joining the center to the point is φ (Figure 2–16).

If $P = (x, y)$ and the center is (h, k) (note that $k = r$), the statement "the circle has rolled without slipping" means that h equals the arc from $(h, 0)$ to (x, y), that is, $h = r\varphi$.

The point of tangency B must be $(h, 0)$. Since φ is the angle PAB, we have $h = r\varphi + 2n\pi r$. (See Figure 2–16.) From the figure, we also have $x + r \sin \varphi = h$. (Note that this formula is correct even when $x > h$.) Hence

$$x + r \sin \varphi = r(\varphi + 2n\pi)$$

or

$$x = r(\varphi + 2n\pi) - r \sin \varphi$$

Also from the figure,

$$y + r \cos \varphi = r$$

or

$$y = r - r \cos \varphi$$

(which holds even if $y > r$). Since

$$\sin \varphi = \sin (\varphi + 2n\pi), \qquad \cos \varphi = \cos (\varphi + 2n\pi)$$

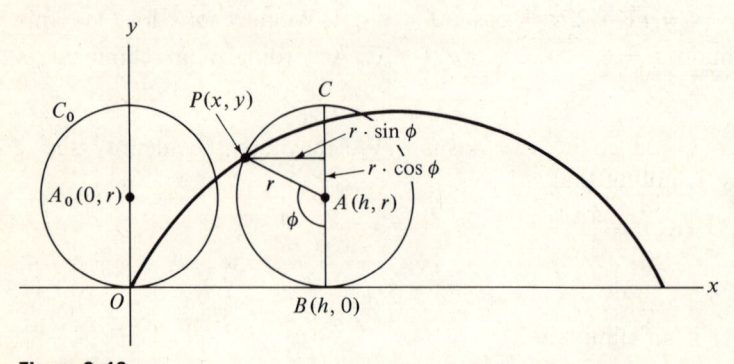

Figure 2–16

it is convenient to let $\theta = \varphi + 2n\pi$ and write

$$\begin{cases} x = r\theta - r\sin\theta = r(\theta - \sin\theta) \\ y = r - r\cos\theta = r(1 - \cos\theta) \end{cases}$$

These, then, are the parametric equations of the cycloid. (See Problem 4.)

EXERCISE 5

1. Eliminate the parameter and sketch the curve. (Unless otherwise stated, the interval is $(-\infty, \infty)$.)

   (a) $\begin{cases} x = t - 2 \\ y = 6t^2 + 7 \end{cases}$

   (b) $\begin{cases} x = 2 - t^2 \\ y = 2 - 3t \end{cases}$

   (c) $\begin{cases} x = 5\cos\theta \\ y = 6\sin\theta + 3 \end{cases}$

   (d) $\begin{cases} x = \csc\theta + 1 \\ y = \cot\theta - 7 \end{cases}$

   (e) $\begin{cases} x = \dfrac{t}{1+t^2} \\ y = \dfrac{-t}{1+t^2} \end{cases} \quad -1 \le t \le 1$

   (f) $\begin{cases} x = t^2 \\ y = t^2 \end{cases}$

   (g) $\begin{cases} x = \cos t \\ y = 3\sin t \end{cases} \quad 0 \le t \le \dfrac{\pi}{2}$

   (h) $\begin{cases} x = t^2 \\ y = t \end{cases} \quad 0 \le t$

   (i) $\begin{cases} x = |t| \\ y = t^2 \end{cases} \quad -1 \le t \le 1$

   (j) $\begin{cases} x = a\sec\theta \\ y = b\tan\theta \end{cases} \quad a, b > 0$

*2. A curve $C: \{x = f(t), y = g(t), t \in [a, b]\}$ is said to be *simple* with respect to the given parametrization if the function $F(t) = (f(t), g(t))$ is one to one on (a, b); that is, if t_1, t_2 are such that $a < t_1 < t_2 < b$, then $F(t_1) \ne F(t_2)$ (that is, $f(t_1) \ne f(t_2)$ or $g(t_1) \ne g(t_2)$). Which of the following are simple curves with respect to the given parametrizations? Plot each curve. A curve (as a point set) may be simple in one parametrization, but not in another. (See Parts (a) and (b) or Parts (c) and (d).)

   (a) $\begin{cases} x = 2 - t \\ y = 1 + t \end{cases} \quad 0 \le t \le 1$

   (b) $\begin{cases} x = 2 - t^2 \\ y = 1 + t^2 \end{cases} \quad -1 \le t \le 1$

(c) $\begin{cases} x = \sin t \\ y = \cos t \end{cases}$ $0 \le t \le 2\pi$

(d) $\begin{cases} x = \sin 2t \\ y = \cos 2t \end{cases}$ $0 \le t \le 2\pi$

(e) $\begin{cases} x = \begin{cases} 0 & 0 \le t \le 1 \\ t-1 & 1 \le t \le 2 \\ 1 & 2 \le t \le 3 \\ 4-t & 3 \le t \le 4 \end{cases} \\ y = \begin{cases} t & 0 \le t \le 1 \\ 1 & 1 \le t \le 2 \\ 3-t & 2 \le t \le 3 \\ 0 & 3 \le t \le 4 \end{cases} \end{cases}$

*3. A simple curve $C: \{x = f(t), y = g(t), t \in [a, b]\}$ is called *closed* with respect to the given parametrization if $f(a) = f(b)$ and $g(a) = g(b)$. Which of the following are closed?

(a) Problem 2(a)
(b) Problem 2(c)
(c) $\begin{cases} x = \sin t \\ y = \cos t \end{cases}$ $(0 \le t \le 3\pi)$
(d) $\begin{cases} x = t^2 \\ y = 1 - t^2 \end{cases}$ $(-2 \le t \le 3)$
(e) Problem 2(e)

4. (See Example 5.)
   (a) Compute points on the cycloid corresponding to $\theta = 0, \frac{1}{2}\pi, \pi, \frac{3}{2}\pi, 2\pi$.
   (b) Eliminate the parameter from the equations for the cycloid.

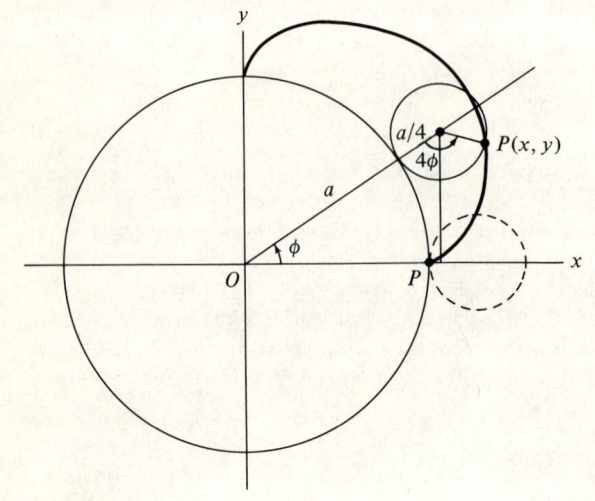

Figure 2-17

5. Find an eliminant for the following equations, and sketch.
$$\begin{cases} x = \dfrac{t}{1+t^3} \\ y = \dfrac{t^2}{1+t^3} \end{cases}$$

*6. A circle of radius $a/4$ rolls on the perimeter outside of a fixed circle of radius a. Find parametric equations for the curve traced by a fixed point on the perimeter of the smaller circle. (Hint: Use as parameter the angle φ shown in Figure 2–17.)

*7. Do Problem 6 if the smaller circle rolls *inside* the larger one (Figure 2–18).

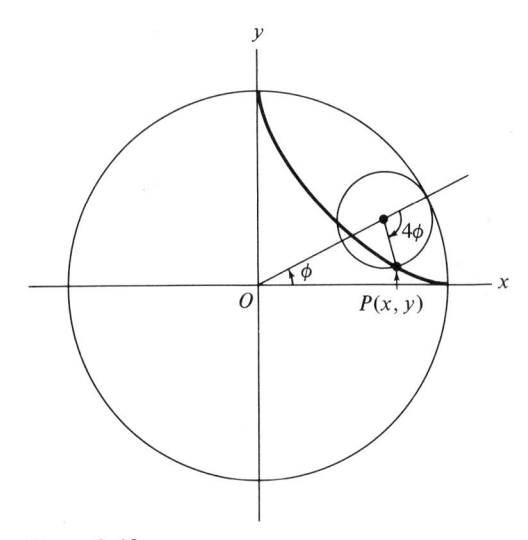

Figure 2-18

chapter 3
Vectors and Line Segments in the Plane

We begin by defining an *arrow* as an ordered pair of points. We then develop the concepts of *length* and *direction* of an arrow, which lead in a natural way to the concept of a *vector*. The definition of *line segment* is then used to define such familiar geometric figures as triangle, parallelogram, and so on. From these definitions, all the theorems of plane Euclidean geometry can be developed. In Section 4, we illustrate how this can be done.

1. ARROWS

In the following $P_1(x_1, y_1)$ and $P_2(x_2, y_2)$ denote equal or distinct points in the plane.

Definition. The *arrow* with *initial point* P_1 and *terminal point* P_2 is the ordered pair $(P_1, P_2) \in R^2 \times R^2$ and is denoted by $\overrightarrow{P_1 P_2}$. We call $x_2 - x_1$ the *x-component* and $y_2 - y_1$ the *y-component* of $\overrightarrow{P_1 P_2}$ (note the *order* of subtraction—coordinate of *terminal* point minus corresponding coordinate of *initial* point).

We represent the arrow $\overrightarrow{P_1 P_2}$ as in Figure 3–1 with the arrowhead placed

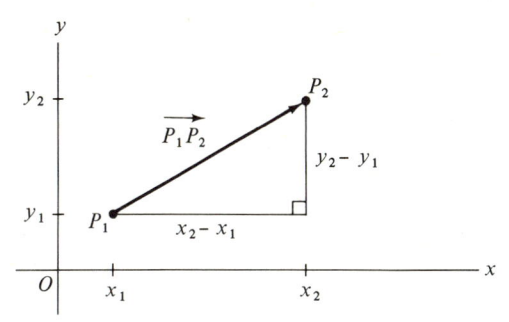

Figure 3-1

at the terminal point P_2. If $P_1 = P_2$, $\overrightarrow{P_1P_2}$ cannot be so represented.

Definition. The *length (magnitude)* of the arrow $\overrightarrow{P_1P_2}$ is the number
$$|\overrightarrow{P_1P_2}| = \sqrt{(x_2 - x_1)^2 + (y_2 - y_1)^2}$$
This number is also called the *distance between the points* P_1 and P_2.

Observe that $|\overrightarrow{P_1P_2}| \geq 0$ and $|\overrightarrow{P_1P_2}| = 0$ iff $P_1 = P_2$. We also have $|\overrightarrow{P_1P_2}| = |\overrightarrow{P_2P_1}|$.

Example 1. $P_1 = (-2, -3)$ and $P_2 = (4, 2)$. The x-component of $\overrightarrow{P_1P_2}$ is $4 - (-2) = 6$. The y-component of $\overrightarrow{P_1P_2}$ is $2 - (-3) = 5$.
$$|\overrightarrow{P_1P_2}| = \sqrt{6^2 + 5^2} = \sqrt{61}$$

Definition. The *scalar product* of $\overrightarrow{P_1P_2}$ by the number a (called a *scalar*), denoted by $a \cdot \overrightarrow{P_1P_2}$ (or simply $a\overrightarrow{P_1P_2}$), is the arrow $\overrightarrow{P_1Q}$ with initial point P_1 and having x-component $a(x_2 - x_1)$ and y-component $a(y_2 - y_1)$. Thus, if $Q = (x, y)$, we have
$$x - x_1 = a(x_2 - x_1) \quad \text{and} \quad y - y_1 = a(y_2 - y_1)$$
so that $x = x_1 + a(x_2 - x_1)$ and $y = y_1 + a(y_2 - y_1)$.

Note that $a\overrightarrow{P_1P_2}$ and $\overrightarrow{P_1P_2}$ have the same initial point. If $a > 0$, we draw $a\overrightarrow{P_1P_2}$ in the *same direction* as $\overrightarrow{P_1P_2}$ (Figure 3-2(a)). If $a < 0$, we draw $a\overrightarrow{P_1P_2}$ in the direction *opposite* to that of $\overrightarrow{P_1P_2}$ (Figure 3-2(b)). The length of $a\overrightarrow{P_1P_2}$ is
$$|a\overrightarrow{P_1P_2}| = \sqrt{[a(x_2 - x_1)]^2 + [a(y_2 - y_1)]^2}$$
$$= |a||\overrightarrow{P_1P_2}|$$

Direction of an Arrow. Let P_1, P_2 be *distinct* points, so that $|\overrightarrow{P_1P_2}| > 0$. Then
$$-1 \leq \frac{x_2 - x_1}{|\overrightarrow{P_1P_2}|} \leq 1, \quad -1 \leq \frac{y_2 - y_1}{|\overrightarrow{P_1P_2}|} \leq 1$$

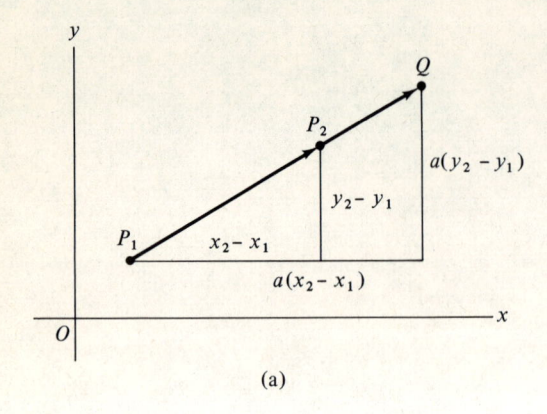

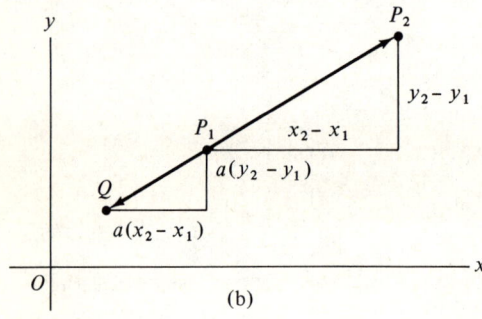

Figure 3-2 (a) $a\overrightarrow{P_1P_2} = \overrightarrow{P_1Q}$ $(a > 0)$. (b) $a\overrightarrow{P_1P_2} = \overrightarrow{P_1Q}$ $(a < 0)$.

and
$$\left(\frac{x_2 - x_1}{|\overrightarrow{P_1P_2}|}\right)^2 + \left(\frac{y_2 - y_1}{|\overrightarrow{P_1P_2}|}\right)^2 = 1$$
Accordingly, there exists a *unique* angle α, $0 \leq \alpha < 2\pi$, such that
$$\cos \alpha = \frac{x_2 - x_1}{|\overrightarrow{P_1P_2}|} \quad \text{and} \quad \sin \alpha = \frac{y_2 - y_1}{|\overrightarrow{P_1P_2}|}$$
α is called the *direction* of the arrow $\overrightarrow{P_1P_2}$ (Figure 3-3). If $\alpha = 0$ or $\alpha = \pi$

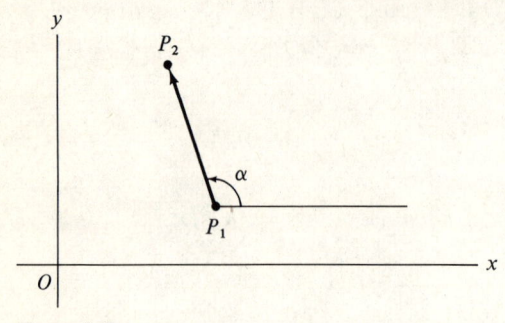

Figure 3-3

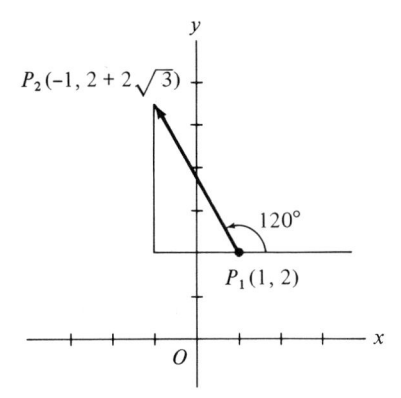

Figure 3-4

(that is, $y_1 = y_2$), we say $\overrightarrow{P_1P_2}$ is *horizontal*. If $\alpha = \frac{1}{2}\pi$ or $\frac{3}{2}\pi$ (that is, $x_1 = x_2$), we say $\overrightarrow{P_1P_2}$ is *vertical*.

If $\overrightarrow{P_1P_2}$ is *not vertical*, we define the *slope* of $\overrightarrow{P_1P_2}$ to be the number

$$m_{\overrightarrow{P_1P_2}} = \frac{y_2 - y_1}{x_2 - x_1} = \tan \alpha$$

If $\overrightarrow{P_1P_2}$ is vertical, we say the slope of $\overrightarrow{P_1P_2}$ is *undefined* or that $\overrightarrow{P_1P_2}$ has *infinite slope*.

Example 2. Let $P_1 = (1, 2)$ and $P_2 = (-1, 2 + 2\sqrt{3})$. Find (a) $|\overrightarrow{P_1P_2}|$, (b) $|(-3)\overrightarrow{P_1P_2}|$, (c) the direction of $\overrightarrow{P_1P_2}$, and (d) the slope of $\overrightarrow{P_1P_2}$ (Figure 3-4).

SOLUTION

(a) $|\overrightarrow{P_1P_2}| = \sqrt{(-2)^2 + (2\sqrt{3})^2} = 4$

(b) $|(-3)\overrightarrow{P_1P_2}| = |-3|\cdot|\overrightarrow{P_1P_2}| = 12$

(c) $\cos \alpha = \frac{-2}{4} = -\frac{1}{2}$; $\sin \alpha = \frac{2\sqrt{3}}{4} = \frac{\sqrt{3}}{2}$. Hence $\alpha = \frac{2}{3}\pi = 120°$

(d) $m_{\overrightarrow{P_1P_2}} = \frac{2\sqrt{3}}{-2} = -\sqrt{3} = \tan \frac{2}{3}\pi$

EXERCISE 1

1. Find the length, direction and slope of the arrow $\overrightarrow{P_1P_2}$.
   (a) $P_1(3, -1)$, $P_2(3 + \sqrt{2}, -1 - \sqrt{2})$
   (b) $P_1(-1, 2)$, $P_2(-1 - 3\sqrt{3}, -1)$
   (c) $P_1(-1, -3)$, $P_2(-5, 1)$
   (d) $P_1(-5, -3)$, $P_2(-1, -2)$
   (e) $P_1(-2, 5)$, $P_2(4, 5)$

2. Let $P_1 = (2, 3)$ and $P_2 = (6, 3 + 4\sqrt{3})$. Find the magnitude and direction of (a) $\overrightarrow{P_1P_2}$, (b) $4\overrightarrow{P_1P_2}$, and (c) $(-4)\overrightarrow{P_1P_2}$.
3. Find the terminal point of the arrow having the given initial point P_1, direction α, and magnitude $|\overrightarrow{P_1P_2}|$.

   (a) $P_1(0, 0), \alpha = 120°, |\overrightarrow{P_1P_2}| = 10$
   (b) $P_1(4, 5), \alpha = 300°, |\overrightarrow{P_1P_2}| = 9$
   (c) $P_1(-1, -4), \alpha = 225°, |\overrightarrow{P_1P_2}| = 5$
   (d) $P_1(-3, 2), \alpha = \frac{7}{6}\pi, |\overrightarrow{P_1P_2}| = 1$

4. (a) Find the terminal point of the arrow with initial point $P_1(3, 2)$, x-component 4, and y-component -6.
   (b) Find the initial point of the arrow with terminal point $P_2(-6, 4)$, x-component -3, and y-component 10.

2. VECTORS

From the definitions given in Section 1, we see that the *length* and *direction* of an arrow $\overrightarrow{P_1P_2}$ are completely determined by the *components* of the arrow. The initial and terminal points *per se* do not play a role in the calculation of these quantities. Accordingly, *any two arrows with the same pair of components have the same length and direction.* Conversely, two arrows with the same length and direction have the same pair of components. In many problems in mathematics, physics, and engineering, one is interested only in the length and direction of an arrow, and so it is convenient to regard as indistinguishable any two arrows having the same pair of components. This is the motivation for the concept of *equivalent arrows.*

Definition. Let $P_1(x_1, y_1), P_2(x_2, y_2), Q_1(u_1, v_1), Q_2(u_2, v_2)$ be (not necessarily distinct) points in the plane. $\overrightarrow{P_1P_2}$ is *equivalent* to $\overrightarrow{Q_1Q_2}$ (written $\overrightarrow{P_1P_2} \sim \overrightarrow{Q_1Q_2}$) iff $x_2 - x_1 = u_2 - u_1$ and $y_2 - y_1 = v_2 - v_1$.

Example 1. Let $P_1 = (-2, -1)$, $P_2 = (2, 4)$, and $Q_1 = (1, -2)$. Find $Q_2 = (u_2, v_2)$ such that $\overrightarrow{P_1P_2} \sim \overrightarrow{Q_1Q_2}$.

SOLUTION. We have $u_2 - 1 = 2 - (-2)$ and $v_2 - (-2) = 4 - (-1)$. Hence $u_2 = 5$ and $v_2 = 3$. So $Q_2 = (5, 3)$ (Figure 3-5).

It is not difficult to show that equivalence of arrows satisfies the following laws. (These are the properties which characterize an *equivalence relation.*)

(1) $\overrightarrow{P_1P_2} \sim \overrightarrow{P_1P_2}$ (reflexive law)
(2) $\overrightarrow{P_1P_2} \sim \overrightarrow{Q_1Q_2}$ implies $\overrightarrow{Q_1Q_2} \sim \overrightarrow{P_1P_2}$ (symmetric law)
(3) $\overrightarrow{P_1P_2} \sim \overrightarrow{Q_1Q_2}$ and $\overrightarrow{Q_1Q_2} \sim \overrightarrow{R_1R_2}$ implies $\overrightarrow{P_1P_2} \sim \overrightarrow{R_1R_2}$
    (transitive law)

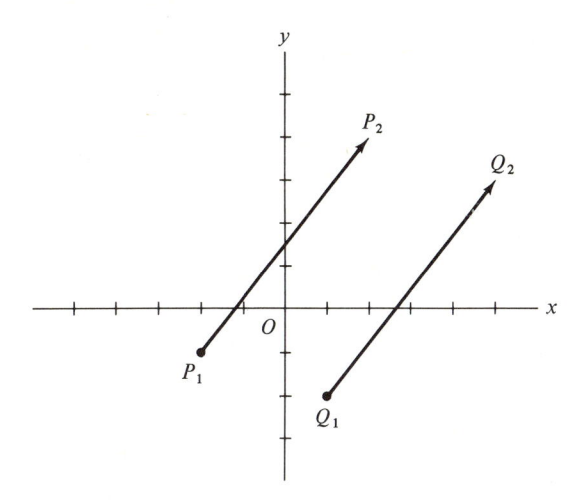

Figure 3-5

Observe that $\overrightarrow{P_1P_2} = \overrightarrow{Q_1Q_2}$ implies $\overrightarrow{P_1P_2} \sim \overrightarrow{Q_1Q_2}$, but not conversely.

We now regard the set of all arrows in the plane as being partitioned into classes, *each class consisting of those, and only those, arrows which are equivalent to one another.* Figure 3-6 shows some arrows belonging to one class. *Any two* (distinct) *classes are disjoint, and every arrow belongs to a class.* Since arrows belong to the same class iff they have the same ordered pair of components, we may denote the *entire class* by the symbol $\mathbf{v} = (a, b)$, where a is the x-component and b the y-component of every arrow belonging to the class $\mathbf{v}$. This suggests the following definition.

Definition. A *plane vector* $\mathbf{v} = (a, b)$ is the class of *all* arrows having x-component a and y-component b. a is called the x-component and b the y-component of $\mathbf{v}$.

Clearly, if $\mathbf{v}_1 = (a_1, b_1)$ and $\mathbf{v}_2 = (a_2, b_2)$, then $\mathbf{v}_1 = \mathbf{v}_2$ iff $a_1 = a_2$ and

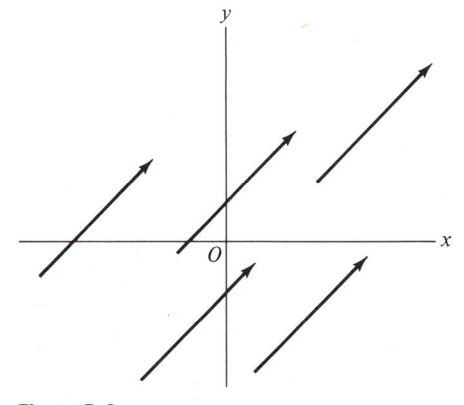

Figure 3-6

$b_1 = b_2$. The *sum* of $\mathbf{v}_1$ and $\mathbf{v}_2$ (in that order) is defined to be the vector

$$\mathbf{v}_1 + \mathbf{v}_2 = (a_1 + a_2, b_1 + b_2)$$

We shall *represent* a vector $\mathbf{v}$ by any arrow belonging to $\mathbf{v}$. Thus, to *plot* a vector $\mathbf{v}$ will mean to plot any arrow belonging to $\mathbf{v}$. Such an arrow is called a *representative* of $\mathbf{v}$, and $\mathbf{v}$ will be called the vector *associated with* the arrow. Since it is convenient at times to identify the vector $\mathbf{v}$ with a representative arrow $\overrightarrow{P_1P_2}$, we often write $\overrightarrow{P_1P_2} = \mathbf{v}$. This simply means that the arrow $\overrightarrow{P_1P_2}$ has the same components as $\mathbf{v}$. We may refer to P_1, P_2 as the initial point and terminal point, respectively, of $\mathbf{v}$. We depict the sum of $\mathbf{v}_1$ and $\mathbf{v}_2$ as follows. Let $\overrightarrow{P_1P_2} = \mathbf{v}_1$ and $\overrightarrow{P_2P_3} = \mathbf{v}_2$. Then $\overrightarrow{P_1P_3} = \mathbf{v}_1 + \mathbf{v}_2$; that is, $\overrightarrow{P_1P_3}$ is a representative of $\mathbf{v}_1 + \mathbf{v}_2$ (Figure 3–7).

To see this, let $P_1 = (x_1, y_1)$, $P_2 = (x_2, y_2)$, $P_3 = (x_3, y_3)$, $\mathbf{v}_1 = (a_1, b_1)$, and $\mathbf{v}_2 = (a_2, b_2)$. Then $x_2 - x_1 = a_1$, $y_2 - y_1 = b_1$, $x_3 - x_2 = a_2$, and $y_3 - y_2 = b_2$. We have

$$x_3 - x_1 = (x_3 - x_2) + (x_2 - x_1) = a_2 + a_1 = a_1 + a_2$$

and

$$y_3 - y_1 = (y_3 - y_2) + (y_2 - y_1) = b_2 + b_1 = b_1 + b_2$$

Thus, $\overrightarrow{P_1P_3} = \mathbf{v}_1 + \mathbf{v}_2$.

If c is a scalar and $\mathbf{v} = (a, b)$, we define the *scalar product* of $\mathbf{v}$ by c to be the vector

$$c\mathbf{v} = (ca, cb)$$

If $\overrightarrow{P_1P_2} = \mathbf{v}$, we shall represent $c\mathbf{v}$ by $c\overrightarrow{P_1P_2}$ (see Section 1). The *zero vector* is the vector $\mathbf{0} = (0, 0)$. The *negative* of $\mathbf{v}$ is the vector $-\mathbf{v} = (-a, -b)$. Clearly, $(-1)\mathbf{v} = -\mathbf{v}$, and $-\mathbf{0} = \mathbf{0}$. The *length (magnitude)* of $\mathbf{v} = (a, b)$ is defined by

$$|\mathbf{v}| = \sqrt{a^2 + b^2}$$

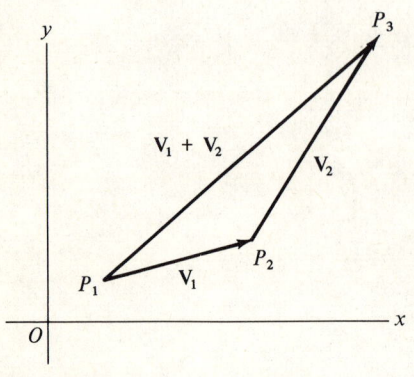

Figure 3-7

It follows that

$$|c\mathbf{v}| = |c| \cdot |\mathbf{v}|$$

for all c and $\mathbf{v}$. We collect the basic properties of vectors in the following theorem.

Theorem 1. *For all vectors $\mathbf{v}, \mathbf{v}_1, \mathbf{v}_2, \mathbf{v}_3$, and scalars c, c_1, c_2, the following laws hold:*

(1) (a) $(\mathbf{v}_1 + \mathbf{v}_2) + \mathbf{v}_3 = \mathbf{v}_1 + (\mathbf{v}_2 + \mathbf{v}_3)$ *(associative law)*
  (b) $\mathbf{v}_1 + \mathbf{v}_2 = \mathbf{v}_2 + \mathbf{v}_1$ *(commutative law)*
  (c) $\mathbf{v} + \mathbf{0} = \mathbf{v}$ *(identity law)*
  (d) $\mathbf{v} + (-\mathbf{v}) = \mathbf{0}$ *(inverse law)*

(2) (a) $c_1(c_2 \mathbf{v}) = (c_1 c_2) \mathbf{v}$
  (b) $c(\mathbf{v}_1 + \mathbf{v}_2) = c\mathbf{v}_1 + c\mathbf{v}_2$
  (c) $(c_1 + c_2)\mathbf{v} = c_1 \mathbf{v} + c_2 \mathbf{v}$
  (d) $1\mathbf{v} = \mathbf{v}$

PROOF. We prove 1(a) and 2(b). The remaining parts are proved in a similar manner. (See Problem 6.)

Let $\mathbf{v}_1 = (a_1, b_1), \mathbf{v}_2 = (a_2, b_2), \mathbf{v}_3 = (a_3, b_3)$. We have

(1a) $(\mathbf{v}_1 + \mathbf{v}_2) + \mathbf{v}_3 = (a_1 + a_2, b_1 + b_2) + (a_3, b_3)$
$= ((a_1 + a_2) + a_3, (b_1 + b_2) + b_3)$
$= (a_1 + (a_2 + a_3), b_1 + (b_2 + b_3))$ (by the associative law of addition for real numbers)
$= (a_1, b_1) + (a_2 + a_3, b_2 + b_3)$
$= (a_1, b_1) + ((a_2, b_2) + (a_3, b_3))$
$= \mathbf{v}_1 + (\mathbf{v}_2 + \mathbf{v}_3)$

(2b) $c(\mathbf{v}_1 + \mathbf{v}_2) = c(a_1 + a_2, b_1 + b_2)$
$= (c(a_1 + a_2), c(b_1 + b_2))$
$= (ca_1 + ca_2, cb_1 + cb_2)$ (by the distributive law for real numbers)
$= (ca_1, cb_1) + (ca_2, cb_2)$
$= c(a_1, b_1) + c(a_2, b_2)$
$= c\mathbf{v}_1 + c\mathbf{v}_2$

The difference of two vectors $\mathbf{v}_1, \mathbf{v}_2$ is defined by

$$\mathbf{v}_1 - \mathbf{v}_2 = \mathbf{v}_1 + (-\mathbf{v}_2)$$

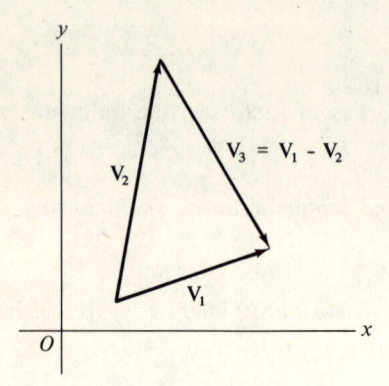

Figure 3-8

Thus, $(a_1, b_1) - (a_2, b_2) = (a_1 - a_2, b_1 - b_2)$. Note that $\mathbf{v}_1 - \mathbf{v}_2 = \mathbf{v}_3$ iff $\mathbf{v}_1 = \mathbf{v}_2 + \mathbf{v}_3$. It follows that if $\mathbf{v}_1, \mathbf{v}_2$ are represented by arrows with the same initial point, then $\mathbf{v}_1 - \mathbf{v}_2$ is that vector represented by *the arrow whose terminal point is that of* $\mathbf{v}_1$ *and whose initial point is the terminal point of* $\mathbf{v}_2$ (Figure 3-8). Some of the immediate consequences of the definitions are contained in the next theorem.

Theorem 2. For all vectors $\mathbf{v}, \mathbf{v}_1, \mathbf{v}_2$, and all scalars c, the following hold:
(1) (a) $0\mathbf{v} = \mathbf{0}$
    (b) $c\mathbf{0} = \mathbf{0}$
    (c) $c\mathbf{v} = \mathbf{0}$ iff $c = 0$ or $\mathbf{v} = \mathbf{0}$
    (d) $(-1)\mathbf{v} = -\mathbf{v}$
    (e) $\mathbf{v}_1 - \mathbf{v}_2 = \mathbf{v}_3$ iff $\mathbf{v}_1 = \mathbf{v}_2 + \mathbf{v}_3$
(2) (a) $|\mathbf{v}_1 - \mathbf{v}_2| = |\mathbf{v}_2 - \mathbf{v}_1|$
    (b) $|\mathbf{v}| = 0$ iff $\mathbf{v} = \mathbf{0}$

PROOF. (See Problem 7.)

Example 2. Given $\mathbf{v}_1 + \mathbf{v}_2 = (2, -3)$ and $(\mathbf{v}_1 - \mathbf{v}_2)/3 = (-1, 4)$. Find $\mathbf{v}_1$ and $\mathbf{v}_2$.

SOLUTION. We have

$\mathbf{v}_1 + \mathbf{v}_2 = (2, -3)$
$\mathbf{v}_1 - \mathbf{v}_2 = (-3, 12)$

Adding the equations, we obtain $2\mathbf{v}_1 = (-1, 9)$. Thus, $\mathbf{v}_1 = (-\frac{1}{2}, \frac{9}{2})$. Subtracting the equations yields $2\mathbf{v}_2 = (5, -15)$, so that $\mathbf{v}_2 = (\frac{5}{2}, -\frac{15}{2})$. (The student should verify that $\mathbf{v}_1, \mathbf{v}_2$ satisfy the given equations.)

EXERCISE 2

1. Let $P_1 = (2, -3)$, $P_2 = (4, 2)$, and $Q_1 = (-1, 2)$.
   (a) Find $Q_2(x, y)$ such that $\overrightarrow{P_1P_2} \sim \overrightarrow{Q_1Q_2}$.
   (b) Find $Q_2(x, y)$ such that $\overrightarrow{P_1P_2} \sim \overrightarrow{Q_2Q_1}$.
2. Determine whether the arrows $\overrightarrow{P_1P_2}$ and $\overrightarrow{Q_1Q_2}$ are equivalent.
   (a) $P_1(1, 4), P_2(6, -1);\quad Q_1(0, 3), Q_2(5, -2)$
   (b) $P_1(0, -10), P_2(-1, 4);\quad Q_1(2, 3), Q_2(1, 6)$
   (c) $P_1(-3, -4), P_2(-1, 3);\quad Q_1(0, 5), Q_2(2, 12)$
3. Given $\mathbf{v}_1 = (2, -3)$ and $\mathbf{v}_2 = (6, 1)$.
   (a) Plot $\mathbf{v}_1, \mathbf{v}_2$ with initial point at the origin.
   (b) Find and plot $\mathbf{v}_1 + \mathbf{v}_2, \mathbf{v}_1 - \mathbf{v}_2$, and $\mathbf{v}_2 - \mathbf{v}_1$.
   (c) Find $|\mathbf{v}_1|, |\mathbf{v}_2|, |\mathbf{v}_1 + \mathbf{v}_2|$, and $|\mathbf{v}_1 - \mathbf{v}_2|$.
4. (a) Given $\mathbf{v}_1 - \mathbf{v}_2 = (4, -6)$ and $\mathbf{v}_1 = (-2, -1)$.
       Find $\mathbf{v}_2$.
   (b) Given $\mathbf{v}_1 + \mathbf{v}_2 = (3, -2)$ and $\mathbf{v}_2 = (-4, 3)$.
       Find $\mathbf{v}_1$.
   (c) Given $(\mathbf{v}_1 - \mathbf{v}_2)/3 = (2, 4)$ and $\mathbf{v}_2/2 = (-1, 3)$.
       Find $2\mathbf{v}_1/3$.
   (d) Given $\mathbf{v}_1 + \mathbf{v}_2 = (-3, 5)$ and $\mathbf{v}_1 - \mathbf{v}_2 = (2, 4)$.
       Find $\mathbf{v}_1$ and $\mathbf{v}_2$.
   (e) Given $2\mathbf{v}_1 - \mathbf{v}_2 = (1, -2)$ and $\mathbf{v}_1 + 3\mathbf{v}_2 = (-2, -4)$.
       Find $\mathbf{v}_1$ and $\mathbf{v}_2$.
5. Prove $|\mathbf{v}_1 + \mathbf{v}_2|^2 + |\mathbf{v}_1 - \mathbf{v}_2|^2 = 2|\mathbf{v}_1|^2 + 2|\mathbf{v}_2|^2$.
6. Prove Theorem 1.
7. Prove Theorem 2.

3. THE ANGLE BETWEEN TWO VECTORS; DOT PRODUCT

If $\mathbf{v} = (a, b)$ is a nonzero vector, we define the *direction* of $\mathbf{v}$ to be the angle α, determined by the following conditions:

$$\cos \alpha = \frac{a}{|\mathbf{v}|}, \quad \sin \alpha = \frac{b}{|\mathbf{v}|} \quad (0 \leq \alpha < 2\pi)$$

Observe that α is well defined (Figure 3-9), since

$$\left(\frac{a}{|\mathbf{v}|}\right)^2 + \left(\frac{b}{|\mathbf{v}|}\right)^2 = 1$$

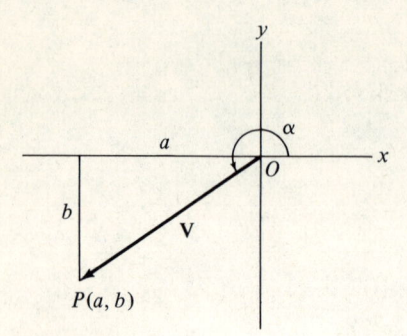

Figure 3-9

The *slope* of **v** is the number

$$m = \tan \alpha = b/a, \text{ provided } a \neq 0. \qquad (\alpha \neq \tfrac{1}{2}\pi, \tfrac{3}{2}\pi)$$

v is said to be *vertical* if $a = 0$, and *horizontal* if $b = 0$. (Compare the above with the corresponding definitions for arrows.)

To define the angle between two nonzero vectors $\mathbf{v}_1 = (a_1, b_1)$ and $\mathbf{v}_2 = (a_2, b_2)$, consider the following: Let $\overrightarrow{P_1P_2} = \mathbf{v}_1$ and $\overrightarrow{P_1P_3} = \mathbf{v}_2$ (Figure 3–10). The angle between $\mathbf{v}_1$ and $\mathbf{v}_2$ should be the angle θ at the vertex P_1 of the triangle $P_1P_2P_3$. From the law of cosines of trigonometry, we have

$$|\overrightarrow{P_2P_3}|^2 = |\overrightarrow{P_1P_2}|^2 + |\overrightarrow{P_1P_3}|^2 - 2|\overrightarrow{P_1P_2}| \cdot |\overrightarrow{P_1P_3}| \cos \theta$$
$$|\mathbf{v}_2 - \mathbf{v}_1|^2 = |\mathbf{v}_1|^2 + |\mathbf{v}_2|^2 - 2|\mathbf{v}_1| \cdot |\mathbf{v}_2| \cos \theta$$
$$(a_2 - a_1)^2 + (b_2 - b_1)^2 = a_1^2 + b_1^2 + a_2^2 + b_2^2 - 2|\mathbf{v}_1| \cdot |\mathbf{v}_2| \cos \theta$$

After simplifying, we obtain

$$\cos \theta = \frac{a_1a_2 + b_1b_2}{|\mathbf{v}_1||\mathbf{v}_2|}, \qquad 0 \leq \theta \leq \pi \tag{1}$$

This formula is taken as the *definition* of the angle θ between $\mathbf{v}_1$ and $\mathbf{v}_2$ (or between any two representative arrows). This is a legitimate definition, since

$$\frac{|a_1a_2 + b_1b_2|}{|\mathbf{v}_1||\mathbf{v}_2|} \leq 1 \qquad \text{(see Problem 10(a))}$$

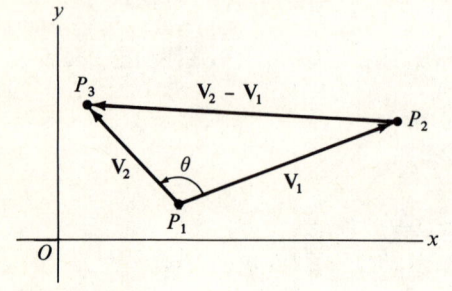

Figure 3-10

We may express θ in terms of the directions α_1, α_2 of $\mathbf{v}_1, \mathbf{v}_2$ as follows:

$$\cos \theta = \frac{a_1}{|\mathbf{v}_1|} \cdot \frac{a_2}{|\mathbf{v}_2|} + \frac{b_1}{|\mathbf{v}_1|} \cdot \frac{b_2}{|\mathbf{v}_2|}$$
$$= \cos \alpha_1 \cos \alpha_2 + \sin \alpha_1 \sin \alpha_2$$

Thus, $\cos \theta = \cos(\alpha_1 - \alpha_2) = \cos(\alpha_2 - \alpha_1)$. Accordingly, $\theta = |\alpha_1 - \alpha_2|$ iff $|\alpha_1 - \alpha_2| \leq \pi$, and $\theta = 2\pi - |\alpha_1 - \alpha_2|$ otherwise.

$\mathbf{v}_1$ and $\mathbf{v}_2$ are said to be *parallel* (written as $\mathbf{v}_1 \| \mathbf{v}_2$ or $\mathbf{v}_2 \| \mathbf{v}_1$) iff $\theta = 0$ or $\theta = \pi$. Thus, $\mathbf{v}_1 \| \mathbf{v}_2$ if $\alpha_1 = \alpha_2$ or $|\alpha_1 - \alpha_2| = \pi$. It is not difficult to show that the (nonzero) vectors $\mathbf{v}_1, \mathbf{v}_2$ are parallel iff $\mathbf{v}_1 = t\mathbf{v}_2$ for some scalar $t \neq 0$. In fact, $\theta = 0$ iff $t > 0$, and $\theta = \pi$ iff $t < 0$. (See Problem 9.) Furthermore, two nonvertical vectors (or arrows) are parallel iff they have the same slope.

The numerator of Formula (1) is called the *dot product* of $\mathbf{v}_1$ and $\mathbf{v}_2$ (or of any two representative arrows), and is denoted by $\mathbf{v}_1 \circ \mathbf{v}_2$. Formula (1) may now be written

$$\cos \theta = \frac{\mathbf{v}_1 \circ \mathbf{v}_2}{|\mathbf{v}_1||\mathbf{v}_2|} \qquad (0 \leq \theta \leq \pi)$$

$\mathbf{v}_1$ and $\mathbf{v}_2$ are said to be *perpendicular* (written $\mathbf{v}_1 \perp \mathbf{v}_2$ or $\mathbf{v}_2 \perp \mathbf{v}_1$) iff $\theta = \pi/2$ (similarly for arrows). It follows that $\mathbf{v}_1 \perp \mathbf{v}_2$ iff $\mathbf{v}_1 \circ \mathbf{v}_2 = 0$. We extend the definition of perpendicularity to include the case where one of the vectors is $\mathbf{0}$. Accordingly, $\mathbf{0}$ is perpendicular to every vector, since $\mathbf{0} \circ \mathbf{v} = \mathbf{v} \circ \mathbf{0} = 0$.

Example 1. Find the angle between the vectors $\mathbf{v}_1 = (-\sqrt{3}, 1)$ and $\mathbf{v}_2 = (2\sqrt{3}, 2)$.

SOLUTION

$$\cos \theta = \frac{\mathbf{v}_1 \circ \mathbf{v}_2}{|\mathbf{v}_1||\mathbf{v}_2|} = \frac{(-\sqrt{3})(2\sqrt{3}) + (1)(2)}{\sqrt{3+1}\sqrt{12+4}} = -\frac{1}{2}$$

Therefore,

$$\theta = 120° = \tfrac{2}{3}\pi$$

As an alternate method, we have

$$\cos \alpha_1 = -\frac{\sqrt{3}}{2}, \quad \sin \alpha_1 = \frac{1}{2}. \qquad \text{Hence } \alpha_1 = 150°.$$

$$\cos \alpha_2 = \frac{\sqrt{3}}{2}, \quad \sin \alpha_2 = \frac{1}{2}. \qquad \text{So } \alpha_2 = 30°.$$

Therefore,

$$\theta = \alpha_1 - \alpha_2 = 120°$$

Example 2. Determine whether the vectors $v_1 = (2, -1)$, $v_2 = (3, 6)$ are perpendicular.

SOLUTION
$v_1 \circ v_2 = (2)(3) + (-1)(6) = 0$

Therefore,

$v_1 \perp v_2$

The important properties of the *dot product* are summarized in the following theorem.

Theorem. *For all vectors* $v, v_1, v_2, v_3,$ *and scalars* c,

(1) $v_1 \circ v_2 = v_2 \circ v_1$
(2) $v_1 \circ (v_2 + v_3) = v_1 \circ v_2 + v_1 \circ v_3$
(3) $v_1 \circ (v_2 - v_3) = v_1 \circ v_2 - v_1 \circ v_3$
(4) $v \circ v = |v|^2$
(5) $c(v_1 \circ v_2) = (cv_1) \circ v_2 = v_1 \circ (cv_2)$
(6) $0 \circ v = 0$
(7) $v \circ v = 0$ iff $v = 0$

PROOF. We prove Part 2 of the theorem and leave the remainder to the student. (See Problem 5.) Let $v_1 = (a_1, b_1)$, $v_2 = (a_2, b_2)$, and $v_3 = (a_3, b_3)$.

$$v_1 \circ (v_2 + v_3) = (a_1, b_1) \circ (a_2 + a_3, b_2 + b_3)$$
$$= a_1(a_2 + a_3) + b_1(b_2 + b_3)$$
$$= (a_1 a_2 + b_1 b_2) + (a_1 a_3 + b_1 b_3)$$
$$= v_1 \circ v_2 + v_1 \circ v_3$$

Example 3. Let $v_1 = (-1, 2)$. Find a vector $v = (a, b)$ such that $v \perp v_1$ and $|v| = \sqrt{5}$.

SOLUTION
$v \circ v_1 = 0$ and $|v|^2 = 5$

Therefore,

$-a + 2b = 0$ and $a^2 + b^2 = 5$

Solving for a and b simultaneously, we obtain the solutions $a = 2, b = 1$ and $a = -2, b = -1$. Thus, both $v = (2, 1)$ and $v = (-2, -1)$ are solution vectors.

Every vector $v = (a, b)$ can be expressed in terms of two *fundamental vectors* (also called *coordinate vectors*). They are $i = (1, 0)$ and $j = (0, 1)$ (Figure 3–11). Thus,

$$v = (a, b) = (a, 0) + (0, b) = a(1, 0) + b(0, 1) = ai + bj$$

In fact, it is this form which is most often used in the application of vectors

The Angle Between Two Vectors; Dot Product

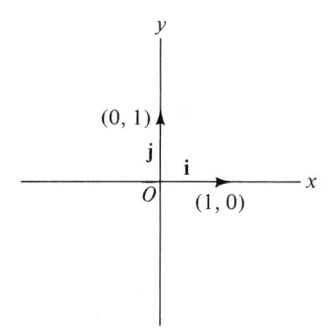

Figure 3-11

to geometry, physics, and engineering. Note that $|\mathbf{i}| = |\mathbf{j}| = 1$ and $\mathbf{i} \circ \mathbf{j} = 0$, so that $\mathbf{i} \perp \mathbf{j}$. a is called the *i*-component and b the *j*-component of $\mathbf{v}$. The vector $\mathbf{v} = 2\mathbf{i} - 3\mathbf{j}$ is shown in Figure 3-12.

Example 4. Given $\mathbf{v}_1 = 2\mathbf{i} - 3\mathbf{j}$, and $\mathbf{v}_2 = -\mathbf{i} - 2\mathbf{j}$. Find (a) $\mathbf{v}_1 + \mathbf{v}_2$, (b) $\mathbf{v}_1 - \mathbf{v}_2$, (c) $\mathbf{v}_1 \circ \mathbf{v}_2$, and (d) $\cos \theta$.

SOLUTION

(a) $\mathbf{v}_1 + \mathbf{v}_2 = \mathbf{i} - 5\mathbf{j}$

(b) $\mathbf{v}_1 - \mathbf{v}_2 = 3\mathbf{i} - \mathbf{j}$

(c) $\mathbf{v}_1 \circ \mathbf{v}_2 = 2(-1) + (-3)(-2) = 4$

(d) $\cos \theta = \dfrac{2(-1) + (-3)(-2)}{\sqrt{4+9}\sqrt{1+4}} = \dfrac{4}{\sqrt{65}}$

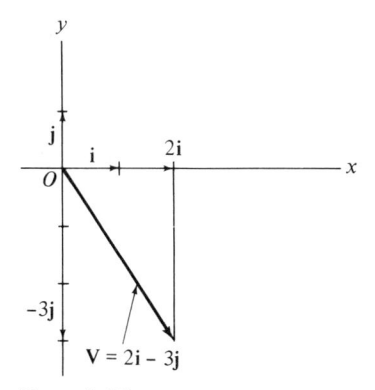

Figure 3-12

EXERCISE 3

For some of the following problems, the student will need a table of trigonometric functions.

1. Find the direction of the following vectors.
    (a) $\mathbf{i} - \mathbf{j}$

Vectors and Line Segments in the Plane

(b) $3\mathbf{j}$
(c) $-2\mathbf{i}$
(d) $3\mathbf{i} - \mathbf{j}$
(e) $-2\mathbf{i} + 3\mathbf{j}$
(f) $\mathbf{i} + \mathbf{j}$

2. Find the angle between the following pairs of vectors.
   (a) $2\mathbf{i} - \mathbf{j}, \mathbf{i} + 2\mathbf{j}$
   (b) $2\mathbf{i} + \mathbf{j}, \mathbf{i} + 3\mathbf{j}$
   (c) $2\mathbf{i} - 4\mathbf{j}, -\mathbf{i} - 3\mathbf{j}$
   (d) $-2\mathbf{i} + \mathbf{j}, -\mathbf{i} - \mathbf{j}$
   (e) $-3\mathbf{i} + \sqrt{3}\mathbf{j}, 3\mathbf{i} - \sqrt{3}\mathbf{j}$

3. Determine whether the following pairs of vectors are perpendicular, parallel, or neither.
   (a) $4\mathbf{i} - \mathbf{j}, 2\mathbf{i} + 8\mathbf{j}$
   (b) $2\mathbf{i} - \mathbf{j}, -6\mathbf{i} + 3\mathbf{j}$
   (c) $\frac{4}{5}\mathbf{i} - \frac{2}{3}\mathbf{j}, 10\mathbf{i} + 12\mathbf{j}$
   (d) $\mathbf{i} + \mathbf{j}, 2\mathbf{i} - 3\mathbf{j}$
   (e) $6\mathbf{i}, -2\mathbf{j}$

4. Find a vector (a) making an angle of 45° with the vector $2\mathbf{i} + \mathbf{j}$, (b) of length $\sqrt{13}$ which is perpendicular to $3\mathbf{i} + 2\mathbf{j}$.

5. (a) Prove the theorem of this section.
   (b) Prove $(\mathbf{v}_1 + \mathbf{v}_2) \circ (\mathbf{v}_1 - \mathbf{v}_2) = (|\mathbf{v}_1| + |\mathbf{v}_2|)(|\mathbf{v}_1| - |\mathbf{v}_2|)$.

6. (a) Prove that if $\mathbf{v}$ is perpendicular to $\mathbf{v}_1$ and $\mathbf{v}_2$, then $\mathbf{v}$ is perpendicular to $c_1\mathbf{v}_1 + c_2\mathbf{v}_2$.
   *(b) Prove that if $\mathbf{v}$ is perpendicular to each of $\mathbf{v}_1, \mathbf{v}_2, \ldots, \mathbf{v}_n$, then
   $$\mathbf{v} \perp c_1\mathbf{v}_1 + c_2\mathbf{v}_2 + \ldots + c_n\mathbf{v}_n$$
   (*Hint*: Use induction and the fact that
   $$\mathbf{v}_1 + \mathbf{v}_2 + \ldots + \mathbf{v}_{n+1} = (\mathbf{v}_1 + \ldots + \mathbf{v}_n) + \mathbf{v}_{n+1}.)$$

†*7. Derive the formula
$$\sin \theta = \frac{|a_1 b_2 - a_2 b_1|}{|\mathbf{v}_1| \cdot |\mathbf{v}_2|}$$
where θ is the angle between $\mathbf{v}_1$ and $\mathbf{v}_2$.

†*8. Show that if m_1, m_2 are the slopes of $\mathbf{v}_1, \mathbf{v}_2$, respectively, and neither m_1 nor m_2 is zero, then $\mathbf{v}_1 \perp \mathbf{v}_2$ iff $m_1 = -1/m_2$.

†*9. Prove that the (nonzero) vectors $\mathbf{v}_1, \mathbf{v}_2$ are parallel iff $\mathbf{v}_1 = t\mathbf{v}_2$ for some scalar $t \neq 0$. Moreover, $t > 0$ iff $\theta = 0$, and $t < 0$ iff $\theta = \pi$. (Use Problem 7.)

†*10. Prove
   (a) $|\mathbf{v}_1 \circ \mathbf{v}_2| \leq |\mathbf{v}_1||\mathbf{v}_2|$ (Cauchy's inequality)
   (*Hint*: Square both sides and simplify.)
   (b) $|\mathbf{v}_1 + \mathbf{v}_2| \leq |\mathbf{v}_1| + |\mathbf{v}_2|$ (triangle inequality)
   (*Hint*: Use Part 4 of the theorem and Problem 10(a).)
   Also show that equality holds iff one of the vectors is zero or the vectors have the same direction. (See Problem 9.)
   (c) $|\mathbf{v}_1 - \mathbf{v}_2| \geq ||\mathbf{v}_1| - |\mathbf{v}_2||$

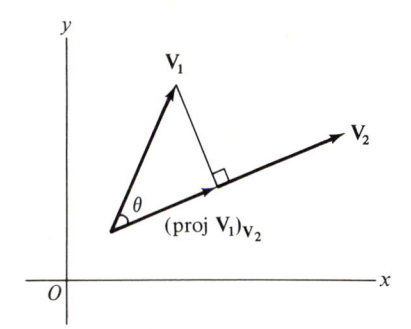

Figure 3-13

*11. Let $\mathbf{v}_1, \mathbf{v}_2$ be nonzero vectors. The *projection of* $\mathbf{v}_1$ *on* $\mathbf{v}_2$ is defined to be the vector (Figure 3-13)

$$(\text{proj } \mathbf{v}_1)_{\mathbf{v}_2} = (|\mathbf{v}_1| \cos \theta) \frac{\mathbf{v}_2}{|\mathbf{v}_2|}$$

Prove

(a) $(\text{proj } \mathbf{v}_1)_{\mathbf{v}_2} = \left(\dfrac{\mathbf{v}_1 \circ \mathbf{v}_2}{|\mathbf{v}_2|^2} \right) \mathbf{v}_2$

(b) $|(\text{proj } \mathbf{v}_1)_{\mathbf{v}_2}| \cdot |\mathbf{v}_2| = |(\text{proj } \mathbf{v}_2)_{\mathbf{v}_1}| \cdot |\mathbf{v}_1|$

(c) $(\text{proj } (\mathbf{v}_1 + \mathbf{v}_2))_{\mathbf{v}_3} = (\text{proj } \mathbf{v}_1)_{\mathbf{v}_3} + (\text{proj } \mathbf{v}_2)_{\mathbf{v}_3}$
(Draw a figure.)

(d) $(\text{proj } \mathbf{v}_1)_{\mathbf{v}_2} = \mathbf{0}$ iff $\mathbf{v}_1 \perp \mathbf{v}_2$

(e) $(\text{proj } \mathbf{v}_1)_{\mathbf{v}_2} = \mathbf{v}_1$ iff $\mathbf{v}_1 \| \mathbf{v}_2$

12. Find $(\text{proj } \mathbf{v}_1)_{\mathbf{v}_2}$. (See Problem 11.)

(a) $\mathbf{v}_1 = a\mathbf{i} + b\mathbf{j}, \mathbf{v}_2 = \mathbf{i}$
(b) $\mathbf{v}_1 = a\mathbf{i} + b\mathbf{j}, \mathbf{v}_2 = \mathbf{j}$
(c) $\mathbf{v}_1 = 3\mathbf{i} - 2\mathbf{j}, \mathbf{v}_2 = 2\mathbf{i} + 3\mathbf{j}$
(d) $\mathbf{v}_1 = \mathbf{i} + \mathbf{j}, \mathbf{v}_2 = \mathbf{i} - \mathbf{j}$
(e) $\mathbf{v}_1 = 3\mathbf{i} - \mathbf{j}, \mathbf{v}_2 = 2\mathbf{i} + \mathbf{j}$
(f) $\mathbf{v}_1 = \mathbf{i} + 2\mathbf{j}, \mathbf{v}_2 = -3\mathbf{i} + \mathbf{j}$
(g) $\mathbf{v}_1 = 4\mathbf{i} + 2\mathbf{j}, \mathbf{v}_2 = 2\mathbf{i} + \mathbf{j}$
(h) $\mathbf{v}_1 = \frac{2}{3}\mathbf{i} - \mathbf{j}, \mathbf{v}_2 = -2\mathbf{i} + 3\mathbf{j}$

4. GEOMETRIC PROBLEMS

In this section, we use the concepts of length and direction to solve certain problems in plane geometry. We begin with the definition of *line segment*.

Line Segment. Let P_1, P_2 be distinct points. The *line segment* (or simply *segment*) joining P_1 and P_2 is the set

$$\overline{P_1 P_2} = \{P \mid \overrightarrow{P_1 P} = t\overrightarrow{P_1 P_2}, 0 \le t \le 1\}$$

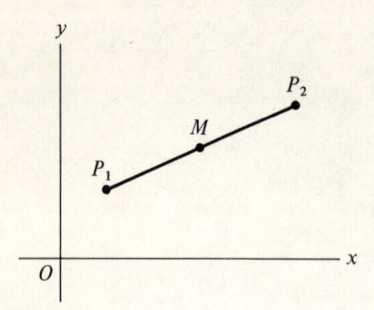

Figure 3-14

Note that $\overline{P_1P_2}$ is a *set of points* in R^2, whereas the arrow $\overrightarrow{P_1P_2}$ is the *ordered pair* of points (P_1, P_2). It is easy to show that $\overline{P_1P_2} = \overline{P_2P_1}$ (see Problem 11). (The segment $\overline{P_1P_2}$ will be pictured in the same way as the arrow $\overrightarrow{P_1P_2}$, except that the arrowhead will be ommitted.)

Observe that $P_1, P_2 \in \overline{P_1P_2}$. We call P_1 and P_2 the *endpoints* of $\overline{P_1P_2}$, and every other point in $\overline{P_1P_2}$ (or *on* $\overline{P_1P_2}$) is said to *lie between* P_1 and P_2. The point M for which $\overrightarrow{P_1M} = \frac{1}{2}\overrightarrow{P_1P_2}$ is called the *midpoint* of $\overline{P_1P_2}$ and is said to be *one half the distance from* P_1 along $\overline{P_1P_2}$ (Figure 3-14). (Note: $|\overrightarrow{P_1M}| = \frac{1}{2}|\overrightarrow{P_1P_2}|$.) The point P for which $\overrightarrow{P_1P} = \frac{1}{3}\overrightarrow{P_1P_2}$ is said to be one third the distance from P_1 along $\overline{P_1P_2}$, and so on. In general, we have $|\overrightarrow{P_1P}| = t|\overrightarrow{P_1P_2}|$, so that $t = |\overrightarrow{P_1P}|/|\overrightarrow{P_1P_2}|$. Thus, *the number t in the definition of* $\overline{P_1P_2}$ *is the ratio of the distance between* P_1 *and* P *to the distance between* P_1 *and* P_2.

Letting $P_1 = (x_1, y_1)$, $P_2 = (x_2, y_2)$, and $P = (x, y)$, we see from the definition of $\overline{P_1P_2}$ that P lies on $\overline{P_1P_2}$ iff x, y satisfy the equations

$$\begin{cases} x = x_1 + t(x_2 - x_1) \\ y = y_1 + t(y_2 - y_1) \end{cases} \text{ for some } t \text{ with } 0 \leq t \leq 1 \quad (1)$$

Equations (1) are the parametric equations of $\overline{P_1P_2}$. Thus, a line segment is an example of a curve. (See Chapter 2, Section 5.)

Example 1. Let $P_1 = (-1, 2)$ and $P_2 = (3, 4)$. Find the point $P(x, y)$ which is $\frac{4}{5}$ of the distance from P_1 along $\overline{P_1P_2}$.

SOLUTION. We have $\overrightarrow{P_1P} = \frac{4}{5}\overrightarrow{P_1P_2}$, so that $t = \frac{4}{5}$. Substituting this value of t in the parametric equations (1), we obtain

$$\begin{cases} x = -1 + \frac{4}{5}(3 - (-1)) \\ y = 2 + \frac{4}{5}(4 - 2) \end{cases}$$

Thus, $x = \frac{11}{5}$, $y = \frac{18}{5}$; so $P = (\frac{11}{5}, \frac{18}{5})$ (Figure 3-15).

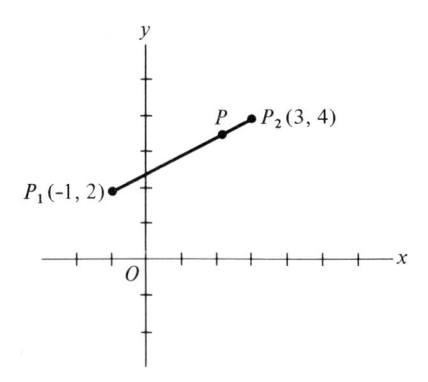

Figure 3-15

CHECK

$$|\overrightarrow{P_1P}| = \sqrt{(\tfrac{11}{5} + 1)^2 + (\tfrac{18}{5} - 2)^2} = \tfrac{8}{5}\sqrt{5}$$

$$|\overrightarrow{P_1P_2}| = \sqrt{(3 + 1)^2 + (4 - 2)^2} = 2\sqrt{5}$$

$$\frac{|\overrightarrow{P_1P}|}{|\overrightarrow{P_1P_2}|} = \frac{4}{5}$$

Observe that P is $\tfrac{1}{5}$ the distance from P_2 along $\overline{P_2P_1}$.

Substituting $t = \tfrac{1}{2}$ in Equations (1), we obtain the *midpoint formulas* (see Problem 10):

$$x = \frac{x_1 + x_2}{2}$$
$$y = \frac{y_1 + y_2}{2}$$
(2)

For segments $\overline{P_1P_2}$, $\overline{P_3P_4}$, the concepts of *length, slope, vertical, horizontal, parallel,* and *perpendicular* are defined as for the arrows $\overrightarrow{P_1P_2}$, $\overrightarrow{P_3P_4}$. If the segments have a common endpoint, say $P_1 = P_3$, then the *angle between the segments* is defined to be the angle between the arrows $\overrightarrow{P_1P_2}$, $\overrightarrow{P_1P_4}$. (This can be extended to intersecting segments without a common endpoint, but this will not be needed for our exercises.) From the law of cosines of trigonometry, we have

$$\overline{P_1P_2} \perp \overline{P_1P_3} \text{ iff } |P_2P_3|^2 = |P_1P_2|^2 + |P_1P_3|^2 \quad \text{(Pythagorean theorem)}.$$

From Problem 8, Exercise 3, we have

$$\overline{P_1P_2} \perp \overline{P_1P_3} \text{ iff } m_{\overline{P_1P_2}} = -\frac{1}{m_{\overline{P_1P_3}}}$$

provided both slopes are defined and neither is zero.

Definition. The (distinct) points $P_1(x_1, y_1)$, $P_2(x_2, y_2)$, and $P_3(x_3, y_3)$ are *collinear* iff one of the following conditions holds:

(1) $m_{\overline{P_1P_2}} = m_{\overline{P_1P_3}}$

that is, $\dfrac{y_2 - y_1}{x_2 - x_1} = \dfrac{y_3 - y_1}{x_3 - x_1}$ $(x_1 \neq x_2, x_1 \neq x_3)$

(2) $x_1 = x_2 = x_3$ (that is, all points lie on the same vertical segment)

REMARK. One can show that if x_1, x_2, x_3 are distinct, then $m_{\overline{P_1P_2}} = m_{\overline{P_1P_3}}$ iff $m_{\overline{P_1P_2}} = m_{\overline{P_2P_3}}$. (See Problem 12.)

Example 2. Determine whether the following points are collinear: $P_1(-1, -2)$, $P_2(1, 2)$, $P_3(3, 5)$.

SOLUTION

$m_{\overline{P_1P_2}} = \dfrac{2 - (-2)}{1 - (-1)} = 2$

$m_{\overline{P_1P_3}} = \dfrac{5 - (-2)}{3 - (-1)} = \dfrac{7}{4}$

Since $m_{\overline{P_1P_2}} \neq m_{\overline{P_1P_3}}$, the points are noncollinear.

Example 3. Given $P_1(2, 1)$, $P_2(4, 5)$, and $P_3(-4, 4)$. Show that $\overline{P_1P_2} \perp \overline{P_1P_3}$.

FIRST SOLUTION

$m_{\overline{P_1P_2}} = \dfrac{5 - 1}{4 - 2} = 2$

$m_{\overline{P_1P_3}} = \dfrac{4 - 1}{-4 - 2} = -\dfrac{1}{2}$

Therefore, $\overline{P_1P_2} \perp \overline{P_1P_3}$, since $m_{\overline{P_1P_3}} = -\dfrac{1}{m_{\overline{P_1P_2}}}$.

SECOND SOLUTION

$|P_1P_2|^2 = (4 - 2)^2 + (5 - 1)^2 = 20$
$|P_1P_3|^2 = (-4 - 2)^2 + (4 - 1)^2 = 45$
$|P_2P_3|^2 = (-4 - 4)^2 + (4 - 5)^2 = 65$

Therefore, $\overline{P_1P_2} \perp \overline{P_1P_3}$, since $|P_1P_2|^2 + |P_1P_3|^2 = |P_2P_3|^2$. Note also that $\overrightarrow{P_1P_2} \circ \overrightarrow{P_1P_3} = 0$.

Let A, B, C, D be distinct points, no three of which are collinear. The *triangle* with *vertices* A, B, C and *sides* $\overline{AB}, \overline{BC}, \overline{CA}$ is the set

$\triangle ABC = \overline{AB} \cup \overline{BC} \cup \overline{CA}$

The *quadrilateral* with *vertices* A, B, C, D and *sides* $\overline{AB}, \overline{BC}, \overline{CD}, \overline{DA}$ is the set quad $ABCD = \overline{AB} \cup \overline{BC} \cup \overline{CD} \cup \overline{DA}$. Special cases of these figures,

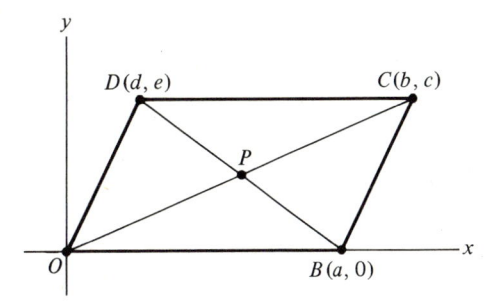

Figure 3-16

such as *isosceles*, *equilateral*, and *right triangle*, *trapezoid*, *parallelogram*, *rhombus*, *rectangle*, and *square* are defined as in plane geometry. Related concepts, such as *opposite* and *adjacent sides*, *hypotenuse*, *diagonal*, *median*, and so on, are also defined as in geometry.

In solving problems involving the above figures, it is convenient to use the origin as a vertex and to locate a second vertex on the positive x-axis. It can be shown that this procedure leads to no loss of generality.

Example 4. Prove that the diagonals of a parallelogram bisect each other.

PROOF. Let the vertices of the parallelogram be $O(0, 0)$, $B(a, 0)$, $C(b, c)$, $D(d, e)$ (Figure 3-16).

Since opposite sides of a parallelogram are parallel, we have

$$\frac{(c - e)}{(b - d)} = \frac{(0 - 0)}{(a - 0)}$$

so $c = e$. Also,

$$\frac{(c - 0)}{(b - a)} = \frac{(e - 0)}{(d - 0)}$$

from which follows $b = a + d$. The midpoints of the diagonals are $(b/2, c/2)$ and $((a + d)/2, e/2)$. Since these points are equal, the diagonals bisect each other.

Example 5. Prove that the medians of a triangle intersect at a point two thirds of the distance from the vertices along the medians.

PROOF. Let the vertices be $O(0, 0)$, $A(a, 0)$, $B(b, c)$, where $a > 0$ and $c > 0$. Then the points $M_1(a/2, 0)$, $M_2(b/2, c/2)$, $M_3((a + b)/2, c/2)$ are the midpoints of the sides of the triangle (Figure 3-17). We next write the parametric equations (1) of the medians, using a different letter for the parameter in each case. We have

$$\overline{OM_3}: \begin{cases} x = t\left(\dfrac{a + b}{2}\right) \\ y = t \cdot \dfrac{c}{2} \end{cases}, \quad 0 \leq t \leq 1$$

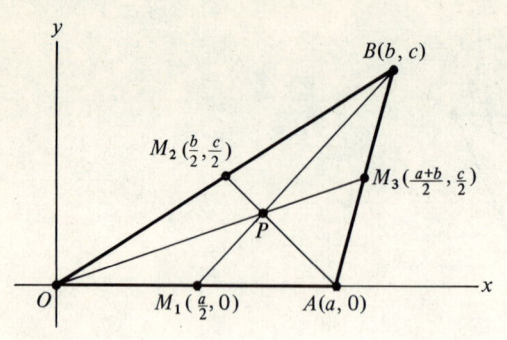

Figure 3-17

$$\overline{AM_2}: \begin{cases} x = a + s\left(\dfrac{b}{2} - a\right) \\ y = s \cdot \dfrac{c}{2} \end{cases}, \quad 0 \leq s \leq 1$$

$$\overline{BM_1}: \begin{cases} x = b + r\left(\dfrac{a}{2} - b\right) \\ y = c - r \cdot c \end{cases}, \quad 0 \leq r \leq 1$$

To find the point of intersection $P(x, y)$ of $\overline{OM_3}$ and $\overline{AM_2}$, we set the x's equal and y's equal and solve simultaneously for s and t. We have

$$t\left(\dfrac{a+b}{2}\right) = a + s\left(\dfrac{b}{2} - a\right)$$

$$t \cdot \dfrac{c}{2} = s \cdot \dfrac{c}{2}$$

The second equation yields $s = t$. Substituting t for s in the first equation yields $t = \frac{2}{3}$. Thus $\overline{OM_3}$ and $\overline{AM_2}$ intersect at the point P two thirds the distance from the vertices along the medians. Similar results can be proved for the medians $\overline{OM_3}$ and $\overline{BM_1}$.

EXERCISE 4

1. Let $P_1 = (4, -1)$ and $P_2 = (-5, 2)$.
   (a) Write the parametric equations of $\overline{P_1 P_2}$.
   (b) Find the midpoint of $\overline{P_1 P_2}$.
   (c) Find the point $P(x, y)$ one fourth the distance from P_1 along $\overline{P_1 P_2}$.
   (d) Find the point $P(x, y)$ three fourths the distance from P_2 along $\overline{P_2 P_1}$. (Compare with (c).)
   (e) Find $m_{\overline{P_1 P_2}}$.

(f) Let $P_3 = (6, 5)$ and $P_4 = (-3, 8)$. Show that $\overline{P_2P_1} \perp \overline{P_1P_3}$ and $\overline{P_2P_1} \parallel \overline{P_3P_4}$. (Draw a figure.)

2. Plot the following sets of points and determine whether they are collinear.
   (a) $P_1(2, 3), P_2(2, -1), P_3(4, 6)$
   (b) $P_1(-1, 4), P_2(-1, 6), P_3(-1, -2)$
   (c) $P_1(1, 3), P_2(2, 5), P_3(4, 9)$
   (d) $P_1(-3, 7), P_2(0, 7), P_3(2, 5)$
   (e) $P_1(6, 2), P_2(-2, 2), P_3(0, 2)$
   (f) $P_1(4, 2), P_2(0, 6), P_3(-4, 7)$
   (g) $P_1(-3, -7), P_2(0, -2), P_3(6, 8)$

3. (a) Find x so that the points $(x, 3), (-4, 2)$, and $(3, 5)$ are collinear.
   (b) Find y so that the points $(4, y), (-2, -3)$, and $(6, 3)$ are collinear.

4. Let $P_1 = (2, -4)$ and $P_2 = (-1, 3)$.
   (a) Find $P_3(0, y)$ such that $\overline{P_1P_2} \perp \overline{P_2P_3}$.
   (b) Find $P_3(x, 0)$ such that $\overline{P_1P_2} \perp \overline{P_2P_3}$.
   (c) Find $P_3(x, y)$ such that $\overline{P_1P_2} \perp \overline{P_2P_3}$ and $|\overline{P_2P_3}| = 5$. (Two solutions; draw a figure.)

5. Find the point C on the line segment joining $A(-3, 2)$ and $B(6, 9)$ such that
   (a) $\dfrac{|\overline{AC}|}{|\overline{CB}|} = \dfrac{2}{3}$
   (b) $\dfrac{|\overline{AC}|}{|\overline{CB}|} = \dfrac{4}{7}$

6. Determine whether the triangles with the following sets of vertices are *right, isosceles, equilateral,* or none of these.
   (a) $A(1, 2), B(-2, 4), C(0, -3)$
   (b) $A(1, 0), B(5, 0), C(3, 4)$
   (c) $A(-1, 0), B(5, 0), C(3, 2)$
   (d) $A(5, 7), B(8, -5), C(0, -7)$
   (e) $A(-1, 0), B(3, 0), C(1, 2\sqrt{3})$

7. Let $P_1(x_1, y_1), P_2(x_2, y_2)$, and $P_3(x_3, y_3)$ be distinct points. Prove that $\overline{P_1P_2} \perp \overline{P_2P_3}$ iff $(x_2 - x_1)(x_3 - x_2) + (y_2 - y_1)(y_3 - y_2) = 0$.

8. Show that if $\triangle ABC$ is a right triangle with $\overline{AC} \perp \overline{CB}$, then $\overline{AB}$ is not perpendicular to either $\overline{AC}$ or $\overline{CB}$.

9. Given $A(3, 7), B(2, 3)$, and $C(5, 6)$. Find $D(x, y)$ so that quad $ABCD$ is a parallelogram.

†10. Prove that if $P_1(x_1, y_1) \neq P_2(x_2, y_2)$, then $M(x, y)$ is the midpoint of $\overline{P_1P_2}$ iff $x = (x_1 + x_2)/2, y = (y_1 + y_2)/2$.

*11. Prove that $\overline{P_1P_2} = \overline{P_2P_1}$.

*12. Let $P_1(x_1, y_1), P_2(x_2, y_2)$, and $P_3(x_3, y_3)$ be points with $x_1 < x_2 < x_3$.
   (a) Prove that $m_{P_1P_2} = m_{P_1P_3}$ iff $m_{P_1P_2} = m_{P_2P_3}$.
   (b) Prove that P_1, P_2, and P_3 are collinear iff
   $|\overline{P_1P_2}| + |\overline{P_2P_3}| = |\overline{P_1P_3}|$

13. Prove that opposite sides of a parallelogram have the same length.

14. (a) The vertices of a triangle are $A(x_1, y_1)$, $B(x_2, y_2)$, and $C(x_3, y_3)$. Show that the medians intersect in the point

$$P\left(\frac{x_1 + x_2 + x_3}{3}, \frac{y_1 + y_2 + y_3}{3}\right)$$

   P is called the *centroid* of the triangle.
   (b) Find the centroid of the triangle whose vertices are $A(3, 5)$, $B(-2, 1)$, and $C(6, -1)$.
   (c) The midpoints of two sides of a triangle are $M_1(1, 2)$, $M_2(-1, 4)$, and the centroid is $(2, 6)$. Find the vertices.

*15. The midpoints of the sides of a triangle are $M_1(3, -2)$, $M_2(-1, 2)$, and $M_3(5, 4)$. Find the vertices.

16. Prove that the diagonals of a rectangle are equal in length.
17. Prove that the diagonals of a square are perpendicular.
18. $\triangle ABC$ (Figure 3-18) is a right triangle; M is the midpoint of the hypotenuse. Prove that $|\overline{MC}| = \frac{1}{2}|\overline{AB}|$.
*19. Prove that in a parallelogram, the sum of the squares of the lengths of the diagonals equals the sum of the squares of the lengths of the sides.
*20. Prove that in a triangle, four times the sum of the squares of the lengths of the medians is equal to three times the sum of the squares of the lengths of the sides.

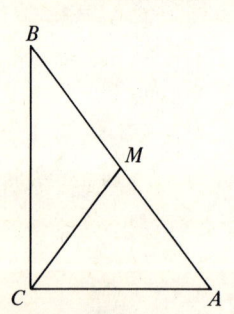

Figure 3-18

chapter 4
Lines, Circles, and Convex Sets in the Plane

1. LINES

In Chapter 3, we defined the *line segment* (or simply *segment*) joining the points $P_1(x_1, y_1)$ and $P_2(x_2, y_2)$ to be the set

$$\overline{P_1P_2} = \{P | \overrightarrow{P_1P} = t\overrightarrow{P_1P_2}, \quad 0 \leq t \leq 1\}$$

$P(x, y)$ is a point on $\overline{P_1P_2}$ iff

$$\begin{cases} x = x_1 + t(x_2 - x_1) \\ y = y_1 + t(y_2 - y_1), \end{cases} \quad \text{for some } t, 0 \leq t \leq 1 \tag{1}$$

We extend this concept to *rays* and *lines* as follows.

Definition. The *ray* with *initial point* P_1 and passing through P_2 with $P_2 \neq P_1$ is the set

$$\text{ray } P_1P_2 = \{P | \overrightarrow{P_1P} = t\overrightarrow{P_1P_2}, \quad t \geq 0\}$$

Definition. The *line* passing through the *distinct* points P_1, P_2 is the set

$$\ell = \{P | \overrightarrow{P_1P} = t\overrightarrow{P_1P_2}, \quad t \in R\}$$

The definitions of segment $\overline{P_1P_2}$, ray P_1P_2, and line ℓ through P_1, P_2 differ

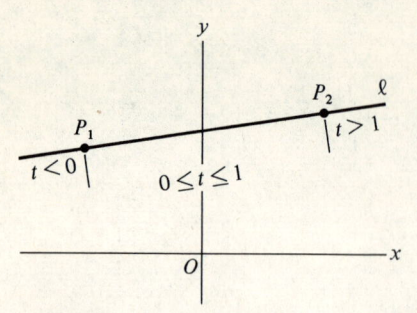

Figure 4-1

only in the range of values of the parameter t. Moreover, Equations (1) are parametric equations for ray P_1P_2 and the line ℓ as well as for the segment $\overline{P_1P_2}$ (Figure 4-1).

Example 1. Find parametric equations for the line ℓ through $P_1(-2, 3)$, $P_2(4, -1)$, and find several points on the line.

SOLUTION. Substituting the coordinates of P_1, P_2 in the Equations (1), we have

$$\ell: \begin{cases} x = -2 + 6t \\ y = 3 - 4t, \end{cases} \quad t \in R$$

$t = 0: P = (-2, 3) = P_1$
$t = 1: P = (4, -1) = P_2$
$t = \frac{1}{2}: P = (1, 1) = M \quad$ (midpoint of $\overline{P_1P_2}$)
$t = 2: P = (10, -5) = P_3$
$t = -1: P = (-8, 7) = P_4 \quad$ (Figure 4-2).

If $x_1 = x_2$ (hence $y_1 \neq y_2$), the equations of ℓ are

$$\begin{cases} x = x_1 \\ y = y_1 + t(y_2 - y_1), \end{cases} \quad t \in R$$

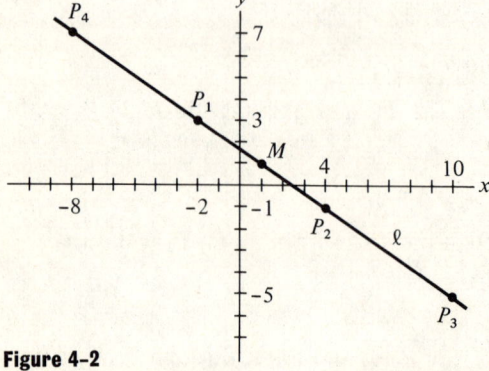

Figure 4-2

Note that for an arbitrary value of y, the point (x_1, y) satisfies the above equations with $t = (y - y_1)/(y_2 - y_1)$. Accordingly, a point $(x, y) \in \ell$ iff $x = x_1$. We may replace the parametric equations by $x = x_1$. In this case, ℓ is said to be *vertical*. Similarly, if $y_1 = y_2$, the equations of ℓ may be replaced by $y = y_1$, and ℓ is called *horizontal*.

Recall that the *slope* of $\overline{P_1 P_2}$ is

$$m_{\overline{P_1 P_2}} = \frac{y_2 - y_1}{x_2 - x_1} \qquad (x_1 \neq x_2)$$

This number is also defined to be the *slope* of ℓ, whenever ℓ is nonvertical. If Q_1, Q_2 are any two points on ℓ, then $m_{\overline{Q_1 Q_2}} = m_{\overline{P_1 P_2}}$. (See Problem 1.)

If ℓ is nonvertical, the first of Equations (1) may be solved for t, yielding

$$t = \frac{x - x_1}{x_2 - x_1}$$

Substituting this value of t in the second equation, we obtain

$$y - y_1 = \frac{y_2 - y_1}{x_2 - x_1}(x - x_1) \tag{2}$$

This is called the *two-point form* of the equation of the line. Letting

$$m = \frac{y_2 - y_1}{x_2 - x_1}$$

we may write Equation (2) as

$$y - y_1 = m(x - x_1) \tag{3}$$

This is called the *point–slope form* of the equation of ℓ. If $m = 0$ (that is, ℓ is horizontal), the equation of ℓ reduces to $y = y_1$.

Assume now that ℓ is neither horizontal nor vertical. Letting $y = 0$ in Equation (3) and solving for x, we find that

$$x = \frac{mx_1 - y_1}{m}$$

Denoting this number by a, we see that the point $(a, 0)$ lies on ℓ. We call $(a, 0)$ or a the *x-intercept* of ℓ. Letting $x = 0$ in Equation (3), we find the *y-intercept* $(0, b)$, where $b = y_1 - mx_1$. Note that a, b are both zero or both nonzero. If a, b are zero, then ℓ passes through the origin. If a, b are nonzero, the slope of ℓ is given by $m = -b/a$. Using the *point–slope form* Equation (3), with $(x_1, y_1) = (0, b)$, we have $y - b = -(b/a)(x - 0)$, which may be written

$$\frac{x}{a} + \frac{y}{b} = 1 \tag{4}$$

This is the *intercept form* of the equation of ℓ.

Using the point $(0, b)$ in the point–slope form Equation (3), we have $y - b = m(x - 0)$, which may be written

$$y = mx + b \tag{5}$$

This is referred to as the *slope–y-intercept form* or simply *slope-intercept form*.

Thus, we see that a line ℓ may be characterized by *two parametric equations* or by a *single equation* in x and y, called the *rectangular equation* of ℓ. The rectangular equation has several forms, each exhibiting special aspects of the line.

Example 2. Write the equation of the line ℓ through $P_1(2, -4)$ and $P_2(-3, 5)$ in (a) two-point form, (b) point–slope form, (c) intercept form, and (d) slope-intercept form.

SOLUTION

(a) $y - (-4) = \left(\dfrac{5 - (-4)}{-3 - 2}\right)(x - 2)$

or

$y - 5 = \left(\dfrac{5 - (-4)}{-3 - 2}\right)(x - (-3))$

(b) $y - (-4) = -\tfrac{9}{5}(x - 2)$

or

$y - 5 = -\tfrac{9}{5}(x - (-3))$

(c) From (b) we find that the intercepts are $a = -\tfrac{2}{9}$, and $b = -\tfrac{2}{5}$. The intercept form is

$$\dfrac{x}{-\tfrac{2}{9}} + \dfrac{y}{-\tfrac{2}{5}} = 1$$

(d) $y = -\tfrac{9}{5}x + (-\tfrac{2}{5})$

Observe that each of the forms in Example 2 reduces to $9x + 5y + 2 = 0$. This equation is of the form $Ax + By + C = 0$ and is called the *general form* of the equation of the line. In the next section, we shall show that every equation of this form represents a line.

Example 3. Write the general form of the equation of the line ℓ with (a) x-intercept 4 and passing through the point $(5, -2)$, (b) y-intercept -3 and slope 7, (c) slope -5 and x-intercept 9, and (d) x-intercept 4 less than y-intercept and slope -3.

SOLUTION

(a) $a = 4$. Using the intercept form, we have $x/4 + y/b = 1$. Substituting $(5, -2)$ in the equation yields $5/4 + -2/b = 1$, whose solution is $b = 8$. The equation of ℓ is, therefore, $x/4 + y/8 = 1$, which reduces to $2x + y - 8 = 0$. An alternate method is to use the two-point form and the fact that ℓ passes through $(4, 0)$ and $(5, -2)$.

(b) $b = -3$ and $m = 7$. The slope-intercept form is $y = 7x - 3$, while the general form is $7x - y - 3 = 0$.

(c) $m = -5$ and $a = 9$. Here we may use the point–slope form, since the slope and the point $(9, 0)$ are given. Thus, $y - 0 = -5(x - 9)$, which reduces to $5x + y - 45 = 0$.

(d) $a = b - 4$ and $m = -b/a = -3$. Solving for a and b, we have $a = 2$ and $b = 6$. The equation of ℓ is, therefore, $x/2 + y/6 = 1$, which in general form is $3x + y - 6 = 0$.

EXERCISE 1

1. Let ℓ be the line through P_1, P_2, and let Q_1, Q_2, be two points on ℓ. Prove that $m_{Q_1Q_2} = m_{P_1P_2}$.

2. Show that if ℓ is neither vertical nor horizontal and passes through $P_1(x_1, y_1)$, $P_2(x_2, y_2)$, then the intercepts are

$$a = \frac{x_1y_2 - x_2y_1}{y_2 - y_1}, \quad b = \frac{y_1x_2 - x_1y_2}{x_2 - x_1}$$

3. Let ℓ be the line through the given pair of points.
    (a) Write the parametric equations for ℓ.
    (b) Write the rectangular equation of ℓ in two-point form, point–slope form, intercept form, and slope-intercept form.
    (c) Sketch the line, showing its intercepts.
    (1) $P_1(-5, 1), P_2(7, -3)$
    (2) $P_1(4, 0), P_2(0, 6)$
    (3) $P_1(-2, -5), P_2(3, 4)$
    (4) $P_1(3, -2), P_2(-4, -2)$
    (5) $P_1(3, 10), P_2(3, -1)$
    (6) $P_1(\frac{2}{3}, \frac{1}{5}), P_2(-\frac{4}{5}, \frac{6}{7})$

4. Write in general form the equation of the line determined by the following conditions. Sketch the line. (As usual, $a = $ x-intercept, $b = $ y-intercept, and $m = $ slope.)
    (a) $m = 11$, through $(-6, 7)$
    (b) $a = 7, b = -4$
    (c) $m = -5, b = 0$
    (d) $m = 0, b = 3$
    (e) $a = -6$, passing through $(-6, -4)$
    (f) $m = 8, a = -7$
    (g) $a = 0, m = -4$
    (h) through $(-1, 8)$ and $(4, -11)$

5. Let $P_1 = (5, 2)$ and $P_2 = (-3, 7)$. Let ℓ be the line through P_1 and P_2.
    (a) Write the equation of ℓ in general form and sketch the line.
    (b) Write the equation of the line ℓ_1 through $(-3, 1)$ and having the same slope as ℓ. Draw a sketch.
    (c) Write the equation of the line ℓ_2 through $(-3, 1)$ having slope $\frac{8}{5}$. Draw a sketch.
    (d) How do the lines ℓ, ℓ_1, and ℓ_2 appear to be related?

6. Write the general form of the equation of the line ℓ satisfying the following conditions.
    (a) The sum of the intercepts is 5, and the slope is 3.

(b) The product of the intercepts is -6, and the line passes through $(3, 4)$ (two solutions).
(c) The slope is three times the y-intercept, and the line passes through the point $(3, 20)$.
(d) ℓ passes through the midpoint of the line segment $(-2, 5)(6, -3)$, and its slope is eual to its x-intercept.

2. THE GENERAL LINEAR EQUATION

In Section 1, we defined the line ℓ passing through the two points $P_1(x_1, y_1)$ and $P_2(x_2, y_2)$ as the set of points $P(x, y)$ satisfying the condition $\overrightarrow{P_1P} = t\overrightarrow{P_1P_2}$ $t \in R$, or, equivalently, the parametric equations

$$\begin{cases} x = x_1 + t(x_2 - x_1) \\ y = y_1 + t(y_2 - y_1) \end{cases} \quad (t \in R)$$

Now the arrow $\overrightarrow{P_1P_2}$ is a representative of the vector

$$\mathbf{v} = (x_2 - x_1)\mathbf{i} + (y_2 - y_1)\mathbf{j} \neq \mathbf{0}$$

and we may write $\mathbf{v} = \overrightarrow{P_1P_2}$. On the other hand, given a point P_1 and a nonzero vector $\mathbf{v} = a\mathbf{i} + b\mathbf{j}$, there exists a unique point $P_2 \neq P_1$ such that $\overrightarrow{P_1P_2} = \mathbf{v}$, and so a unique line ℓ passing through P_1, P_2 is determined. We shall refer to the line ℓ determined by P_1 and $\mathbf{v}$ as *the line through P_1 parallel to $\mathbf{v}$* (Figure 4–3). (In this context, a, b do not denote the x- and y-intercepts of Section 1.) Thus,

$$\ell = \{P(x, y) | \overrightarrow{P_1P} = t\mathbf{v}, \quad t \in R\}$$

and the parametric equations of ℓ are

$$\begin{cases} x = x_1 + at \\ y = y_1 + bt, \quad t \in R \end{cases}$$

It is not difficult to show that if Q is any point on ℓ and $\mathbf{v}_1$ is any (nonzero)

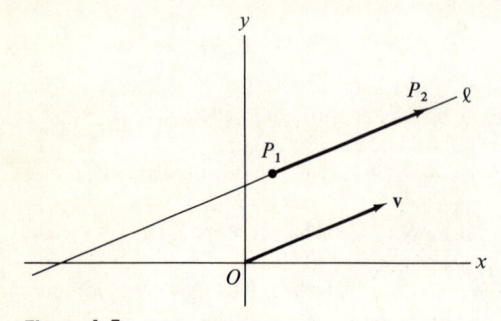

Figure 4–3

The General Linear Equation

vector parallel to **v** (that is, $\mathbf{v}_1 = k\mathbf{v}$ for some $k \neq 0$), then ℓ is equal to the line ℓ_1 through Q parallel to $\mathbf{v}_1$. (See Problem 4.)

Since $\mathbf{v} = a\mathbf{i} + b\mathbf{j} \neq \mathbf{0}$, either $a \neq 0$ or $b \neq 0$. If $a \neq 0$, we may solve for t in terms of x to obtain $t = (x - x_1)/a$. Substituting this expression for t in the second equation, we have $y = y_1 + (b/a)(x - x_1)$, which may be written

$$bx - ay + ay_1 - bx_1 = 0$$

The same equation results if we assume $b \neq 0$ and solve for t in terms of y. We see then that the line ℓ consists of all points $P(x, y)$ satisfying an equation of the form

$$Ax + By + C = 0 \tag{1}$$

where $A \neq 0$ or $B \neq 0$ ($A = b$, $B = -a$, $C = ay_1 - bx_1$). Equation (1) is called the *general linear equation* in two variables. Thus, every line is represented by a *linear equation*. In the following theorem, we show that every linear equation represents a line.

Theorem. *Let A, B, and C be real numbers with $A \neq 0$ or $B \neq 0$. Then the set of points*

$$S = \{P(x, y) | Ax + By + C = 0\}$$

is a line.

PROOF. We first show that S is a nonvoid set. If $A \neq 0$, let y_1 be any real number. Substituting y_1 for y and solving for x yields $x = x_1$, where

$$x_1 = \frac{-(By_1 + C)}{A}$$

Accordingly, $P_1(x_1, y_1) \in S$. The case $B \neq 0$ is treated similarly. Thus,

$$Ax_1 + By_1 + C = 0$$

and so $S \neq \emptyset$.

Now let $P(x, y)$ be any point in S so that

$$Ax + By + C = 0$$

Subtracting the last two equations and simplifying, we obtain

$$A(x - x_1) + B(y - y_1) = 0$$

Assuming $A \neq 0$ (the case $B \neq 0$ is similar), we have

$$x - x_1 = \frac{-B(y - y_1)}{A}$$

Letting $t = (y - y_1)/A$, we may write

$$\begin{cases} x - x_1 = -Bt \\ y - y_1 = At \end{cases}$$

Lines, Circles, and Convex Sets in the Plane

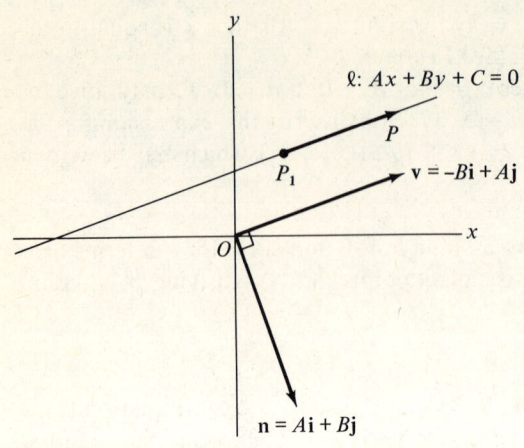

Figure 4-4

Thus, if $P(x, y) \in S$, there exists $t \in R$ such that
$$\begin{cases} x = x_1 + (-B)t \\ y = y_1 + At \end{cases}$$

That is, $P(x, y)$ lies on the line ℓ through (x_1, y_1) parallel to $\mathbf{v} = -B\mathbf{i} + A\mathbf{j}$. Conversely, if $P(x, y)$ lies on this line, we may reverse our steps to obtain $A(x - x_1) + B(y - y_1) = 0$. Adding this equation to $Ax_1 + By_1 + C = 0$, we have $Ax + By + C = 0$, so that $P(x, y) \in S$. Thus, S is equal to the line ℓ.

REMARK. The line ℓ with equation $Ax + By + C = 0$ is parallel to the vector $\mathbf{v} = -B\mathbf{i} + A\mathbf{j}$. If we let $\mathbf{n} = A\mathbf{i} + B\mathbf{j}$, we see that

$$\mathbf{n} \circ \mathbf{v} = -AB + AB = 0$$

so that $\mathbf{n} \perp \mathbf{v}$. $\mathbf{n}$ is said to be perpendicular to ℓ ($\mathbf{n} \perp \ell$) and is called a *normal vector* (or simply a *normal*) to ℓ (Figure 4-4). Thus, if $P_1(x_1, y_1) \in \ell$, then $P(x, y) \in \ell$ iff $\overrightarrow{P_1P} \perp \mathbf{n}$, that is, $\overrightarrow{P_1P} \circ \mathbf{n} = 0$, which leads to the very useful form for the equation of ℓ

$$A(x - x_1) + B(y - y_1) = 0$$

Example 1. Find the equation of the line ℓ passing through $P_1(-2, 3)$ and having normal vector $\mathbf{n} = 5\mathbf{i} + 4\mathbf{j}$.

SOLUTION. $P(x, y)$ lies on ℓ iff $\mathbf{n} \circ \overrightarrow{P_1P} = 0$ (Figure 4-5). Therefore,

$$5(x + 2) + 4(y - 3) = 0$$

which reduces to $5x + 4y - 2 = 0$. Since ℓ is parallel to $\mathbf{v} = -4\mathbf{i} + 5\mathbf{j}$, parametric equations for ℓ are

$$\ell: \begin{cases} x = -2 - 4t \\ y = 3 + 5t, \end{cases} \quad t \in R$$

The General Linear Equation

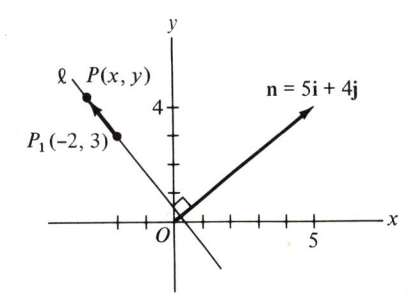

Figure 4-5

Example 2. Given $\ell: 2x - 3y + 6 = 0$. (a) Find a vector perpendicular to ℓ. (b) Find a vector parallel to ℓ. (c) Write a set of parametric equations for ℓ. (d) Find the slope and intercepts of ℓ. (e) Sketch the graph of ℓ.

SOLUTION. $A = 2$ and $B = -3$.
(a) $\mathbf{v}_1 = 2\mathbf{i} - 3\mathbf{j} \perp \ell$
(b) $\mathbf{v}_2 = 3\mathbf{i} + 2\mathbf{j} \| \ell$.   (Note that $\mathbf{v}_1 \perp \mathbf{v}_2$.)
(c) To find a point on ℓ, substitute any number for x, say $x = 1$, then $y = \frac{8}{3}$; so $(1, \frac{8}{3})$ lies on ℓ. Using $\mathbf{v}_2 = 3\mathbf{i} + 2\mathbf{j} \| \ell$, we obtain parametric equations

$$\ell: \begin{cases} x = 1 + 3t \\ y = \frac{8}{3} + 2t, \end{cases} \quad t \in R$$

CHECK. Eliminating the parameter t, we have $t = (x - 1)/3$ from the first equation, and so $y = \frac{8}{3} + \frac{2}{3}(x - 1)$, which reduces to

$$2x - 3y + 6 = 0$$

(d) From the given equation, we have $y = \frac{2}{3}x + 2$. Therefore, $m = \frac{2}{3}$ and $b = 2$. The x-intercept is $a = -3$.
(e) See Figure 4-6.

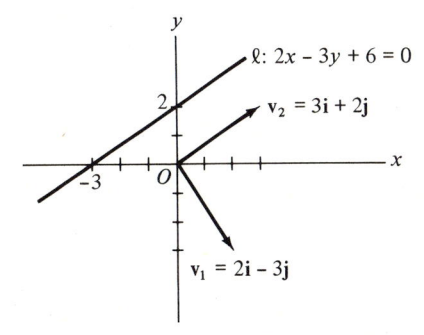

Figure 4-6

Lines, Circles, and Convex Sets in the Plane

EXERCISE 2

1. Write parametric equations and the general equation for the following lines and sketch the graphs.
   (a) through $(5, 0)$ parallel to $\mathbf{v} = 2\mathbf{i} + 6\mathbf{j}$
   (b) through $(-2, -1)$ parallel to $\mathbf{v} = -\mathbf{i} - 4\mathbf{j}$
   (c) through $(0, 0)$ perpendicular to $\mathbf{v} = -\mathbf{i} + \mathbf{j}$
   (d) through $(-3, -4)$ perpendicular to $\mathbf{v} = 2\mathbf{i}$

2. Find vectors $\mathbf{v}_1, \mathbf{v}_2$ which are perpendicular and parallel, respectively, to the following lines. Sketch the lines and plot the vectors.
   (a) $x + y - 3 = 0$
   (b) $2x - 7y + 5 = 0$
   (c) $4x + 3y = 6$
   (d) $y = 5x - 3$
   (e) $2x - 3 = 0$
   (f) $4y + 2 = 0$
   (g) $x = 6y - 3$

3. Find the point of intersection (if it exists) for the following pairs of lines. Sketch the lines.
   (a) $\begin{cases} 2x - y + 1 = 0 \\ 4x - y - 1 = 0 \end{cases}$
   (b) $\begin{cases} 3x + y = 9 \\ 6x - y = 27 \end{cases}$
   (c) $\begin{cases} 3x - 4y + 1 = 0 \\ 2x + 3y - 5 = 0 \end{cases}$

   Solve (d) and (e) by two methods:
   1. Eliminate the parameters and reduce to general form.
   2. Find a pair of values (s, t) of the parameters for which the corresponding point (x, y) lies on both lines.

   (d) $\ell_1 : \begin{cases} x = 1 + 3t \\ y = 4 + t \end{cases}$
   $\ell_2 : \begin{cases} x = -\frac{1}{2} - 3s \\ y = 2 + 2s \end{cases}$

   (e) $\ell_1 : \begin{cases} x = 3 + 6t \\ y = -2 + 3t \end{cases}$
   $\ell_2 : \begin{cases} x = 3 + 2s \\ y = 5 + s \end{cases}$

*4. Let ℓ be the line through P_1 parallel to $\mathbf{v}$. Let Q be a point on ℓ and $\mathbf{v}_1 = k\mathbf{v}$ ($k \neq 0$). Prove that ℓ is equal to the line ℓ_1 through Q parallel to $\mathbf{v}_1$.

3. ANGLE OF INCLINATION; ANGLE BETWEEN TWO LINES; PARALLEL AND PERPENDICULAR LINES

Let ℓ be a nonvertical line passing through the two points $P_1(x_1, y_1)$ and $P_2(x_2, y_2)$. The *angle of inclination* of ℓ is the angle α defined by

Angle of Inclination; Angle Between Two Lines; Parallel and Perpendicular Lines

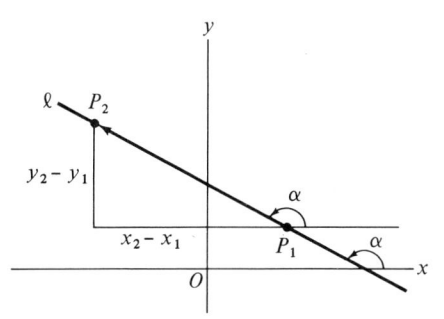

Figure 4-7

$$\tan \alpha = m = \frac{y_2 - y_1}{x_2 - x_1}, \qquad 0 \leq \alpha \leq \pi$$

and is pictured as shown in Figure 4-7. Notice that the angle of inclination is always a first or second quadrant angle. If ℓ is vertical, α is defined to be $\frac{1}{2}\pi$. If $\alpha = 0$ or $\alpha = \pi$, ℓ is horizontal (and so $m = 0$). We may obtain the angle of inclination for a nonvertical line from the general form as follows:

$$\ell = Ax + By + C = 0 \qquad (B \neq 0)$$

Reducing to slope-intercept form, we have

$$y = -\frac{A}{B}x + \left(\frac{-C}{B}\right) = mx + b$$

Thus, $m = \tan \alpha = -A/B$.

Example 1. Find the angle of inclination of the line $2x + 2\sqrt{3}\,y + 5 = 0$.

SOLUTION. Reducing to slope-intercept form,

$$y = -\frac{1}{\sqrt{3}}x - \frac{5}{2\sqrt{3}}$$

Therefore, $m = \tan \alpha = -1/\sqrt{3}$, and so $\alpha = \frac{5}{6}\pi = 150°$.

The Angle Between Two Lines. Let ℓ_1, ℓ_2 have equations

$$A_1x + B_1y + C_1 = 0 \quad \text{and} \quad A_2x + B_2y + C_2 = 0$$

respectively. Then $\mathbf{v}_1 = -B_1\mathbf{i} + A_1\mathbf{j} \| \ell_1$ and $\mathbf{v}_2 = -B_2\mathbf{i} + A_2\mathbf{j} \| \ell_2$. Recall that the angle θ between $\mathbf{v}_1$ and $\mathbf{v}_2$ is given by (see Figure 4-8)

$$\cos \theta = \frac{\mathbf{v}_1 \circ \mathbf{v}_2}{|\mathbf{v}_1| \cdot |\mathbf{v}_2|}, \qquad 0 \leq \theta \leq \pi$$

We define the angle between ℓ_1 and ℓ_2 to be the angle φ given by

$$\cos \varphi = |\cos \theta| = \frac{|\mathbf{v}_1 \circ \mathbf{v}_2|}{|\mathbf{v}_1| \cdot |\mathbf{v}_2|}, \qquad 0 \leq \varphi \leq \frac{\pi}{2}$$

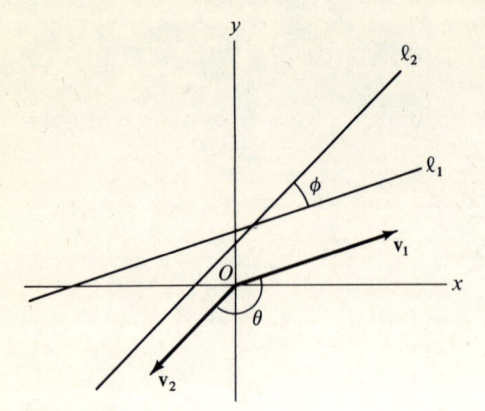

Figure 4-8

Thus, $\varphi = \theta$ if $0 \leq \theta \leq \frac{1}{2}\pi$, whereas $\varphi = \pi - \theta$ if $\frac{1}{2}\pi < \theta \leq \pi$. We sometimes refer to the angle $\pi - \varphi$ as *an* angle between ℓ_1 and ℓ_2.

Observe that

$$\cos \theta = \frac{A_1 A_2 + B_1 B_2}{\sqrt{A_1^2 + B_1^2}\sqrt{A_2^2 + B_2^2}}$$

so that θ is also the angle between the vectors $\mathbf{n}_1 = A_1 \mathbf{i} + B_1 \mathbf{j}$ and $\mathbf{n}_2 = A_2 \mathbf{i} + B_2 \mathbf{j}$, which are normal to ℓ_1 and ℓ_2, respectively. From the definition of φ, we have

$$\cos \varphi = \frac{|A_1 A_2 + B_1 B_2|}{\sqrt{A_1^2 + B_1^2}\sqrt{A_2^2 + B_2^2}}$$

Using this formula and the identity $\sin^2 \varphi + \cos^2 \varphi = 1$, we obtain the formula

$$\sin \varphi = \frac{|A_1 B_2 - A_2 B_1|}{\sqrt{A_1^2 + B_1^2}\sqrt{A_2^2 + B_2^2}}$$

If $\varphi = 0$, we say ℓ_1 and ℓ_2 are *parallel* and write $\ell_1 \| \ell_2$. From the formula for $\sin \varphi$, we see that ℓ_1 and ℓ_2 are parallel iff $A_1 B_2 - A_2 B_1 = 0$. The expression $A_1 B_2 - A_2 B_1$ is written

$$\begin{vmatrix} A_1 & B_1 \\ A_2 & B_2 \end{vmatrix}$$

and is called a (second-order) *determinant*. (See the Appendix for a further discussion of determinants.) The equation $A_1 B_2 - A_2 B_1 = 0$ may be written $A_1/B_1 = A_2/B_2$, provided B_1 and B_2 are nonzero. Thus, two lines are parallel iff *they are both vertical or have the same slope;* that is, they have *the same angle of inclination*. If ℓ_1 and ℓ_2 are parallel and have the same y-intercept, they

Angle of Inclination; Angle Between Two Lines; Parallel and Perpendicular Lines

coincide. Now ℓ_1 and ℓ_2 have the same y-intercept iff $C_1/B_1 = C_2/B_2$, or equivalently,

$$\begin{vmatrix} B_1 & C_1 \\ B_2 & C_2 \end{vmatrix} = 0$$

Thus, the condition that the nonvertical lines ℓ_1 and ℓ_2 be equal is

$$\begin{vmatrix} A_1 & B_1 \\ A_2 & B_2 \end{vmatrix} = \begin{vmatrix} B_1 & C_1 \\ B_2 & C_2 \end{vmatrix} = 0$$

that is,

$$A_1/A_2 = B_1/B_2 = C_1/C_2$$

provided A_2, B_2, and C_2 are nonzero. The vertical lines ℓ_1 and ℓ_2 are equal iff $C_1/A_1 = C_2/A_2$.

If $\varphi = \frac{1}{2}\pi$, we say ℓ_1 and ℓ_2 are perpendicular (or orthogonal) and write $\ell_1 \perp \ell_2$. From the formula for $\cos \varphi$, we see that $\ell_1 \perp \ell_2$ iff $A_1 A_2 + B_1 B_2 = 0$. If neither ℓ_1 nor ℓ_2 is vertical, this may be written $A_1/B_1 \cdot A_2/B_2 = -1$. Letting m_1 and m_2 denote the slope of ℓ_1 and ℓ_2, respectively, we see that $\ell_1 \perp \ell_2$ iff $m_1 = -1/m_2$. Thus, *two (nonvertical) lines are perpendicular iff the slope of one is equal to the negative reciprocal of the slope of the other.*

Example 2. Find the angle between

$$\ell_1 : x + 3y - 6 = 0 \quad \text{and} \quad \ell_2 : 6x - 2y + 3 = 0$$

SOLUTION

$$\cos \varphi = \frac{|1 \cdot 6 + 3(-2)|}{\sqrt{1+9} \cdot \sqrt{36+4}} = 0$$

Therefore, $\varphi = \pi/2$ and so $\ell_1 \perp \ell_2$. Note that $m_1 = -\frac{1}{3}$, whereas $m_2 = 3$; that is, $m_1 = -1/m_2$.

Example 3. Find the angle between

$$\ell_1 : 3x + 2y - 4 = 0 \quad \text{and} \quad \ell_2 : x + 5y + 2 = 0$$

SOLUTION

$$\cos \varphi = \frac{|3 \cdot 1 + 2 \cdot 5|}{\sqrt{9+4}\sqrt{1+25}} = \frac{13}{\sqrt{13}\sqrt{26}} = \frac{1}{\sqrt{2}}$$

Therefore $\varphi = \dfrac{\pi}{4} = 45°$.

The *perpendicular bisector* of the line segment $\overline{P_1 P_2}$ is the line ℓ passing through the midpoint M of $\overline{P_1 P_2}$ and perpendicular to $\overline{P_1 P_2}$. It is not difficult to prove that this is precisely the set of all points P equidistant from P_1 and P_2. (See Problem 8.)

Example 4. Find the perpendicular bisector of $(3, 2)(-1, -4)$.

SOLUTION. The midpoint is $M(1, -1)$. The slope of the line segment is $\frac{3}{2}$. Accordingly, the equation of the perpendicular bisector is

$$y + 1 = -\tfrac{2}{3}(x - 1)$$

that is,

$$2x + 3y + 1 = 0$$

It is convenient to have a formula for the angle φ in terms of the slopes of ℓ_1 and ℓ_2. Assuming $\varphi \neq \tfrac{1}{2}\pi$,

$$\tan \varphi = \frac{\sin \varphi}{\cos \varphi} = \left|\frac{A_1 B_2 - A_2 B_1}{A_1 A_2 + B_1 B_2}\right|$$

If neither ℓ_1 nor ℓ_2 is vertical, we may divide numerator and denominator by $B_1 B_2$ to obtain

$$\tan \varphi = \left|\frac{m_1 - m_2}{1 + m_1 m_2}\right|$$

EXERCISE 3

(*Note:* The student will need a table of trigonometric functions for certain problems in this exercise.)

1. Find the angle of inclination of each of the following lines. Sketch the lines.
    (a) $x - \sqrt{3}y + 2 = 0$
    (b) $2x - 2y + 3 = 0$
    (c) $x + y + 7 = 0$
    (d) $2x + 3y - 6 = 0$
    (e) $x - 4y + 8 = 0$
    (f) $2x = 3$
    (g) $3y = 7$
    (h) $2x + 3y = 0$

2. Find the angle φ between the lines ℓ_1 and ℓ_2 having slopes m_1 and m_2, respectively.
    (a) $m_1 = -\tfrac{2}{3}, m_2 = \tfrac{3}{5}$
    (b) $m_1 = -\tfrac{5}{4}, m_2 = \tfrac{4}{5}$
    (c) $m_1 = 7, m_2 = -3$
    (d) $m_1 = 0, m_2 = -4$

3. Determine whether the following pairs of lines are parallel or perpendicular. If neither, find the angle between the lines. Sketch the lines.
    (a) $\begin{cases} 3x - 2y = 6 \\ -9x + 6y = 10 \end{cases}$
    (b) $\begin{cases} \tfrac{2}{3}x + 4y = 2 \\ \dfrac{x}{2} + 3y = -6 \end{cases}$

(c) $\begin{cases} y = \frac{2}{3}x + 4 \\ 3x + 2y = 7 \end{cases}$

(d) $\begin{cases} x = \frac{4}{5}y + 3 \\ 8x + 10y = 9 \end{cases}$

(e) $\begin{cases} 4x - 5y = 20 \\ -12x + 15y = -60 \end{cases}$

(f) $\begin{cases} y = 3x + 6 \\ 2x - 4y = 9 \end{cases}$

(g) $\begin{cases} y - 2\sqrt{3}x = 4 \\ 7y - \sqrt{3}x = 14 \end{cases}$

(h) $\begin{cases} x + y = 5 \\ 2x - 2y = 5 \end{cases}$

*4. Let $\ell_1: A_1x + B_1y + C_1 = 0$ be neither vertical nor horizontal and let $\ell_2: A_2x + C_2 = 0$ be vertical. Prove that the angle φ between ℓ_1 and ℓ_2 is given by $\tan \varphi = 1/|m_1|$, where m_1 is the slope of ℓ_1.

5. Find the equation of the line ℓ
   (a) parallel to the line $2x - 5y = 10$, and passing through the point $(1, -4)$
   (b) perpendicular to the line $x = 2y - 4$, and passing through the point $(-4, 2)$.
   (c) which is the perpendicular bisector of the segment $\overline{(-5, 6)(3, 2)}$
   (d) passing through the point of intersection of the lines $2x - y = 7$ and $5x + y = 7$, and perpendicular to the line $4x - 3y = 12$
   (e) making an angle of $45°$ with the line $2x - y + 1 = 0$ and having y-intercept -3.

*6. Prove that if ℓ_1 and ℓ_2 have slopes m_1 and m_2, respectively, and the angle between ℓ_1 and ℓ_2 is $\frac{1}{4}\pi$, then

$$m_2 = \frac{1 + m_1}{1 - m_1} \quad \text{or} \quad m_2 = \frac{m_1 - 1}{m_1 + 1}$$

*7. Let ℓ_1 and ℓ_2 be two parallel lines; let $P_1, Q_1 \in \ell_1$ and $P_2, Q_2 \in \ell_2$ such that $\overline{P_1P_2} \perp \ell_1$ and $\overline{Q_1Q_2} \perp \ell_1$. Prove that $|\overline{P_1P_2}| = |\overline{Q_1Q_2}|$.

8. Prove that the perpendicular bisector of the segment $\overline{P_1P_2}$ is equal to the set of all points $P(x, y)$ equidistant from P_1 and P_2.

4. DISTANCE FROM A POINT TO A LINE; THE NORMAL FORM; FAMILIES OF LINES

To determine the distance from a point P_1 to a line ℓ, we proceed as follows.

Recall that the vector $\mathbf{n} = A\mathbf{i} + B\mathbf{j}$ is perpendicular to the line $\ell: Ax + By + C = 0$. Consequently, the line ℓ_1 through $P_1(x_1, y_1)$ perpendicular to ℓ (and hence parallel to $\mathbf{n}$) has parametric equations (see Figure 4–9)

$$\ell_1: \begin{cases} x = x_1 + At \\ y = y_1 + Bt, \end{cases} \quad t \in R$$

Lines, Circles, and Convex Sets in the Plane

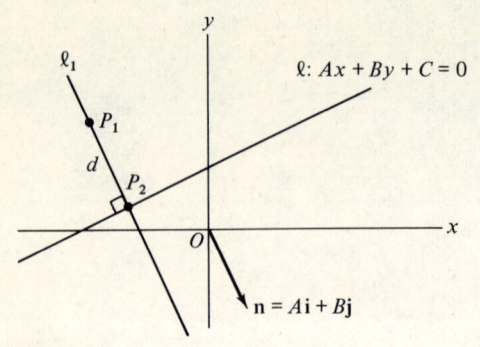

Figure 4-9

The lines ℓ and ℓ_1 intersect in a point $P_2(x_2, y_2)$ which is determined as follows: Substituting x, y from the parametric equations of ℓ_1 in the equation of ℓ, we obtain $A(x_1 + At) + B(y_1 + Bt) + C = 0$. Solving for t,

$$t = -\frac{Ax_1 + By_1 + C}{A^2 + B^2}$$

Denoting this value of t by t_1, the coordinates of P_2 are

$x_2 = x_1 + At_1$
$y_2 = y_1 + Bt_1$

We define the *distance d from P_1 to ℓ* to be the length of the segment $\overline{P_1 P_2}$. Thus,

$$\begin{aligned} d^2 &= (x_2 - x_1)^2 + (y_2 - y_1)^2 \\ &= (A^2 + B^2)t_1^2 \\ &= \frac{(Ax_1 + By_1 + C)^2}{A^2 + B^2} \end{aligned}$$

Therefore,

$$d = \frac{|Ax_1 + By_1 + C|}{\sqrt{A^2 + B^2}}$$

Observe that $d = 0$ iff $P_1(x_1, y_1)$ lies on ℓ.

Example 1. Find the distance from the point $P_1(4, -3)$ to the line $\ell: 3x = 2y + 6$.

SOLUTION. We first write the equation of ℓ in general form: $3x - 2y - 6 = 0$. Then

$$d = \frac{|3(4) - 2(-3) - 6|}{\sqrt{3^2 + (-2)^2}}$$

$$= \frac{12}{\sqrt{13}}$$

Distance from a Point to a Line; the Normal Form; Families of Lines

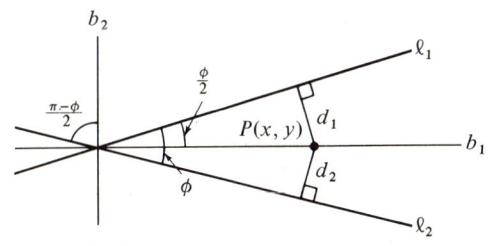

Figure 4–10

Let ℓ_1 and ℓ_2 be nonparallel lines intersecting in the angle φ. An *angle bisector* for the lines ℓ_1 and ℓ_2 is a line b such that every point of b is equidistant from ℓ_1 and ℓ_2 (Figure 4–10). The method used in the next example can be used to show that for every pair of nonparallel lines ℓ_1 and ℓ_2,

(1) there are two angle bisectors;
(2) the two angle bisectors are perpendicular;
(3) one of the angle bisectors makes an angle $\varphi/2$ with ℓ_1 and ℓ_2, whereas the other makes an angle $(\pi - \varphi)/2$ with ℓ_1 and ℓ_2. (See Problem 7.)

Example 2. Find the equations of the angle bisectors for the lines $\ell_1: \sqrt{3}x - y + \sqrt{3} + 1 = 0$ and $\ell_2: x - \sqrt{3}y = 0$.

SOLUTION. Let $P(x, y)$ be any point on an angle bisector. From the definition, we have

$$\frac{|\sqrt{3}x - y + \sqrt{3} + 1|}{\sqrt{3 + 1}} = \frac{|x - \sqrt{3}y|}{\sqrt{1 + 3}}$$

Since both sides are nonnegative, we may square both sides to obtain the equivalent equation

$$(\sqrt{3}x - y + \sqrt{3} + 1)^2 = (x - \sqrt{3}y)^2$$

Transposing and factoring,

$$[(\sqrt{3}x - y + \sqrt{3} + 1) - (x - \sqrt{3}y)]$$
$$\cdot [(\sqrt{3}x - y + \sqrt{3} + 1) + (x - \sqrt{3}y)] = 0$$

Setting each factor equal to zero, we obtain the two bisectors:

$$b_1: (\sqrt{3} - 1)x + (\sqrt{3} - 1)y + \sqrt{3} + 1 = 0$$
$$b_2: (\sqrt{3} + 1)x - (\sqrt{3} + 1)y + \sqrt{3} + 1 = 0$$

The student may verify that if φ is the angle between ℓ_1 and ℓ_2 and φ' is the angle between ℓ_1 and b_2, then $\tan 2\varphi' = \tan \varphi$; that is, $\varphi' = \frac{1}{2}\varphi$. (See Problem 6.)

The Normal Form of the Equation of a Line. Let ℓ have equation $Ax + By + C = 0$. The vector $\mathbf{n} = A\mathbf{i} + B\mathbf{j}$ is normal to ℓ and so the line

Lines, Circles, and Convex Sets in the Plane

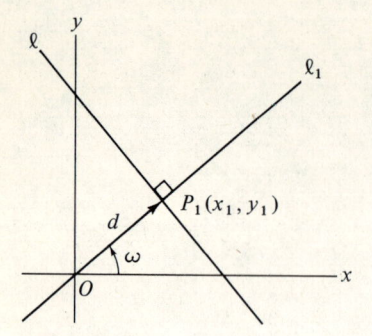

Figure 4-11

ℓ_1 through the origin and perpendicular to ℓ has parametric equations $x = At$, $y = Bt$ (Figure 4-11). The value of t corresponding to the point of intersection $P_1(x_1, y_1)$ of ℓ and ℓ_1 is $t_1 = -C/(A^2 + B^2)$. The coordinates of P_1 are

$$x_1 = \frac{-AC}{A^2 + B^2}, \quad y_1 = \frac{-BC}{A^2 + B^2}$$

The vector $\overrightarrow{OP_1}$ is normal to ℓ and is given by

$$\overrightarrow{OP_1} = \left(\frac{-AC}{A^2 + B^2}\right)\mathbf{i} + \left(\frac{-BC}{A^2 + B^2}\right)\mathbf{j}$$

The length of $\overrightarrow{OP_1}$ is

$$d = |\overrightarrow{OP_1}| = \frac{|C|}{\sqrt{A^2 + B^2}}$$

Assuming $C \neq 0$ (so that ℓ does not pass through the origin), the direction of $\overrightarrow{OP_1}$ is the angle ω (omega) given by

$$\cos \omega = \frac{x_1}{|\overrightarrow{OP_1}|} = -\frac{C}{|C|} \cdot \frac{A}{\sqrt{A^2 + B^2}}$$

$$\sin \omega = \frac{y_1}{|\overrightarrow{OP_1}|} = -\frac{C}{|C|} \cdot \frac{B}{\sqrt{A^2 + B^2}}$$

ω is called the *normal angle* for ℓ. The equation of ℓ may now be written

$$\frac{A}{\sqrt{A^2 + B^2}} x + \frac{B}{\sqrt{A^2 + B^2}} y + \frac{C}{\sqrt{A^2 + B^2}} = 0$$

$$\left(-\frac{|C|}{C} \cos \omega\right) x + \left(-\frac{|C|}{C} \sin \omega\right) y + \frac{C}{\sqrt{A^2 + B^2}} = 0$$

$$x \cos \omega + y \sin \omega - \frac{|C|}{\sqrt{A^2 + B^2}} = 0$$

since $C^2/|C| = |C|$. Since the distance from the origin to ℓ is

$$d = \frac{|C|}{\sqrt{A^2 + B^2}} > 0$$

the equation of ℓ may now be written in the *normal form*

$$x \cos \omega + y \sin \omega - d = 0$$

that is,

$$\frac{A}{-\sqrt{A^2 + B^2}} x + \frac{B}{-\sqrt{A^2 + B^2}} y + \frac{C}{-\sqrt{A^2 + B^2}} = 0 \qquad \text{if } C > 0$$

or

$$\frac{A}{\sqrt{A^2 + B^2}} x + \frac{B}{\sqrt{A^2 + B^2}} y + \frac{C}{\sqrt{A^2 + B^2}} = 0 \qquad \text{if } C < 0$$

The normal form exhibits the direction ω of the normal vector $\overrightarrow{OP_1}$ and the distance d from the origin to ℓ. If $C = 0$, we adopt the convention that $0 \leq \omega \leq \pi$, so that $\sin \omega \geq 0$. In this case, we reduce to normal form by dividing by $\sqrt{A^2 + B^2}$ or $-\sqrt{A^2 + B^2}$, according as $B > 0$ or $B < 0$, respectively. (*Note:* To find the distance from a point $P_1(x_1, y_1)$ to the line ℓ, reduce to normal form, and substitute the coordinates of P_1 in the absolute value of the left member.)

Example 3. Reduce the equation to normal form and find d and ω. Draw a figure. $\sqrt{3}x + y + 4 = 0$.

SOLUTION

$$-\frac{\sqrt{3}x}{2} - \frac{1}{2}y - 2 = 0$$

Therefore, $\cos \omega = -\dfrac{\sqrt{3}}{2}$, $\sin \omega = -\dfrac{1}{2}$, and $d = 2$.

Therefore, $\omega = \frac{7}{6}\pi = 210°$ (see Figure 4–12).

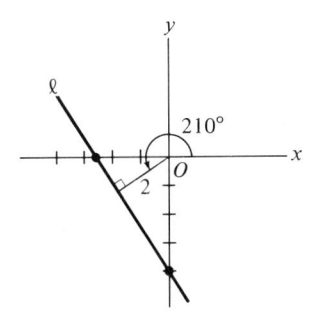

Figure 4–12

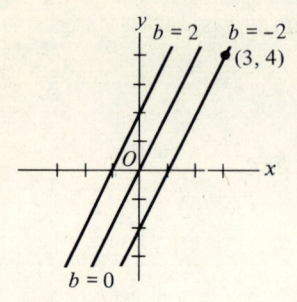

Figure 4-13

Families of Lines. A *family* of lines is simply a set of lines. The equation of a family of lines differs from the equation of a specific line only in that certain constants are unspecified. We give several examples.

Example 4. Find the equation of the family of lines having slope 2. Find that member of the family passing through (3, 4).

SOLUTION. Using the slope-intercept form with $m = 2$, we have $y = 2x + b$. Several members of this family are shown in Figure 4-13. Substituting (3, 4) in the equation of the family of lines, we obtain $b = -2$. The required member is $y = 2x - 2$.

Example 5. Find the equation of the family of lines for which the sum of the intercepts is 6. Find the member of the family whose slope is -3.

SOLUTION. Using the intercept form $x/a + y/b = 1$ and the fact that $a + b = 6$, the equation of the family of lines may be written $x/a + y/(6 - a) = 1$. The slope of any member of this family is $m = -(6 - a)/a$. When $m = -3$, we find that $a = \frac{3}{2}$, so that the desired member is

$$\frac{x}{\frac{3}{2}} + \frac{y}{\frac{9}{2}} = 1$$

Example 6. Write the equation of the family of lines whose distance from the origin is 2. Find the member whose slope is $\frac{2}{3}$.

SOLUTION. Using the normal form, we find the equation of the family of lines to be $x \cos \omega + y \sin \omega - 2 = 0$. The slope of any member is $m = -(\cos \omega)/(\sin \omega)$ (if $\sin \omega \neq 0$). When $m = \frac{2}{3}$, we have $\cot \omega = -\frac{2}{3}$. Hence $\sin \omega = \pm 3/\sqrt{13}$, and $\cos \omega = \mp 2/\sqrt{13}$. Accordingly, there are two solutions:

$$\ell_1: -\frac{2}{\sqrt{13}}x + \frac{3}{\sqrt{13}}y - 2 = 0$$

$$\ell_2: \frac{2}{\sqrt{13}}x - \frac{3}{\sqrt{13}}y - 2 = 0$$

Example 7. Given the lines $\ell_1: x - 2y - 3 = 0$ and $\ell_2: x + 3y - 8 = 0$. (a) Find the equation of the family of lines passing through the point of intersection of ℓ_1 and ℓ_2. (b) Find that member of the family whose y-intercept is 2.

SOLUTION. We use a technique which avoids the calculation of the point of intersection, and is applicable to other curves, for example, circles.

(a) We first observe that for arbitrary constants h and k (not both zero), the equation

$$h(x - 2y - 3) + k(x + 3y - 8) = 0 \tag{1}$$

represents the family of all lines through the point of intersection $P_1(x_1, y_1)$ of ℓ_1 and ℓ_2 (assuming $\ell_1 \not\parallel \ell_2$). This follows from the fact that $x_1 - 2y_1 - 3 = 0$ and $x_1 + 3y_1 - 8 = 0$. Since h and k are not both zero and $\ell_1 \not\parallel \ell_2$, Equation (1) is a linear equation, hence represents a line for all h and k. Furthermore, the equation of every line through $P_1(x_1, y_1)$ is obtainable from Equation (1) by appropriate choice of h and k. (See Problem 8.)

(b) Writing Equation (1) in general form, we have

$$(h + k)x + (-2h + 3k)y + (-3h - 8k) = 0$$

Setting the y-intercept equal to 2,

$$\frac{3h + 8k}{-2h + 3k} = 2$$

This equation reduces to $7h + 2k = 0$, so that $h/k = -2/7$. Thus, we may take $h = -2$ and $k = 7$ (or any proportional pairs) in Equation (1) to obtain the equation

$$x + 5y - 10 = 0$$

EXERCISE 4

1. Find the distance from the given point to the given line.
    (a) $P_1(3, -2); \ell: 2x - 4y = 7$
    (b) $P_1(0, 0); \ell: y = 4x - 3$
    (c) $P_1(6, -1); \ell: 4y - 6x = 9$
    (d) $P_1(0, -5); \ell: 6x = 5$
    (e) $P_1(6, 0); \ell: 3y = 7$

2. Find the angle bisectors for the following pairs of lines. As a partial check, show that the angle bisectors are perpendicular.

    (a) $\begin{cases} x + y = 6 \\ x - y = 4 \end{cases}$

    (b) $\begin{cases} 3x + y = 9 \\ x + 3y = 9 \end{cases}$

(c) $\begin{cases} 3x + 4y = 12 \\ x + \sqrt{3}y = \sqrt{3} \end{cases}$

(d) $\begin{cases} 3y = 5 \\ x + 3y = 6 \end{cases}$

(e) $\begin{cases} \sqrt{3}x + y + 2\sqrt{3} - 3 = 0 \\ \sqrt{3}x - y + 4 - \sqrt{3} = 0 \end{cases}$

(f) $\begin{cases} 3x - 4y = 12 \\ 5x + 12y = 60 \end{cases}$

3. Reduce the following equations to normal form. Find ω and d, and sketch.
   (a) $\sqrt{3}x + y = 8$
   (b) $x + y - 2\sqrt{2} = 0$
   (c) $3y - 4 = 0$
   (d) $x - \sqrt{3}y + 6 = 0$
   (e) $x + y + 2 = 0$
   (f) $x - \sqrt{3}y - 6 = 0$
   (g) $x + 3y + 4 = 0$
   (h) $3x - 4y = 12$
   (i) $2x = 5$
   (j) $x - \sqrt{3}y = 0$
   (k) $x + y = 0$

4. Write the equation of the family of lines having the given property. Find the indicated member(s). Draw a figure.
   (a) The product of the intercepts is 1; the member passing through $(2, -3)$.
   (b) y-intercept equals 4; the member whose distance from the origin is 2.
   (c) The distance from the origin is 5; the member whose slope is $\frac{1}{2}$.
   (d) The normal angle is $210°$; the member having x-intercept -3.
   (e) Passing through $(4, -2)$; the member having equal intercepts.

5. Write the equation of the family of lines passing through the point of intersection of the given pair of lines. Find the indicated member of the family. Solve each problem by two methods (see Example 7 for one method).
   (a) $2x - y - 4 = 0$, $3x + y - 11 = 0$; the member having slope $\frac{3}{4}$.
   (b) $3x - y = 7$, $4x - y = 9$; the member having y-intercept -2.
   (c) $2x + y = -1$, $x + y = 2$; the member whose distance from the origin is $6/\sqrt{58}$.
   (d) $x - 2y - 1 = 0$, $3x - y + 7 = 0$; the member passing through $(2, 1)$.

*6. Complete Example 2 by showing that $\tan 2\varphi' = \tan \varphi$.

*7. Given $\ell_1 : y = m_1 x$ and $\ell_2 : y = m_2 x (m_1 \neq m_2)$. Let φ be the angle of intersection of ℓ_1 and ℓ_2. Prove that
   (a) there are two angle bisectors for the lines ℓ_1 and ℓ_2;
   (b) the two angle bisectors are perpendicular and pass through the point of intersection of ℓ_1 and ℓ_2;
   (c) one of the angle bisectors makes an angle $\frac{1}{2}\varphi$ with ℓ_1 and ℓ_2, whereas the other makes an angle $(\pi - \varphi)/2$ with ℓ_1 and ℓ_2.

*8. (a) Show that if $\ell_1: A_1x + B_1y + C_1 = 0$ and $\ell_2: A_2x + B_2y + C_2 = 0$ are nonparallel and h, k are not both zero, then the equation

$$h(A_1x + B_1y + C_1) + k(A_2x + B_2y + C_2) = 0$$

represents a line. (You must show that $hA_1 + kA_2$ and $hB_1 + kB_2$ are not both zero.)

(b) Show that if $\ell: Ax + By + C = 0$ is a line through the point of intersection $P_1(x_1, y_1)$ of ℓ_1 and ℓ_2, then there exist h and k (not both zero) such that

$$h(A_1x + B_1y + C_1) + k(A_2x + B_2y + C_2) = Ax + By + C$$

5. CIRCLES AND TANGENT LINES

Definition. The *circle* with center $A(h, k)$ and radius $r \geq 0$ is the set (see Figure 4-14)

$$C_r(A) = \{P(x, y) \, | \, \overline{AP} = r\}$$

A point $P(x, y)$ lies on the circle iff

$$\sqrt{(x - h)^2 + (y - k)^2} = r$$

or equivalently,

$$(x - h)^2 + (y - k)^2 = r^2 \tag{1}$$

This is called the *standard form* of the equation of the circle. If $r = 0$, the equation represents a "point-circle" $A(h, k)$.

Example 1. The equation of the circle with center $A(2, -3)$ and radius 4 is

$$(x - 2)^2 + (y + 3)^2 = 16$$

If the binomials in Equation (1) are squared and the terms rearranged, the equation may be written

$$x^2 + y^2 - 2hx - 2ky + h^2 + k^2 - r^2 = 0$$

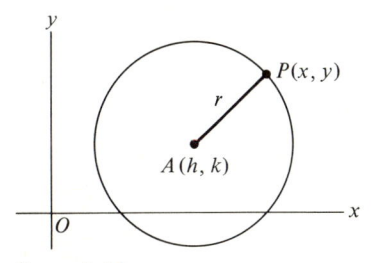

Figure 4-14

which is of the form

$$Ax^2 + Ay^2 + Bx + Cy + D = 0 \tag{2}$$

where A, B, C and D are constants and $A \neq 0$. That every equation of this form represents a circle or no locus may be shown by completing the square in x and y:

$$Ax^2 + Ay^2 + Bx + Cy + D = 0$$

$$\left(x^2 + \frac{Bx}{A} + \left(\frac{B}{2A}\right)^2\right) + \left(y^2 + \frac{Cy}{A} + \left(\frac{C}{2A}\right)^2\right) = \frac{B^2}{4A^2} + \frac{C^2}{4A^2} - \frac{D}{A}$$

$$\left(x + \frac{B}{2A}\right)^2 + \left(y + \frac{C}{2A}\right)^2 = \frac{B^2 + C^2 - 4AD}{4A^2}$$

Denoting the right member by Q, we conclude that if $Q \geq 0$, the equation represents a circle with center $(-B/2A, -C/2A)$ and radius $\sqrt{Q}$. If $Q < 0$, there is no locus, since the sum of squares cannot be negative. Equation (2) is called the *general form* of the equation of the circle.

Example 2. Reduce to standard form and find the center and radius of the following circle:

$$4x^2 + 4y^2 - 4x + 32y + 1 = 0$$

SOLUTION. Dividing through by 4 and rearranging terms, we have

$$\left(x^2 - x + \left(-\tfrac{1}{2}\right)^2\right) + (y^2 + 8y + 4^2) = \tfrac{1}{4} + 16 - \tfrac{1}{4}$$

$$(x - \tfrac{1}{2})^2 + (y + 4)^2 = 16$$

This is the circle with center $A(\tfrac{1}{2}, -4)$ and radius $r = 4$.

Tangent Line to a Circle. Let $C_r(A)$ be the circle with equation $(x - h)^2 + (y - k)^2 = r^2$, so that the center is $A(h, k)$ and the radius is r. We assume that $r > 0$. Now let $P_0(x_0, y_0)$ be a point on the circle so that

$$(x_0 - h)^2 + (y_0 - k)^2 = r^2$$

The line T through P_0 and perpendicular to $\overline{AP_0}$ is called the *tangent* to $C_r(A)$ at P_0 (Figure 4–15). Its equation is (see Section 2)

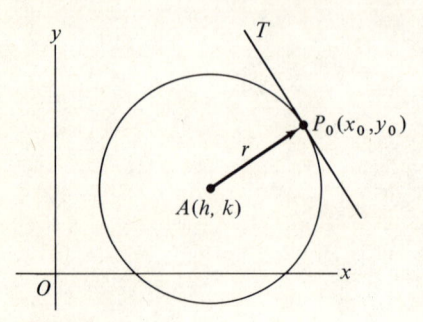

Figure 4–15

$T: (x_0 - h)(x - x_0) + (y_0 - k)(y - y_0) = 0$

It can be shown that P_0 is the only point common to T and $C_r(A)$. Moreover, if ℓ is a line which intersects $C_r(A)$ only in the point $P_0(x_0, y_0)$, then ℓ is perpendicular to $\overrightarrow{AP_0}$, and is therefore the tangent to $C_r(A)$ at P_0. (See Problem 7.)

Clearly, there is only one tangent line to a given circle at a given point on the circle. Moreover, the distance from the center to the tangent line is the radius r of the circle. Conversely, if the distance from the center $A(h, k)$ to a line ℓ is r, then ℓ is tangent to the circle.

Example 3. Find the equation of the line T tangent to the circle $x^2 + y^2 - 2x + 2y - 23 = 0$ at the point $P_0(4, 3)$.

SOLUTION. The standard form of the equation of the circle is

$(x - 1)^2 + (y + 1)^2 = 25$

Thus, the center is $A(1, -1)$ and the radius is $r = 5$. The slope of $\overrightarrow{AP_0}$ is $\frac{4}{3}$. The slope of the tangent line is, therefore, $-\frac{3}{4}$. The equation of T is

$T: y - 3 = -\frac{3}{4}(x - 4)$

which reduces to

$3x + 4y - 24 = 0$

Example 4. Given the circle $(x - 1)^2 + (y + 1)^2 = 25$ and the point $Q(0, 6)$. Find all the lines through Q and tangent to the circle.

SOLUTION. The family of all lines through Q (except the vertical one) is

$y - 6 = m(x - 0)$

or

$mx - y + 6 = 0$

Since the radius of the circle is 5, we seek those members of the family whose distance d from the center $(1, -1)$ is 5. Thus,

$d = \dfrac{|m \cdot 1 - (-1) + 6|}{\sqrt{m^2 + 1}} = 5$

$(m + 7)^2 = 25(m^2 + 1)$
$12m^2 - 7m - 12 = 0$
$(3m - 4)(4m + 3) = 0$

The roots are $m_1 = \frac{4}{3}$ and $m_2 = -\frac{3}{4}$. The two tangent lines are (see Figure 4-16)

$T_1: y - 6 = \frac{4}{3}(x - 0)$

or

$4x - 3y + 18 = 0$

Lines, Circles, and Convex Sets in the Plane

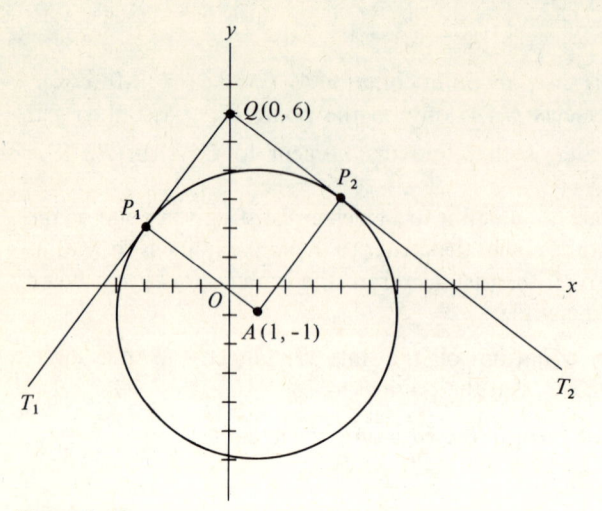

Figure 4-16

and $T_2: y - 6 = -\frac{3}{4}(x - 0)$

or

$3x + 4y - 24 = 0$

Note: Example 4 can be solved by first finding the points of tangency. (See Problem 8.)

If the distance from $P_1(x_1, y_1)$ to the center $A(h, k)$ of $C_r(A)$ is greater than $r > 0$, then there are exactly two lines through P_1 tangent to the circle. (See Problem 9.) The distance between P_1 and the point of tangency $P_0(x_0, y_0)$ is called the *tangential distance* from P_1 to the circle and can be obtained as follows. Since $\overline{P_0P_1} \perp \overline{P_0A}$, we have (see Figure 4-17)

$|\overline{P_0P_1}|^2 = |\overline{AP_1}|^2 - |\overline{AP_0}|^2$
$|\overline{P_0P_1}|^2 = (x_1 - h)^2 + (y_1 - k)^2 - r^2$
$|\overline{P_0P_1}|^2 = x_1^2 + y_1^2 + ax_1 + by_1 + c$

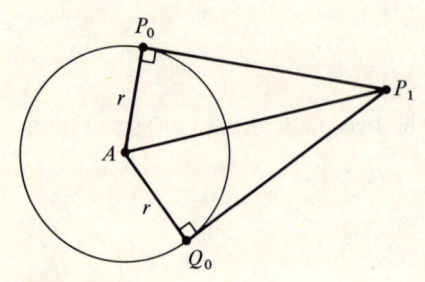

Figure 4-17

Circles and Tangent Lines

Example 5. Find the tangential distance from $P_1(-2, 4)$ to the circle $3x^2 + 3y^2 - 18x + 12y + 12 = 0$.

SOLUTION. We must first divide by 3 to obtain

$$x^2 + y^2 - 6x + 4y + 4 = 0$$

Letting d denote the tangential distance, we have

$$d^2 = (-2)^2 + 4^2 - 6(-2) + 4(4) + 4 = 52$$

Thus, $d = \sqrt{52}$.

EXERCISE 5

1. Write the equation of the circle with the given center A and radius r. Give both the standard form and the general form.
    (a) $A(0, 0); r = 4$
    (b) $A(-2, -2); r = 0$
    (c) $A(4, -\tfrac{1}{2}); r = \sqrt{2}$
    (d) $A(-\tfrac{3}{2}, \tfrac{2}{3}); r = \sqrt[3]{4}$
    (e) $A(5, \tfrac{3}{4}); r = \tfrac{1}{5}$

2. Reduce the following equations to standard form. If the equation represents a circle, give the center and radius.
    (a) $x^2 + y^2 - 6x + 8y + 20 = 0$
    (b) $36x^2 + 36y^2 - 48x + 36y = 551$
    (c) $9x^2 + 9y^2 + 90x - 12y + 274 = 0$
    (d) $16x^2 + 16y^2 - 32x - 8y + 65 = 0$
    (e) $25x^2 + 25y^2 - 250x - 10y + 626 = 0$
    (f) $16x^2 + 16y^2 - 24x + 32y + 25 = 0$

3. Find the equation of the line T tangent to the given circle at the given point.
    (a) $(x - 1)^2 + (y - 1)^2 = 20; (3, 5)$
    (b) $(x + 2)^2 + (y - 3)^2 = 32; (2, -1)$
    (c) $x^2 + y^2 + 4x - 2y - 20 = 0; (-2, -4)$
    (d) $4x^2 - 24x + 4y^2 + 20y = 12; (-1, -1)$

4. Find the equations of the lines (if they exist) passing through the given point and tangent to the given circle. Where appropriate, find the tangential distance from the point to the circle. Draw a figure.
    (a) $(\tfrac{13}{2}, 0); x^2 + y^2 = 13$
    (b) $(1, 1 + \sqrt{3}); (x - 2)^2 + (y - 1)^2 = 1$
    (c) $(-2, 2); x^2 + y^2 + 2x - 2y - 2 = 0$
    (d) $(2, 1); x^2 + y^2 - 6x + 4y - 12 = 0$
    (e) $(-1, 2); x^2 + y^2 + 4x + 2y = 0$
    (f) $(-7, -2); (x + 1)^2 + (y - 2)^2 = 26$

5. Write the equation of the circle with the given point as center and tangent to the given line. Draw a figure.
   (a) $A(3, 1); 3x - 2y + 6 = 0$
   (b) $A(0, 0); x + 4y = 8$
   (c) $A(-2, -3); x + y - 2 = 0$
   (d) $A(-1, 2); 4x + 3y + 12 = 0$
6. A *chord* of a circle $C_r(A)$ $(r > 0)$ is a line segment whose endpoints lie on the circle. A *diameter* is a chord containing the center A of the circle. In parts (a) through (d) assume the center of the circle is at the origin.
   (a) Prove that if d is the length of a diameter, then $d = 2r$. (Thus, A is the midpoint of any diameter and all diameters have the same length.)
   (b) Prove that a line joining the center of a circle to the midpoint of a chord is perpendicular to the chord.
   (c) Prove that a line through the center of a circle perpendicular to a chord passes through the midpoint of the chord.
   (d) Prove that if $\overline{P_1 P_2}$ is a diameter and P_3 is any other point on the circle, then $\overline{P_1 P_3} \perp \overline{P_3 P_2}$. (Thus, an angle inscribed in a semicircle is a right angle.)
   (e) Write the equation of the circle, one of whose diameters is the segment $(-2, 3)(4, -1)$.
   (f) Write the equation of the circle with center $A(4, 3)$ and which has the segment $(1, 5)(3, 2)$ as a chord.
*7. Let $C_r(A)$ be the circle with center $A(h, k)$ and radius $r > 0$. Consider the line $\ell : y = mx + b$.
   (a) Prove that ℓ intersects $C_r(A)$ in at most two points.
   (b) Prove that if ℓ intersects $C_r(A)$ in exactly one point $P_0(x_0, y_0)$, then ℓ is perpendicular to $\overrightarrow{AP_0}$. (For simplicity, let $A = (0, 0)$. Solve simultaneously the equations of ℓ and $C_r(A)$, and use the fact that the resulting quadratic equation can have only one solution.)
*8. Solve Example 4 by first finding the points of tangency.
*9. Let $P_1(x_1, y_1)$ be a point and $C_r(A)$ a circle with positive radius r.
   (a) Show that if $|\overline{AP_1}| > r$, there are exactly two lines through P_1 tangent to the circle.
   (b) Show that if $|\overline{AP_1}| < r$, there are no lines through P_1 tangent to the circle.

6. FAMILIES OF CIRCLES

Observe that in both the general form $x^2 + y^2 + ax + by + c = 0$ and the standard form $(x - h)^2 + (y - k)^2 = r^2$, the equation of a circle involves an ordered triple of constants (a, b, c) or (h, k, r) called *parameters*. A particular circle is determined when, and only when, either of these triples is known. When conditions which yield equations relating a, b, c or h, k, r are imposed, so that one or more of the constants may be eliminated, the result is the equation of the family of circles satisfying the conditions.

Example 1. Find the equation of the family of circles whose centers lie on the line $\ell : 2x - 3y = 6$. Sketch a few members of the family.

Families of Circles

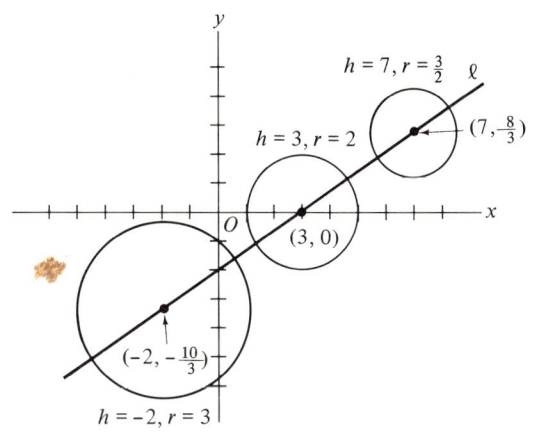

Figure 4-18

SOLUTION. The center (h, k) must satisfy the equation of ℓ; thus

$$2h - 3k = 6$$

Hence $k = (2h - 6)/3$. Substituting this expression for k in the standard form, we have the equation of the required family of circles (see Figure 4-18)

$$(x - h)^2 + \left(y - \frac{2h - 6}{3}\right)^2 = r^2$$

Example 2. Write the equation of the family of circles whose centers lie on the line $\ell: y = x$ and which are tangent to the x- and y-axes. Find the member (or members) of the family passing through $(4, 2)$.

SOLUTION. The center (h, k) must satisfy $k = h$. Moreover, since the members of the family are tangent to the coordinate axes, we must have $r = |h| = |k|$. The equation of this family is, accordingly,

$$(x - h)^2 + (y - h)^2 = h^2$$

For those members passing through $(4, 2)$, we must have

$$(4 - h)^2 + (2 - h)^2 = h^2$$

or

$$h^2 - 12h + 20 = 0$$

The roots are $h = 2$ and $h = 10$. Thus, there are two members of the family passing through $(4, 2)$. They are

$$C_1: (x - 2)^2 + (y - 2)^2 = 4$$
$$C_2: (x - 10)^2 + (y - 10)^2 = 100$$

Example 3. Find the equation of the family of circles passing through the points $P_1(-1, 3)$ and $P_2(2, 1)$. Find the member(s) of the family passing through $P_3(0, -2)$ (Figure 4-19).

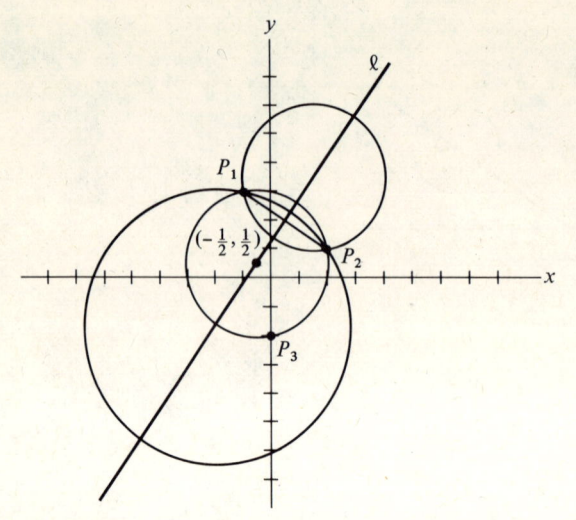

Figure 4-19

SOLUTION. We use the fact that the center $A(h, k)$ must lie on the perpendicular bisector of $\overline{P_1 P_2}$, since it is equidistant from P_1 and P_2. The equation of the perpendicular bisector is

$\ell : 6x - 4y + 5 = 0$

Therefore,

$6h - 4k + 5 = 0$

Moreover, the radius r is given by

$r^2 = |\overline{AP_1}|^2 = (h + 1)^2 + (k - 3)^2$

Thus,

$k = \dfrac{6h + 5}{4}$

and

$r^2 = (h + 1)^2 + \left(\dfrac{6h - 7}{4}\right)^2$

Substituting in the standard form, we obtain

$(x - h)^2 + \left(y - \dfrac{6h + 5}{4}\right)^2 = (h + 1)^2 + \left(\dfrac{6h - 7}{4}\right)^2$

To find the member passing through $P_3(0, -2)$, we substitute $(0, -2)$ in the equation of the family and solve for h. We find that $h = -\frac{1}{2}$. The required member is, accordingly,

$(x + \frac{1}{2})^2 + (y - \frac{1}{2})^2 = \frac{13}{2}$

Families of Circles

A more direct method for obtaining the circle passing through three *noncollinear* points is as follows. Denoting the three points by P_1, P_2, and P_3, we obtain the equations of the perpendicular bisectors of $\overline{P_1P_2}$ and $\overline{P_2P_3}$. These must intersect, since $\overline{P_1P_2} \not\parallel \overline{P_2P_3}$. The point of intersection is the center $A(h, k)$ of the required circle, and the radius is

$$r = |\overline{AP_1}| = |\overline{AP_2}| = |\overline{AP_3}|$$

This method shows that every three noncollinear points determine a unique circle.

A third method consists in solving the three equations in a, b, and c obtained by substituting the coordinates of the three points in the equation

$$x^2 + y^2 + ax + by + c = 0$$

Family of Circles Through Intersections of Two Circles; Radical Axis. Consider the two circles

$$C_1 : x^2 + y^2 + a_1 x + b_1 y + c_1 = 0$$
$$C_2 : x^2 + y^2 + a_2 x + b_2 y + c_2 = 0$$

Let s and t be any numbers, not both zero, and consider the equation

$$s(x^2 + y^2 + a_1 x + b_1 y + c_1) + t(x^2 + y^2 + a_2 x + b_2 y + c_2) = 0 \quad (1)$$

or equivalently,

$$(s + t)x^2 + (s + t)y^2 + (sa_1 + ta_2)x + (sb_1 + tb_2)y + sc_1 + tc_2 = 0$$

If $P_1(x_1, y_1)$ lies on both C_1 and C_2, (x_1, y_1) clearly satisfies Equation (1). Accordingly, Equation (1) represents a curve passing through the points of intersection of C_1 and C_2.

If $s + t \neq 0$, Equation (1) represents a *circle* passing through the points of intersection of C_1 and C_2.

If $s + t = 0$, $t = -s$, and Equation (1) reduces to

$$(a_1 - a_2)x + (b_1 - b_2)y + c_1 - c_2 = 0 \quad (2)$$

since s and t are nonzero. This is the equation of a line, called the *radical axis* of C_1 and C_2. It may be obtained by subtracting the equations of C_1 and C_2. The *line of centers* of C_1 and C_2 is the line joining the centers of the two circles. It is not difficult to show that the radical axis is perpendicular to the line of centers (Figure 4–20).

Equation (1) is the equation of the family of *all* circles passing through the points of intersection of C_1 and C_2, including the radical axis. (See Problem 4.)

Example 4. Given the circles

$$C_1 : x^2 + y^2 - 6x + 8y - 25 = 0$$
$$C_2 : x^2 + y^2 + 2x - 16y + 15 = 0$$

Lines, Circles, and Convex Sets in the Plane

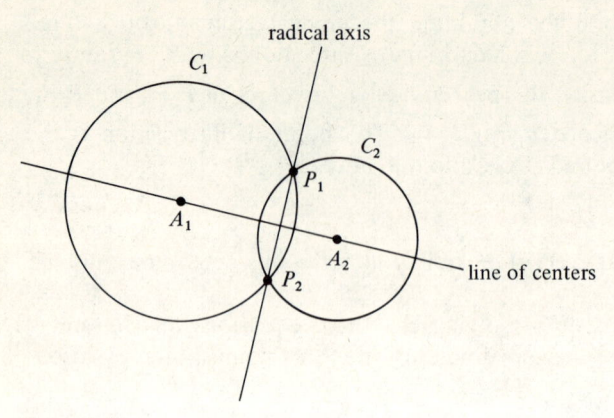

Figure 4-20

(a) Find the circle passing through the points of intersection of C_1 and C_2 and passing through the origin.
(b) Find the radical axis.

SOLUTION

(a) The required circle belongs to the family of circles having equation

$$s(x^2 + y^2 - 6x + 8y - 25) + t(x^2 + y^2 + 2x - 16y + 15) = 0$$

Substituting $(0, 0)$ in the equation of the family, we have

$$s(-25) + t(15) = 0$$

so that

$$\frac{s}{t} = \frac{3}{5}$$

Thus, we may take $s = 3$ and $t = 5$ (or any proportional pairs) in the equation of the family to obtain the equation

$$8x^2 + 8y^2 - 8x - 56y = 0$$

(b) Subtracting the equations of C_1 and C_2 yields the radical axis $x - 3y + 5 = 0$.

EXERCISE 6

1. Write the equation of the family of circles satisfying the given conditions. Find the indicated member(s) of the family. Use either the standard form or the general form, whichever seems appropriate. Draw a figure.
   (a) Centers lie on the line $x + y = 5$; the member(s) of radius 3 tangent to the y-axis.
   (b) Tangent to the coordinate axes and lying in the 2nd or 4th quadrants; the member(s) passing through $(-4, 4)$.

(c) Tangent to the x-axis and the line $y = 2x$; the member(s) having radius 4.
(d) Tangent to the lines $3x - 2y + 4 = 0$ and $3x - 2y + 8 = 0$; the member(s) tangent to the y-axis.
(e) Passing through the points (2, 2) and (−1, 1); the member(s) passing through the origin.

2. Write the equation of the circle (if it exists) passing through the three given points.
    (a) $P_1(-1, 1), P_2(2, 3), P_3(3, -1)$
    (b) $P_1(0, 0), P_2(1, 1), P_3(0, 4)$
    (c) $P_1(-2, -2), P_2(0, 5), P_3(4, 0)$
    (d) $P_1(1, 3), P_2(-2, 1), P_3(1, -4)$
    (e) $P_1(0, -2), P_2(1, 1), P_3(2, 4)$

3. (a) Find the equation of the family of circles passing through the points of intersection of the given pair of circles C_1 and C_2.
   (b) Find the indicated member of the family.
   (c) Find the radical axis and verify that it is perpendicular to the line of centers of the two given circles. Draw a figure.
   (1) $C_1: 4x^2 + 4y^2 - 4x + 12y - 16 = 0$
       $C_2: x^2 + y^2 - 8x - 4y + 10 = 0$
       The member passing through $(1, -7)$.
   (2) $C_1: x^2 + y^2 + 8x - 8y + 22 = 0$
       $C_2: x^2 + y^2 - x + y - 14 = 0$
       The member passing through $(-1, 1)$.
   (3) $C_1: 2x^2 + 2y^2 + 12x - 5y = 0$
       $C_2: 2x^2 + 2y^2 + 16x - 11y = 0$
       The member whose center is on the y-axis.

*4. Let $C_1: x^2 + y^2 + a_1 x + b_1 y + c_1 = 0$ and $C_2: x^2 + y^2 + a_2 x + b_2 y + c_2 = 0$ be distinct circles intersecting in the distinct points $P_1(x_1, y_1)$ and $P_2(x_2, y_2)$ (Figure 4–20).
   (a) Prove that the line of centers is the perpendicular bisector of $\overline{P_1 P_2}$.
   (b) Prove that if $C: x^2 + y^2 + ax + by + c = 0$ is any circle passing through P_1, and P_2, the center of C lies on the line of centers.
   (c) Prove that if $C: x^2 + y^2 + ax + by + c = 0$ is any circle passing through P_1 and P_2, then the equation of C is given by Equation (1) for suitable choice of the ratio $\lambda = t/s$ (or s/t). (Hint: For such s and t to exist, we must have $s + t \neq 0$. Write Equation (1) as

$$x^2 + y^2 + \frac{sa_1 + ta_2}{s + t} x + \frac{sb_1 + tb_2}{s + t} y + \frac{sc_1 + tc_2}{s + t} = 0$$

Find $\lambda = t/s$ such that

$$x^2 + y^2 + \frac{a_1 + \lambda a_2}{1 + \lambda} x + \frac{b_1 + \lambda b_2}{1 + \lambda} y + \frac{c_1 + \lambda c_2}{1 + \lambda} = x^2 + y^2 + ax + by + c$$

*7. CONVEX SETS

The concept of *convexity* of a set plays an important role in much of mathematics. In Chapter 7, we shall relate convexity to the graphs of the conic sections. Convex sets in space will be discussed in Chapter 9. In this section, we show that a half plane and the interior of a circle are convex sets.

Definition. A set S in the plane is *convex* iff for all points $P_1, P_2 \in S$, $\overline{P_1 P_2} \subseteq S$.

Thus, a set is convex iff it contains the line segment joining any two of its points (Figure 4–21). It follows that the empty set as well as sets consisting of a single point, line segments, rays, and lines are all convex sets. The following theorem is quite easy to prove from the definition.

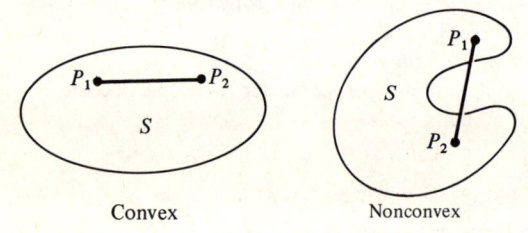

Convex   Nonconvex

Figure 4-21

Theorem 1. *The intersection of two convex sets is convex.*

PROOF. Let S_1 and S_2 be convex sets and let $S = S_1 \cap S_2$. Let P_1 and P_2 be points in S. Since P_1, P_2 are in S_1 and S_1 is convex, $\overline{P_1 P_2} \subseteq S_1$. Similarly, $\overline{P_1 P_2} \subseteq S_2$. Thus, $\overline{P_1 P_2} \subseteq S_1 \cap S_2 = S$. Therefore, S is convex. (Similarly, we can prove that the intersection of any number of convex sets is convex.)

It is clear that the union of two convex sets is not, in general, convex. (See Problem 1.)

Recall that a line is a set of points (x, y) satisfying an equation of the form

$$\ell: Ax + By + C = 0$$

where A and B are not both zero. If we let $f(x, y) = Ax + By + C$, then

$$\ell = \{(x, y) | f(x, y) = 0\}$$

Now for any point (x, y) in the plane, one and only one of the following holds: $f(x, y) = 0, f(x, y) > 0$, or $f(x, y) < 0$. Let

$$S_1 = \{(x, y) | f(x, y) > 0\}$$
$$S_2 = \{(x, y) | f(x, y) < 0\}$$

S_1 and S_2 are called the "two sides" of the line or the two *half planes* determined by ℓ. In the next theorem, we prove that ℓ, S_1, and S_2 are convex sets. The following lemma is used in the proof of the theorem.

Lemma. Let $P_1(x_1, y_1)$ and $P_2(x_2, y_2)$ be points, $P(x, y) \in \overline{P_1 P_2}$, and $f(x, y) = Ax + By + C$. Then there exists t with $0 \leq t \leq 1$ such that

$$f(x, y) = (1 - t)f(x_1, y_1) + tf(x_2, y_2)$$

PROOF. Since $P(x, y) \in \overline{P_1 P_2}$, there exists t with $0 \leq t \leq 1$ such that (see Section 1)

$$\begin{cases} x = x_1 + t(x_2 - x_1) \\ y = y_1 + t(y_2 - y_1) \end{cases}$$

Therefore,

$$\begin{aligned} f(x, y) &= A[x_1 + t(x_2 - x_1)] + B[y_1 + t(y_2 - y_1)] + C \\ &= Ax_1 + By_1 + C + t[(Ax_2 + By_2 + C) - (Ax_1 + By_1 + C)] \\ &= f(x_1, y_1) + t[f(x_2, y_2) - f(x_1, y_1)] \\ &= (1 - t)f(x_1, y_1) + tf(x_2, y_2) \end{aligned}$$

Theorem 2. *The sets ℓ, S_1, and S_2 defined above are nonvoid convex sets.*

PROOF. We shall consider only S_1 and leave the rest of the proof as an exercise. (See Problem 2.) To prove that S_1 is nonvoid, we must show that there exists (x, y) such that $f(x, y) > 0$; that is, $Ax + By + C > 0$. We assume $B \neq 0$. (The case $A \neq 0$ is similar.)

Choose x_0 arbitrarily, and determine y so that $By > -Ax_0 - C$; that is, choose $y > (-Ax_0 - C)/B$ if $B > 0$, and $y < (-Ax_0 - C)/B$ if $B < 0$. Then $f(x_0, y) > 0$, so that $(x_0, y) \in S_1$.

To prove S_1 convex, let $P_1(x_1, y_1)$ and $P_2(x_2, y_2)$ be points in S_1 and let $P(x, y) \in \overline{P_1 P_2}$. We show that $P(x, y) \in S$. By the lemma there exists t with $0 \leq t \leq 1$ such that

$$f(x, y) = (1 - t)f(x_1, y_1) + tf(x_2, y_2)$$

Now by hypothesis $f(x_1, y_1) > 0$ and $f(x_2, y_2) > 0$. Furthermore, at least one of t and $1 - t$ must be positive. It follows that

$$(1 - t)f(x_1, y_1) + tf(x_2, y_2) > 0$$

and so $f(x, y) > 0$. Thus, $P(x, y) \in S_1$ and so $\overline{P_1 P_2} \subseteq S_1$; that is, S_1 is convex.

The essence of the next theorem is that a line segment passing from S_1 to S_2 must intersect ℓ (Figure 4-22).

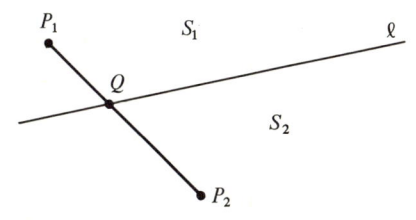

Figure 4-22

Theorem 3. Let $P_1(x_1, y_1) \in S_1$ and $P_2(x_2, y_2) \in S_2$. Then $\overline{P_1P_2}$ intersects ℓ in exactly one point.

PROOF. We seek a point $Q(x, y)$ such that $Q \in \ell \cap \overline{P_1P_2}$; that is, for some t with $0 \leq t \leq 1$,

$$x = x_1 + t(x_2 - x_1)$$
$$y = y_1 + t(y_2 - y_1)$$

and $f(x, y) = 0$. By the lemma we must have

$$f(x, y) = (1 - t)f(x_1, y_1) + tf(x_2, y_2) = 0$$

Solving for t, we obtain

$$t = \frac{f(x_1, y_1)}{f(x_1, y_1) - f(x_2, y_2)}$$

Since $f(x_1, y_1) > 0$ and $f(x_2, y_2) < 0$, it follows that $f(x_1, y_1) - f(x_2, y_2) > f(x_1, y_1) > 0$, and so

$$0 < \frac{f(x_1, y_1)}{f(x_1, y_1) - f(x_2, y_2)} < 1$$

The point $Q(x, y)$ on $\overline{P_1P_2}$ corresponding to this value of t has the required properties.

For a given line ℓ, the related half planes S_1 and S_2 may be represented graphically as in the following example.

Example. Given the line $\ell: 3x - 2y + 6 = 0$. Sketch the line and the sets S_1 and S_2.

SOLUTION

$$S_1 = \{(x, y) | 3x - 2y + 6 > 0\}$$

that is, $y < \frac{3}{2}x + 3$. Now let x_1 be any number and let $y_1 = \frac{3}{2}x_1 + 3$. Then (x_1, y_1) lies on ℓ (Figure 4-23). Accordingly, if $y < y_1 = \frac{3}{2}x_1 + 3$, the point

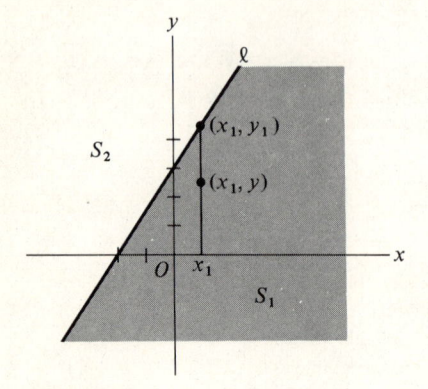

Figure 4-23

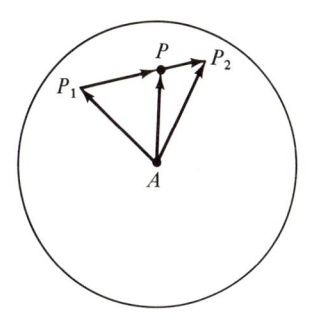

Figure 4-24

(x_1, y) is plotted below (x_1, y_1). Therefore, the set of all points (x, y) such that $y < \frac{3}{2}x + 3$ (that is, the set S_1) lies below the line ℓ (Figure 4–23). Similarly, S_2 consists of all points above the line ℓ. In Exercise 7 are some examples in which S_1 lies above the line, whereas S_2 lies below.

In the next theorem, we show how vectors may be used to prove that a set is convex. Recall that $C_r(A)$ is the circle with center A and radius r. If $r > 0$, we define the *interior* of the circle to be the set

$$\text{int } C_r(A) = \{P(x,y) \,|\, |\overrightarrow{AP}| < r\}$$

Theorem 4. *int $C_r(A)$ is a convex set.*

PROOF. Let $P_1 P_2 \in \text{int } C_r(A)$ and $P \in \overline{P_1 P_2}$ (Figure 4–24). Then $\overrightarrow{AP_1} + \overrightarrow{P_1 P} = \overrightarrow{AP}$. But $\overrightarrow{P_1 P} = t\overrightarrow{P_1 P_2}$ for some t with $0 \leq t \leq 1$, and $\overrightarrow{P_1 P_2} = \overrightarrow{AP_2} - \overrightarrow{AP_1}$. Therefore,

$$\overrightarrow{AP} = \overrightarrow{AP_1} + t(\overrightarrow{AP_2} - \overrightarrow{AP_1})$$
$$= (1 - t)\overrightarrow{AP_1} + t\overrightarrow{AP_2}$$

By the triangle inequality,

$$|\overrightarrow{AP}| \leq |(1 - t)\overrightarrow{AP_1}| + |t\overrightarrow{AP_2}|$$
$$= (1 - t)|\overrightarrow{AP_1}| + t|\overrightarrow{AP_2}| \quad \text{(since } 1 - t \geq 0 \text{ and } t \geq 0\text{)}$$
$$< (1 - t)r + tr \quad \text{(since } P_1, P_2 \in \text{int } C_r(A)\text{)}$$
$$= r$$

Therefore, $|\overrightarrow{AP}| < r$, and so $P \in \text{int } C_r(A)$. Hence $\overline{P_1 P_2} \subseteq \text{int } C_r(A)$, and int $C_r(A)$ is convex.

EXERCISE 7

1. Draw a figure showing two convex sets whose union is nonconvex.
2. Complete the proof of Theorem 2.
3. Use the definition of convex set to prove that the following sets are convex.
   (a) The "solid" triangle with vertices $A(0, 0)$, $B(1, 0)$, $C(0, 1)$; that is,
   $\{(x, y) \,|\, x \geq 0, y \geq 0, \text{ and } y \leq 1 - x\}$

(b) The "solid" rectangle with vertices $A(0, 0)$, $B(3, 0)$, $C(3, 2)$, $D(0, 2)$; that is
$\{(x, y) | 0 \le x \le 3 \text{ and } 0 \le y \le 2\}$

(c) The "solid" triangle with vertices $A(-2, 0)$, $B(2, 0)$, $C(0, 4)$; that is
$\{(x, y) | -2 \le x \le 2, y \ge 0, y \le 2x + 4, y \le -2x + 4\}$

*(d) The "interior" of the parabola $y^2 = x$; that is, $\{(x, y) | x > 0 \text{ and } y^2 < x\}$.

4. Draw a figure showing the following point sets:
   (a) $\{(x, y) | 2x + y - 4 \ge 0 \text{ and } x - 3y + 3 < 0\}$
   (b) $\{(x, y) | y > 2 \text{ and } 3x - y + 6 > 0\}$
   (c) $\{(x, y) | y \ge -3 \text{ and } x \le 2\}$
   (d) $\{(x, y) | -1 \le x \le 1\}$
   (e) $\left\{(x, y) \Big| \dfrac{x}{-8} + \dfrac{y}{5} \le 1, \dfrac{x}{3} + \dfrac{y}{4} \le 1, \text{ and } \dfrac{x}{-6} + \dfrac{y}{1} \ge 1\right\}$
   (f) $\{(x, y) | x^2 + y^2 < 9 \text{ and } y \ge x\}$

5. For the following lines and line segments,
   (a) indicate in a figure the sets S_1 and S_2.
   (b) write the parametric equations of the given line segment and find the value of the parameter t (if it exists) for which the corresponding point lies on ℓ. Draw a figure.

   (1) $\ell: 3x - 5y + 15 = 0; \overline{(-5, 2)(3, -1)}$
   (2) $\ell: x - 5y + 5 = 0; \overline{(2, 1)(3, -1)}$
   (3) $\ell: y = 3; \overline{(-4, 5)(3, 1)}$
   (4) $\ell: 2x + 3y + 6 = 0; \overline{(-1, 2)(-5, 1)}$
   (5) $\ell: x - y = 4; \overline{(1, -1)(2, 2)}$

*6. Let $A(3, 2)$, $B(-3, 4)$, and $C(-1, 1)$ be the vertices of a triangle. Express the solid triangle ABC as the intersection of three half planes. Conclude that $\triangle ABC$ is convex.

*7. Let $C_r(A)$ and int $C_r(A)$ be as in Theorem 4. The *exterior* of $C_r(A)$ is the set
$$\text{ext } C_r(A) = \{P \mid |\overrightarrow{AP}| > r\}$$
Prove that if $P_1 \in \text{int } C_r(A)$ and $P_2 \in \text{ext } C_r(A)$, the segment $\overline{P_1 P_2}$ intersects $C_r(A)$ in exactly one point.

chapter 5
Polar Coordinates

1. DEFINITION OF POLAR COORDINATES

In many instances, curves in the x–y plane which are represented by a rather complicated equation in x and y have a relatively simple representation in "polar coordinates." These are defined as follows.

Recall that associated with the point $P(x, y)$ is its position vector $\overrightarrow{OP}$, where O is the origin (Figure 5–1). Let r denote the length of $\overrightarrow{OP}$, and θ an angle which $\overrightarrow{OP}$ makes with the positive x-axis. Then

$$\begin{cases} x = r \cos \theta \\ y = r \sin \theta \end{cases} \tag{1}$$

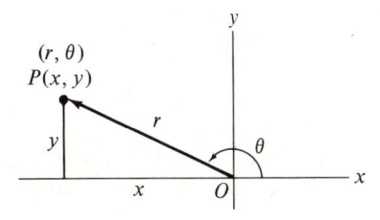

Figure 5-1

The pair (r, θ) is called "a polar coordinate pair" for the point (x, y). In fact, any numbers r, θ (positive, negative, or zero) satisfying Equations (1) are called *polar coordinates* for the point (x, y). Observe that if r, θ are polar coordinates of P, then so are $-r, \theta + \pi$, since $\cos(\theta + \pi) = -\cos\theta$ and $\sin(\theta + \pi) = -\sin\theta$. Furthermore, since $\cos\theta$ and $\sin\theta$ are periodic functions with period 2π (that is, $\cos(\theta + 2\pi) = \cos\theta$ and $\sin(\theta + 2\pi) = \sin\theta$), each point $P(x, y)$ has associated with it infinitely many polar coordinate pairs. However, it is clear that each polar coordinate pair determines exactly one point (x, y) satisfying Equations (1); that is Equations (1) determine a function which maps (r, θ) into (x, y). When using polar coordinates (r, θ) to represent a point (x, y), we usually refer to the origin as the *pole* and the positive x-axis as the *polar axis*. We call r the *radius vector* and θ the *polar angle*. To plot the point (x, y) having polar coordinates (r, θ), we draw the angle θ with vertex at the pole and initial side along the polar axis (counter clockwise if $\theta > 0$, clockwise if $\theta < 0$). An analysis of Equations (1) shows that if $r > 0$, $P(x, y)$ is located on the terminal side of θ at a distance r from the pole; if $r < 0$, $P(x, y)$ is located on the *extension of the terminal side through the pole* at a distance $|r|$ from the pole. The same "dot" in Figure 5–1 depicts both the pair (x, y) and any associated polar coordinate pair (r, θ). The angle θ may be given in radian measure or degree measure. However, when degree measure is intended, the degree symbol "°" shall be used. For instance, in the polar coordinate pair $(5, 2)$, $\theta = 2$ radians; whereas in the polar pair $(5, 2°)$, $\theta = 2$ degrees.

Example 1. Plot the points in the x–y plane having the following polar coordinates: $(2, 30°), (-3, 45°), (3, -60°), (1, 120°), (-2, -150°), (-3, 120°), (\frac{3}{2}, 225°), (4, \frac{1}{2}\pi), (-2, \frac{2}{3}\pi)$.

SOLUTION. See Figure 5–2.

Note that the pairs $(2, 30°)$ and $(-2, -150°)$ represent the same (rectangular) point, $(\sqrt{3}, \frac{1}{2})$. Similarly, both $(3, -60°)$ and $(-3, 120°)$ represent the point $(3/2, -3\sqrt{3}/2)$ in rectangular coordinates.

Given the polar coordinates r, θ of a point P, the rectangular coordinates x, y are determined uniquely by Equations (1). Conversely, given a point $P(x, y)$ in rectangular coordinates, we may use Equations (1) to determine any number of polar coordinate pairs (r, θ) corresponding to (x, y). In fact, by squaring Equations (1) and adding, we obtain $r^2 = x^2 + y^2$. If $x \neq 0$, we may divide the second by the first of Equations (1) to obtain $\tan\theta = y/x$. Thus, to convert from rectangular to polar coordinates, we use

$$\begin{cases} r = \sqrt{x^2 + y^2} \\ \tan\theta = \dfrac{y}{x} \end{cases} \quad (2)$$

Care must be exercised in using the second of Equations (2) to assure that θ is chosen in the proper quadrant. (Recall that $\tan(\theta + \pi) = \tan\theta$.) If $x = 0$ and $y \neq 0$, we may take $r = |y|$, and $\theta = \frac{1}{2}\pi$ if $y > 0$, $\theta = -\frac{1}{2}\pi$ if $y < 0$. For the origin $(0, 0)$, we take $r = 0$ and θ arbitrary.

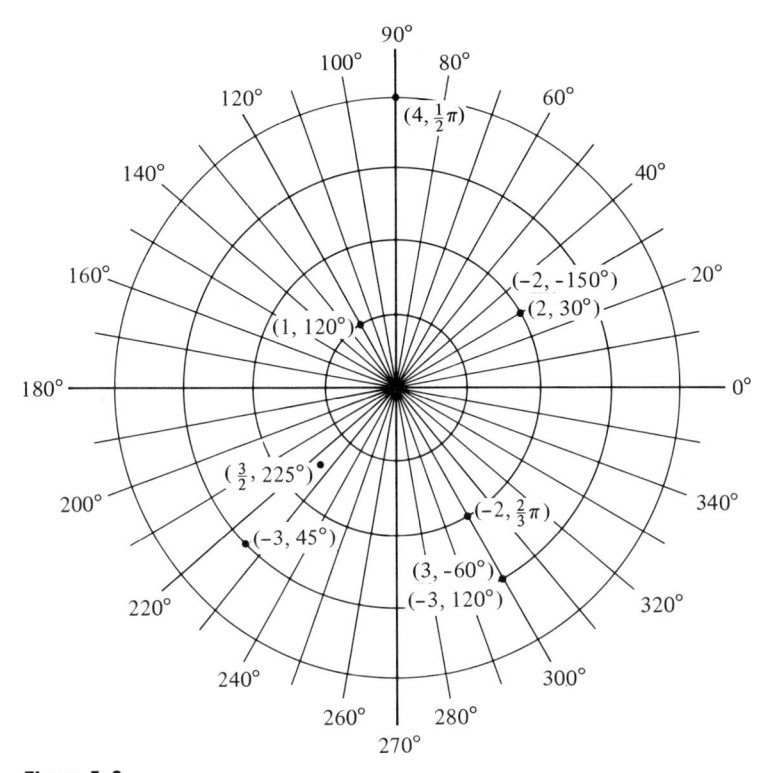

Figure 5-2

Example 2
(a) Find the point (x, y) having polar coordinates $(-3, 30°)$.
(b) Find two pairs of polar coordinates for the (rectangular) point $(-1, -\sqrt{3})$.

SOLUTION
(a) $r = -3, \theta = 30°$

$x = (-3)\cos 30° = -\dfrac{3\sqrt{3}}{2}$

$y = (-3)\sin 30° = -\tfrac{3}{2}$

Therefore, the required point is $(-3\sqrt{3}/2, -3/2)$.

(b) $x = -1, y = -\sqrt{3}$

$r = \sqrt{(-1)^2 + (-\sqrt{3})^2} = \sqrt{4} = 2$

$\tan \theta = \dfrac{-\sqrt{3}}{-1} = \sqrt{3}$

Therefore, $\theta = 60°$ or $240°$.

However, since $(-1, -\sqrt{3})$ lies in the third quadrant, and we have chosen

r positive, we must choose $\theta = 240°$. Thus, a pair of polar coordinates for the given point is $(2, 240°)$. If we take $r = -2$, then we must choose $\theta = 60°$. Thus, $(-2, 60°)$ is another polar coordinate pair representing the given point.

Notationally, (r, θ) always denotes a polar coordinate pair and (x, y) a rectangular pair. When a particular pair such as $(2, 1)$ is given, one must specify whether polar or rectangular coordinates are intended.

Two polar coordinate pairs (r_1, θ_1), (r_2, θ_2), corresponding to the same rectangular pair (x, y), are said to be *equivalent*. From Equations (1), we see that any pair of the form $(r, \theta + 2n\pi)$ or $(-r, \theta + (2n + 1)\pi)$, n an integer, is equivalent to (r, θ). Moreover, it can be shown that for $r \neq 0$ these are the only pairs equivalent to (r, θ). (See Problem 3.)

EXERCISE 1

1. Find the rectangular coordinates of the points corresponding to the following polar coordinate pairs. Plot the points in the x–y plane.
   (a) $(2, -\frac{1}{6}\pi)$
   (b) $(-1, \pi)$
   (c) $(1, 0)$
   (d) $(3, \frac{7}{4}\pi)$
   (e) $(0, 1)$
   (f) $(-2, -\frac{1}{4}\pi)$
   (g) $(2, \frac{3}{4}\pi)$
   (h) $(-\frac{5}{2}, 300°)$
   (i) $(2, -15°)$

2. Find three polar coordinate pairs corresponding to each of the following points given in rectangular coordinates. Plot the points in the x–y plane.
   (a) $(1, 1)$
   (b) $(\sqrt{3}, -1)$
   (c) $(-1, -\sqrt{3})$
   (d) $(0, 3)$
   (e) $(2, 0)$
   (f) $(-1, 0)$
   (g) $(-1, 1)$
   (h) $(-2, -3)$

*3. Prove that if (r_1, θ_1) and (r_2, θ_2) are equivalent, then

   (i) $r_1 = r_2 = 0$ (θ_1, θ_2 arbitrary)

   or

   (ii) $r_1 = r_2$ and $\theta_1 = \theta_2 + 2n\pi$ for some integer n

   or

   (iii) $r_1 = -r_2$ and $\theta_1 = \theta_2 + (2n + 1)\pi$ for some integer n

   (*Hint:* Multiply $r_1 \cos \theta_1 = r_2 \cos \theta_2$ and $r_2 \sin \theta_2 = r_1 \sin \theta_1$ to obtain $r_1 r_2 (\sin \theta_1 \cos \theta_2 - \cos \theta_1 \sin \theta_2) = 0$.)

2. POLAR CURVES

We are concerned here with graphs of equations of the form

$$F(r, \theta) = 0 \tag{1}$$

where r, θ are polar coordinates and F is a function of two variables. The fact that r, θ are polar coordinates becomes significant when we interpret the graph of Equation (1) as a *set of points in the x–y plane*.

Definition. The graph of $F(r, \theta) = 0$ (in the x–y plane) is the set of all points (x, y) such that for some polar coordinate pair (r, θ) corresponding to (x, y), $F(r, \theta) = 0$. The graph of $F(r, \theta) = 0$ is called a *polar curve*, while the equation $F(r, \theta) = 0$ is called a *polar equation*.

REMARK. All that is required for a point (x, y) to be on a polar curve is that there exist *at least one* polar coordinate pair (r, θ) corresponding to (x, y) and satisfying Equation (1). Since a point (x, y) is associated with infinitely many (equivalent) polar pairs (r, θ), it often happens that while *some* of these pairs satisfy Equation (1), others do not. They, in fact, may satisfy another equation $G(r, \theta) = 0$. We are, thus, faced with the (sometimes disturbing) fact that *a graph in the x–y plane may correspond to more than one polar equation*. (See Example 1.)

In sketching the graph of a polar equation, we plot the points (x, y) directly by measuring the polar angle θ from the polar axis and then marking off the radius vector r along the terminal side of θ (or the extension of the terminal side).

Example 1. Sketch the graphs of the polar equations (a) $r = a (a > 0,$ constant), and (b) $\theta = c (c > 0,$ constant).

SOLUTION

(a) The graph of $r = a$ consists of all points (x, y) which have radius vector $r = a$. θ is arbitrary. The graph is the circle with center at O and radius a. (Note that this is also the graph of the equation $r = -a,\ a > 0$.) (See Figure 5–3(a).)

(b) The graph of $\theta = c$ consists of all points (x, y) which have polar angle $\theta = c$. r is arbitrary. The graph is the straight line through the pole, making angle c with the polar axis. (This graph also has polar equation $\theta = c + n\pi$, n an integer.) (See Figure 5–3(b).)

The equation $F(r, \theta) = 0$ often has the form $r = f(\theta)$, where f is a function of one variable.

Example 2. Sketch the polar curve $r = \cos \theta$.

DISCUSSION. We first make a table of pairs (r, θ) satisfying the given equation, tentatively letting θ vary from 0 to 2π, although the range needed to complete

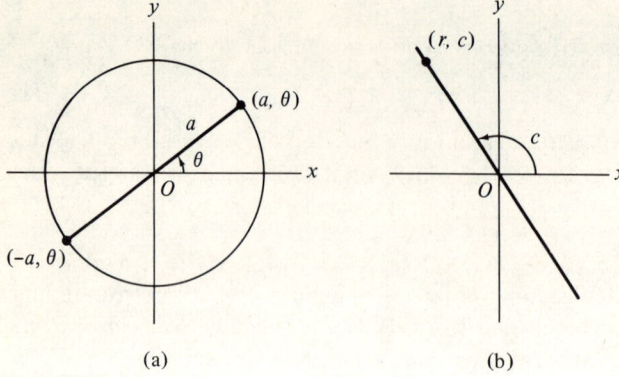

Figure 5-3 (a) $r = a$; $r = -a$. (b) $\theta = c$; $\theta = c+\pi$.

the graph may be different from this. The table and graph are shown in Figure 5-4. Note that the points (x, y) corresponding to θ in the range $\pi \leq \theta \leq 2\pi$ are precisely those points obtained with θ in the range $0 \leq \theta \leq \pi$.

We can save time in sketching polar curves if we take advantage of any symmetries the curve may have. The tests for symmetry which follow are sufficient but not necessary. That is to say, if the polar equation satisfies the test, the curve possesses the indicated symmetry. However, if the equation does not "pass the test," the curve may or may not have the indicated symmetry. (This is due to the fact that of two equivalent pairs, one may satisfy the equation whereas the other may not.)

Tests for Symmetry. (See Figure 5-5.)
(1) If $F(r, \theta) = 0$ implies $F(-r, \theta) = 0$, then the curve is symmetric with respect to the origin (Figure 5-5(a)).
(2) If $F(r, \theta) = 0$ implies $F(r, -\theta) = 0$, the curve is symmetric with respect to the x-axis (Figure 5-5(b)).

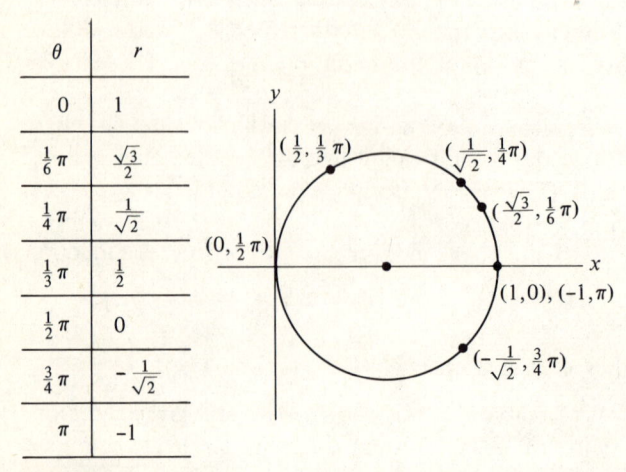

θ	r
0	1
$\frac{1}{6}\pi$	$\frac{\sqrt{3}}{2}$
$\frac{1}{4}\pi$	$\frac{1}{\sqrt{2}}$
$\frac{1}{3}\pi$	$\frac{1}{2}$
$\frac{1}{2}\pi$	0
$\frac{3}{4}\pi$	$-\frac{1}{\sqrt{2}}$
π	-1

Figure 5-4 $r = \cos\theta$.

Polar Curves

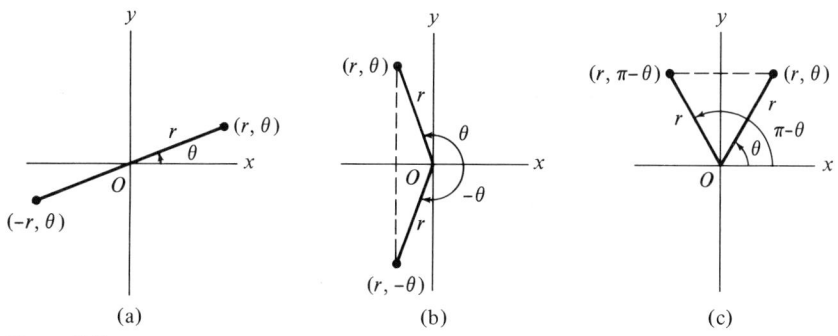

Figure 5-5

(3) If $F(r, \theta) = 0$ implies $F(r, \pi - \theta) = 0$, the curve is symmetric with respect to the y-axis (Figure 5-5(c)).

Recall that any two of the above symmetries imply the third. In Example 2 above, $r = \cos \theta$ implies $r = \cos(-\theta)$, since $\cos(-\theta) = \cos \theta$. Therefore, the curve is symmetric about the x-axis. Consequently, having sketched the graph for θ in the range $0 \le \theta \le \frac{1}{2}\pi$, we obtain the graph for θ in the range $-\frac{1}{2}\pi \le \theta \le 0$ by reflection about the x-axis.

Example 3. Test for symmetry and sketch the curve $r = \sin 3\theta$.

DISCUSSION. Since $r = \sin 3\theta$ does not imply $-r = \sin 3\theta$, the test for symmetry about the origin fails. Also, $r = \sin 3\theta$ does not imply $r = \sin 3(-\theta)$, since $\sin 3(-\theta) = -\sin 3\theta$. Thus, the test for symmetry about the x-axis fails. We now consider symmetry about the y-axis. Replacing

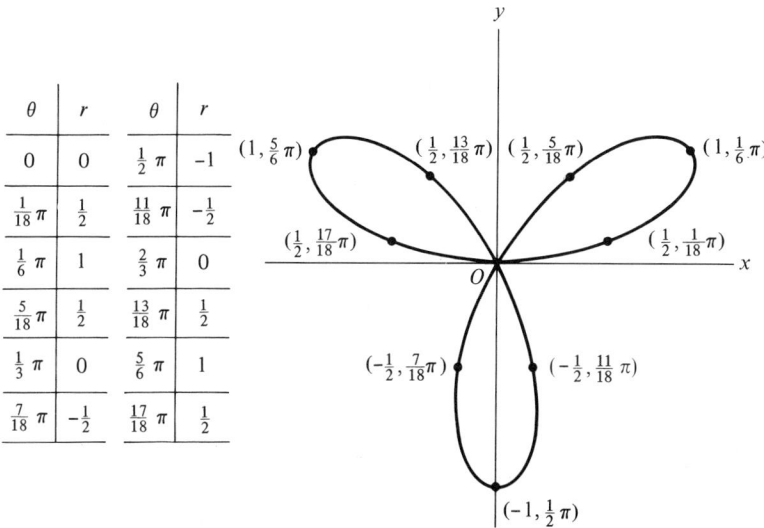

θ	r	θ	r
0	0	$\frac{1}{2}\pi$	-1
$\frac{1}{18}\pi$	$\frac{1}{2}$	$\frac{11}{18}\pi$	$-\frac{1}{2}$
$\frac{1}{6}\pi$	1	$\frac{2}{3}\pi$	0
$\frac{5}{18}\pi$	$\frac{1}{2}$	$\frac{13}{18}\pi$	$\frac{1}{2}$
$\frac{1}{3}\pi$	0	$\frac{5}{6}\pi$	1
$\frac{7}{18}\pi$	$-\frac{1}{2}$	$\frac{17}{18}\pi$	$\frac{1}{2}$

Figure 5-6 $r = \sin 3\theta$.

θ by $\pi - \theta$, we have

$$\sin 3(\pi - \theta) = \sin(3\pi - 3\theta)$$
$$= \sin 3\pi \cos 3\theta - \cos 3\pi \sin 3\theta$$
$$= 0 - (-1)\sin 3\theta$$
$$= \sin 3\theta$$

Therefore, $r = \sin 3\theta$ implies $r = \sin 3(\pi - \theta)$; that is, the curve is symmetric about the y-axis (Figure 5–6).

Example 4. Sketch the polar curve $r^2 = \sin \theta$.

DISCUSSION. Since $r^2 = \sin \theta$ implies $(-r)^2 = \sin \theta$, the curve is symmetric about the origin.

Since $\sin(-\theta) = -\sin \theta$, $r^2 = \sin \theta$ does not imply $r^2 = \sin(-\theta)$. The test for symmetry about the x-axis fails.

Because $\sin(\pi - \theta) = \sin \theta$, $r^2 = \sin \theta$ implies $r^2 = \sin(\pi - \theta)$; so the curve is symmetric about the y-axis.

Observe that since $\sin \theta$ is negative for $\pi < \theta < 2\pi$, there are no points corresponding to θ in this range (Figure 5–7). Note also that the graph *is* symmetric about the x-axis. This is due to the fact that $(r, -\theta)$ is equivalent to $(-r, \pi - \theta)$, and $r^2 = \sin \theta$ implies $(-r)^2 = \sin(\pi - \theta)$.

Example 5. Sketch the graph of $r = \theta$ ($-\pi \leq \theta \leq 2\pi$).

DISCUSSION. Here θ must be measured in radians. The length of the radius vector is equal to the numerical value of the radian measure of θ (Figure 5–8).

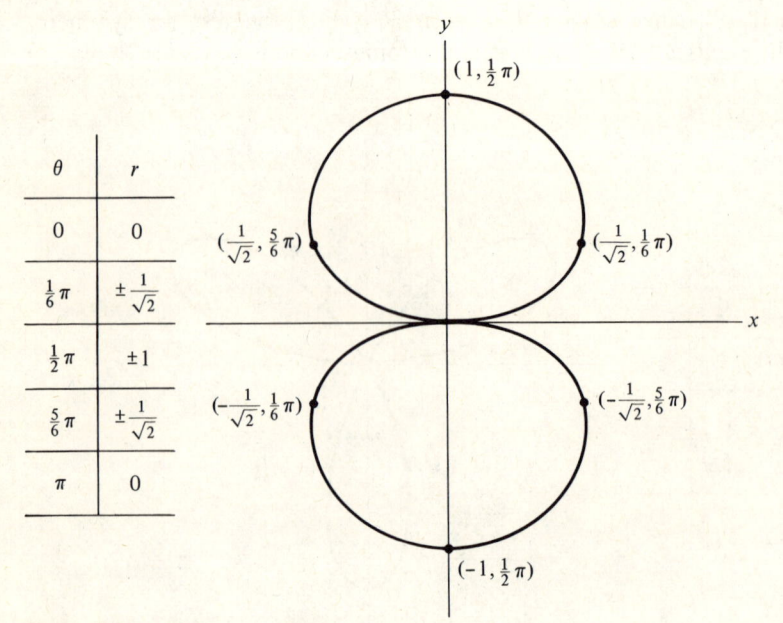

θ	r
0	0
$\frac{1}{6}\pi$	$\pm\frac{1}{\sqrt{2}}$
$\frac{1}{2}\pi$	± 1
$\frac{5}{6}\pi$	$\pm\frac{1}{\sqrt{2}}$
π	0

Figure 5–7 $r^2 = \sin \theta$.

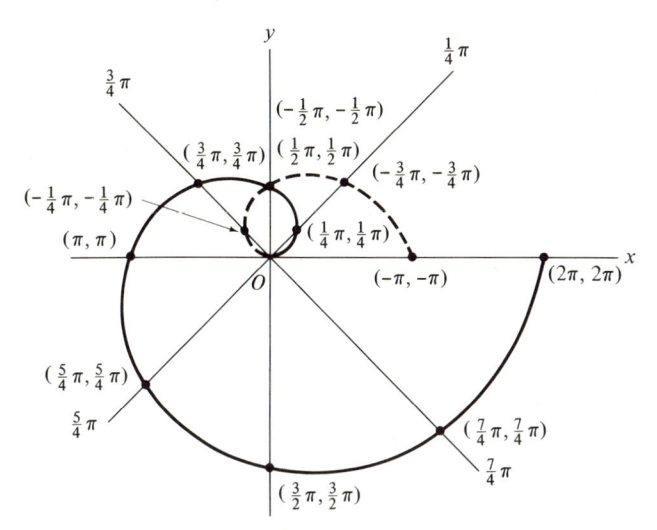

Figure 5-8 $r = \theta \, (-\pi \le \theta \le 2\pi)$.

EXERCISE 2

In Problems 1 to 22, test for symmetry and sketch the graphs of the polar equations.
1. $r = \cos 3\theta$
2. $r = 2(1 + \sin \theta)$
3. $r = 4 \sin 5\theta$
4. $r = 2 \sin 4\theta$
5. $r = 2 + 3 \cos \theta$
6. $r = a \cos \theta \quad (a > 0)$
7. $r^2 = 4 \cos 2\theta$
8. $r = a \cos \frac{1}{3}\theta \quad (a > 0)$
9. $r = a(1 - \cos \theta) \quad (a > 0)$
10. $r = 1 + 2 \sin \theta$
11. $r\theta = 2$
12. $r = 2 \cos (\theta + \frac{1}{4}\pi)$
13. $r \cos \theta = 3$
14. $r \sin \theta = -2$
15. $r^2 = 2 \cos \theta$
16. $r^2 = 4 \sin 2\theta$
17. $r = 4 \sin \frac{1}{2}\theta$
18. $r = \cos 2\theta$
19. $r = 4 \sin^2 \theta$
20. $r = \theta \quad (0 \le \theta)$
21. $r = -\theta \quad (0 \le \theta)$
22. $r \cos (\theta - \frac{1}{4}\pi) = 6$

In Problems 23 and 24, show that the two equations have the same graph.

23. $\begin{cases} r = 2 + \cos \theta \\ r = \cos \theta - 2 \end{cases}$

24. $\begin{cases} r = 1 + 2 \sin \theta \\ r = 2 \sin \theta - 1 \end{cases}$

3. TRANSFORMING EQUATIONS FROM RECTANGULAR TO POLAR COORDINATES AND VICE VERSA; EQUATIONS OF LINES AND CIRCLES IN POLAR COORDINATES

To obtain a polar equation for a curve given by the rectangular equation

$$F(x, y) = 0 \tag{1}$$

one merely makes the substitution $x = r \cos \theta$ and $y = r \sin \theta$ in Equation (1). The result is an equation of the form

$$G(r, \theta) = 0 \tag{2}$$

To transform an equation of the form of Equation (2) to that of Equation (1), we make substitutions using the equations $x = r \cos \theta$, $y = r \sin \theta$, $r^2 = x^2 + y^2$, and $\tan \theta = y/x$. The following examples illustrate the technique.

Example 1. Convert the equation $x^2 + y^2 = a^2$ $(a > 0)$ to polar coordinates.

SOLUTION. Since $x^2 + y^2 = r^2$, the given equation reduces to

$$r^2 = a^2$$

or

$$r = a$$

(Another solution is $r = -a$.)

Example 2. Find a polar equation for the straight line $ax + by + c = 0$.

SOLUTION. Substituting $x = r \cos \theta$ and $y = r \sin \theta$ in the equation, we obtain

$$ar \cos \theta + br \sin \theta + c = 0$$

Hence

$$r = \frac{-c}{a \cos \theta + b \sin \theta}$$

Example 3. Convert the following equation to rectangular coordinates:

$$r = \frac{a}{1 + \cos \theta}$$

SOLUTION. The given equation is equivalent to

$$r + r \cos \theta = a$$

Therefore,

$$r = a - x$$
$$r^2 = (a - x)^2$$

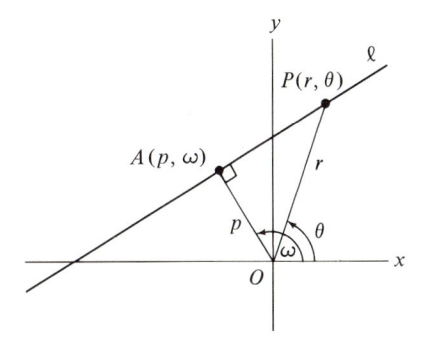

Figure 5-9 $r\cos(\theta-\omega) = p$ $(p>0)$.

which reduces to

$$y^2 = a^2 - 2ax$$

Note: Squaring both sides of an equation may introduce extraneous solutions. The graph of the resulting equation always contains the graph of the original equation. In the exercises, it is often necessary to square in order to express the final equation as a polynomial in x and y (that is, free of radicals).

One may derive the equation of a straight line and of a circle directly in polar coordinates as shown in the following example.

Example 4. Find the equation of the line ℓ passing through the point A with polar coordinates (p, ω), $p > 0$, and perpendicular to $\overline{OA}$ (Figure 5-9).

SOLUTION. Let (r, θ) be polar coordinates of an arbitrary point P on ℓ. Since $\triangle OAP$ is a right triangle, we have

$$r\cos(\theta - \omega) = p \tag{3}$$

(This may also be written as $r\cos(\omega - \theta) = p$.) Observe that the point $A(p, \omega)$ satisfies Equation (3).

Example 5. Find the equation of the circle whose center (in polar coordinates) is $C(c, \alpha)$ and whose radius is $a > 0$.

SOLUTION. Let (r, θ) be polar coordinates of an arbitrary point P on the circle (Figure 5-10). By the law of cosines of trigonometry,

$$a^2 = c^2 + r^2 - 2cr\cos(\theta - \alpha)$$

Therefore,

$$r^2 - 2cr\cos(\theta - \alpha) + c^2 = a^2 \tag{4}$$

(This may also be written with $\alpha - \theta$ in place of $\theta - \alpha$.) In the special case in which the circle passes through the origin, we have $c^2 = a^2$ and Equation (4) reduces to

$$r = 2c\cos(\theta - \alpha) \tag{5}$$

We see that the origin satisfies Equation (5) by taking $r = 0$ and $\theta = \alpha + \tfrac{1}{2}\pi$.

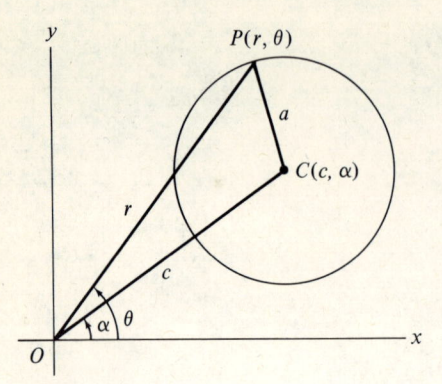

Figure 5-10

If the center is at O, then $c = 0$ and Equation (4) reduces to $r = a$ (or $r = -a$).

As a concluding remark, we note that if a polar curve has an equation of the form $r = f(\theta)$, where f is a function of one variable, we may obtain (rectangular) parametric equations for the curve from the equations $x = r \cos \theta$ and $y = r \sin \theta$. Substituting $f(\theta)$ for r, we have

$$\begin{cases} x = f(\theta) \cos \theta \\ y = f(\theta) \sin \theta, \quad a \leq \theta \leq b \end{cases} \tag{6}$$

These are parametric equations for the curve with θ as parameter.

EXERCISE 3

1. Transform the following rectangular equations to polar form.
   - (a) $x + y = a$
   - (b) $x^2 + xy + y^2 = 1$
   - (c) $x = 2 \cos t, y = 3 \sin t$
   - (d) $xy = a^2$
   - (e) $x^2 + y^2 - 12y = 0$
   - (f) $y = 3x$
   - (g) $y^2 - x^2 = a^2$
   - (h) $x^2 + y^2 = 9$
   - (i) $(x^2 + y^2)^3 = 4x^2 y^2$
   - (j) $\dfrac{xy}{x^2 + y^2} = 4$
   - (k) $\dfrac{x^2 \cdot y^2}{x^2 + y^2} = 1$

2. Transform the following polar equations to rectangular form. Simplify your answer, removing all radicals.
   - (a) $r = 10$
   - (b) $\tan \theta = -1$

(c) $r = a \sec \theta$

(d) $r = \dfrac{1}{1 - \sin \theta}$

(e) $r^2 = \sin 2\theta$

(f) $r^2 \sin 2\theta = 2$

(g) $r(2 - \cos \theta) = 2$

3. Find a polar equation for the line passing through the given point $A(r, \theta)$ and perpendicular to $\overline{OA}$. Draw a figure.
   (a) $A(4, \frac{1}{6}\pi)$
   (b) $A(3, \pi)$
   (c) $A(-2, \frac{1}{4}\pi)$
   (d) $A(3, 0)$
   (e) $A(6, -150°)$

4. Find a polar equation for the circle with given center $C(c, \alpha)$ and radius a. Draw a figure.
   (a) $C(4, \frac{1}{3}\pi), a = 4$
   (b) $C(-2, 150°), a = 3$
   (c) $C(0, \frac{1}{4}\pi), a = 5$
   (d) $C(2, 0), a = 5$

5. Derive the normal form of the equation of a line from Equation (3). (See Chapter 4, Section 4.)

6. Show that the formula in polar coordinates for the distance between the points $P_1(r_1, \theta_1)$ and $P_2(r_2, \theta_2)$ is

$$\overline{|P_1 P_2|} = \sqrt{r_1^2 + r_2^2 - 2r_1 r_2 \cos(\theta_1 - \theta_2)}$$

7. Find the distance between the points A and B given in polar coordinates. Do not convert to rectangular coordinates.
   (a) $A(4, 60°), B(3, 0°)$
   (b) $A(3, 170°), B(5, 20°)$
   (c) $A(4, -45°), B(7, 45°)$
   (d) $A(-3, 150°), B(5, -75°)$
   (e) $A(6, 210°), B(4, 300°)$

8. Find a pair of rectangular parametric equations for the curve with the given polar equation.
   (a) $r = 1 - \cos \theta$
   (b) $r \cos \theta = 1$
   (c) $r = 5$
   (d) $\theta = \frac{1}{3}\pi$
   (e) $1/r = \sin 2\theta$

9. Show that the parametric equations
$$\begin{cases} x = 2(t + 1) \\ y = t(t + 2) \end{cases}$$
and the polar equation $r(1 - \sin \theta) = 2$ represent the same curve.

*10. A line segment of constant length $2a$ moves so that its endpoints lie on the x- and y-axes.
   (a) Find a polar equation of the path of its midpoint.
   (b) Find a polar equation of the path of the foot of the perpendicular drawn from the origin to the line segment.

4. INTERSECTIONS OF POLAR CURVES

We find points of intersection of the polar curves

$$\begin{cases} C_1: F_1(r, \theta) = 0 \\ C_2: F_2(r, \theta) = 0 \end{cases}$$

by solving these equations simultaneously for r, θ. The pairs (r, θ) satisfying both equations simultaneously will, of course, represent points lying on both curves; hence, these will be points of intersection. However, as we shall see, there are examples of polar curves which have a point of intersection (x, y), but for which no corresponding polar coordinate pair simultaneously satisfies the equations. For instance, $(2, 30°)$ may satisfy $F_1(r, \theta) = 0$ and not $F_2(r, \theta) = 0$, while $(-2, 210°)$ may satisfy $F_2(r, \theta) = 0$ but not $F_1(r, \theta) = 0$. However, $(2, 30°)$ and $(-2, 210°)$ are equivalent pairs, since they both represent $(\sqrt{3}, 1)$ in the x–y plane, and $(\sqrt{3}, 1)$ may be a point of intersection of the two curves. Furthermore, no simultaneous solution may correspond to $(\sqrt{3}, 1)$. In order to find such points, we sketch the curves and proceed by inspection.

Example 1. Find the points of intersection of the curves $r = 2\cos\theta - 1$ and $\theta = \frac{1}{6}\pi$ (Figure 5–11).

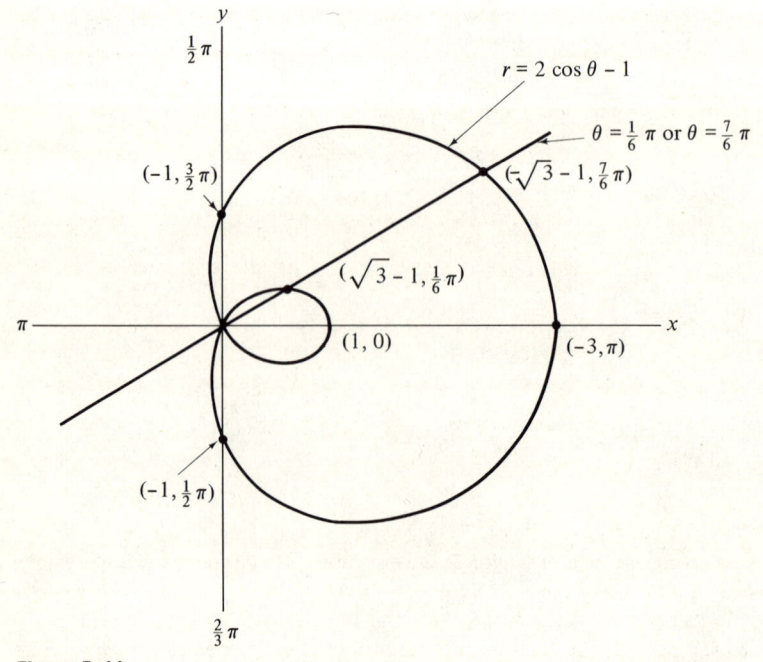

Figure 5-11

SOLUTION. The only simultaneous solution is $\theta = \frac{1}{6}\pi$,

$$r = 2\cos\tfrac{1}{6}\pi - 1 = \sqrt{3} - 1 \approx 0.73$$

that is $(\sqrt{3} - 1, \frac{1}{6}\pi)$.

Now the graph of $\theta = \frac{7}{6}\pi$ is the same as that of $\theta = \frac{1}{6}\pi$. Therefore, another point of intersection is $\theta = \frac{7}{6}\pi$, $r = 2\cos\frac{7}{6}\pi - 1 = -\sqrt{3} - 1$, that is $(-(\sqrt{3} + 1), \frac{7}{6}\pi)$. (See Figure 5-11.) Note that this pair does not satisfy the equation $\theta = \frac{1}{6}\pi$.

Example 2. Find the points of intersection of the curves $r = \sin 2\theta$ and $r = \cos\theta$. Sketch the graphs.

SOLUTION. Solving simultaneously, we have

$$\sin 2\theta = \cos\theta$$
$$2\sin\theta\cos\theta = \cos\theta \quad \text{(the double angle formula)}$$
$$\cos\theta(2\sin\theta - 1) = 0$$

Therefore,

$$\cos\theta = 0; \theta = \tfrac{1}{2}\pi, \tfrac{3}{2}\pi, \ldots, (2n+1)\tfrac{1}{2}\pi.$$
$$\sin\theta = \tfrac{1}{2}; \theta = \tfrac{1}{6}\pi, \tfrac{5}{6}\pi, \ldots, \tfrac{1}{6}\pi + 2n\pi, \tfrac{5}{6}\pi + 2n\pi$$

Substituting these values of θ in one of the given equations, say $r = \cos\theta$, we obtain the simultaneous solutions: $(0, \tfrac{1}{2}\pi)$, $(\sqrt{3}/2, \tfrac{1}{6}\pi)$, $(-\sqrt{3}/2, \tfrac{5}{6}\pi)$. (Note that the other solutions are equivalent to one of these.) From Figure 5-12, we see that these are the only points of intersection. (Note that the tests for symmetry fail when applied to the curve $r = \sin 2\theta$, even though the curve possesses all three symmetries.)

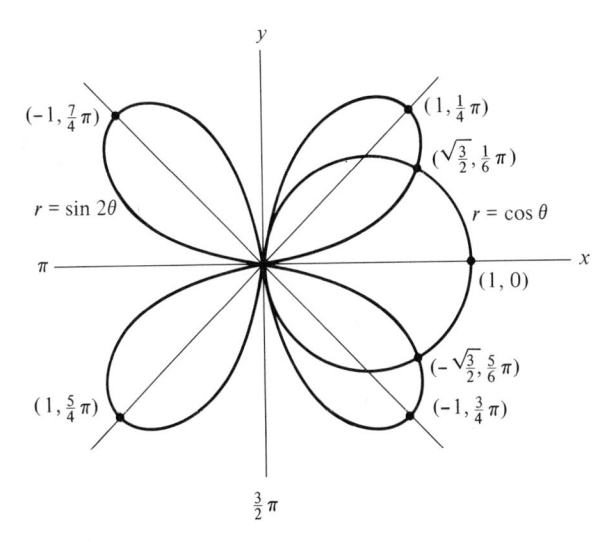

Figure 5-12

EXERCISE 4

Find the points of intersection of the following polar curves. Sketch the curves and plot all points of intersection.

1. $r = \cos 2\theta,\ r = \cos \theta$
2. $r = 2(\cos \theta - 1),\ r = 4 \sin \theta$
3. $r = 1 - \cos \theta,\ r = 1 - \sin \theta$
4. $r = \sin 2\theta,\ \theta = \frac{1}{3}\pi$
5. $r = \sin \theta,\ r = 1 - \sin \theta$
6. $r = 2 \cos \theta,\ r = 2 \sin \theta$
7. $r = 2 \sin \theta - 3,\ r = 2$
8. $2r \cos \theta = 3,\ r = 6 \cos \theta$
9. $r = 4(1 + \sin \theta),\ r \sin \theta = 3$
10. $r = 2(1 + \cos \theta),\ r(1 - \cos \theta) = 1$
11. $r^2 = \sin 2\theta,\ r^2 = \cos 2\theta$
12. $r = 2 \cos \frac{1}{2}\theta,\ r = 1$

chapter 6
Translation and Rotation of Axes; Isometries of the Plane

1. TRANSLATION AND ROTATION OF AXES

To facilitate the investigation of certain curves, one often makes a "change of variables" in the equations of the curves in order to simplify them. The most elementary changes of variable are translations and rotations. These are special one-to-one correspondences (mappings) from R^2 to itself.

Definition. Let (h, k) be a given point. The mapping of R^2 to itself which maps an arbitrary point (x, y) into (x', y'), where

$$\begin{cases} x' = x - h \\ y' = y - k \end{cases} \qquad (1)$$

is called the *translation of axes to the point* (h, k).

We justify the term "translation of axes" by constructing a pictorial representation of the mapping as follows. Label the lines $x = h$ and $y = k$ the y'-axis and the x'-axis, respectively (Figure 6–1). For an arbitrary point $P(x, y)$, the numbers $|x - h|$ and $|y - k|$ are the distances from the y'-axis and

Translation and Rotation of Axes; Isometries of the Plane

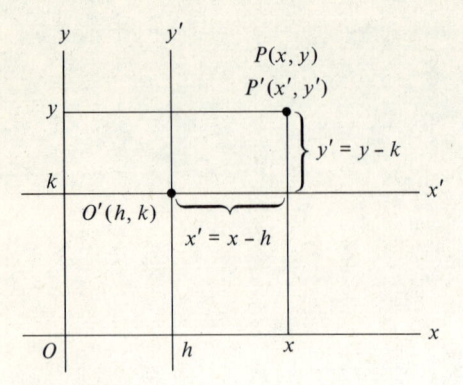

Figure 6-1

x'-axis, respectively, to the point $P(x, y)$. We say that the point $P(x, y)$, referred to the x'- and y'-axes, has coordinates $x' = x - h$ and $y' = y - k$. Note that Figure 6-1 shows two diagrams of R^2, one called "the x–y plane" and the other "the x'–y' plane." The same "dot" represents both the point $P(x, y)$ and its image $P'(x', y')$, under the translation. (h, k) in the x–y plane is the origin in the x'–y' plane, and we denote it by O'.

Example 1. Translate axes to the point $(-3, 4)$ and find the coordinates of $P(5, -2)$ in the x'–y' plane.

SOLUTION. The equations for the translation of axes (Equations (1)) are $x' = x + 3$ and $y' = y - 4$. For the point $P(5, -2)$, $x' = 5 + 3 = 8$ and $y' = -2 - 4 = -6$. The coordinates of P in the x'–y' plane are $(8, -6)$ (Figure 6-2).

Given the graph of the equation $f(x, y) = 0$, we may find the equation of the graph of its image under a translation of axes by solving Equations (1) for x, y in terms of x', y' and substituting in the equation, the result being an equation of the form $g(x', y') = 0$, called the *transform* of $f(x, y) = 0$. The same picture is used to represent both the graph of $f(x, y) = 0$ and the graph of $g(x', y') = 0$. In some cases, by an appropriate choice of (h, k), one can

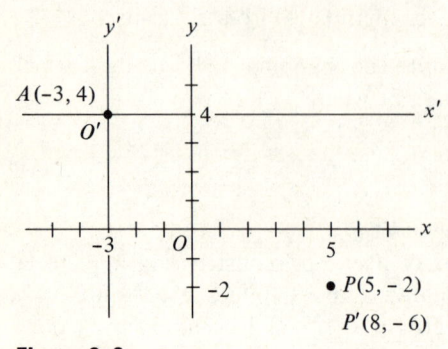

Figure 6-2

obtain an equation in x', y' which is simpler than the one in x, y, thus facilitating the sketching of its graph.

Example 2. Given the equation $y^2 - 4y - 4x - 8 = 0$, determine a translation of axes (if possible) which transforms the equation to one without linear terms.

SOLUTION. Substituting $x = x' + h$ and $y = y' + k$ in the equation, we obtain

$$(y' + k)^2 - 4(y' + k) - 4(x' + h) - 8 = 0$$
$$y'^2 + (2k - 4)y' - 4x' + (k^2 - 4k - 4h - 8) = 0$$

We may now eliminate the linear terms in y' by setting $2k - 4 = 0$ and solving for k. Thus, $k = 2$. However, there is no way to eliminate the linear term in x'. If, in addition, we wish to eliminate the constant term, we set $k^2 - 4k - 4h - 8 = 0$ and solve for h, using $k = 2$. Thus, $h = -3$. Accordingly, a translation of axes to the point $O'(-3, 2)$ yields the transform $y'^2 = 4x'$ (Figure 6–3).

An alternate method is to complete the square in y and rearrange the equation as follows:

$$y^2 - 4y - 4x - 8 = 0$$
$$y^2 - 4y \qquad = 4x + 8$$
$$y^2 - 4y \qquad + 4 = 4x + 12 \qquad \text{(adding the square of one half the coefficient of } y \text{ to both sides)}$$
$$(y - 2)^2 = 4(x + 3)$$

Now let $y' = y - 2$ and $x' = x + 3 = x - (-3)$. These, then, are the translation of axes equations which yield $y'^2 = 4x'$.

Definition. Let φ be a given angle with $0 \leq \varphi < \pi$. The rotation of axes through the angle φ is the mapping of R^2 to itself defined as follows: An arbitrary point $P(x, y)$ with polar coordinates (r, θ), $0 \leq \theta < 2\pi$, is mapped

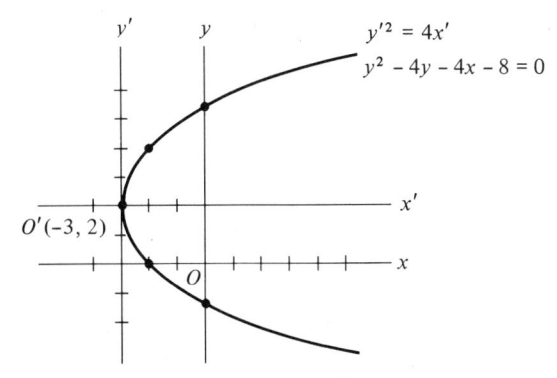

Figure 6–3

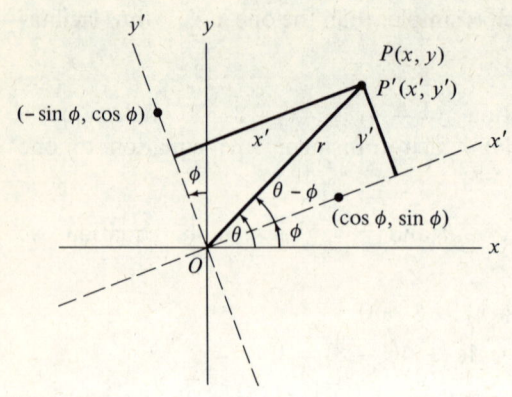

Figure 6-4

into the point $P'(x', y')$ with polar coordinates $(r, \theta - \varphi)$. (Thus, both P and P' lie on the circle of radius r and center at O. The difference between their polar angles is φ.) (See Figure 6-4.)

Formulas expressing x', y' in terms of x, y and φ may be derived as follows:

$$x' = r\cos(\theta - \varphi) = r(\cos\theta \cos\varphi + \sin\theta \sin\varphi)$$
$$= (r\cos\theta)\cos\varphi + (r\sin\theta)\sin\varphi$$
$$= x\cos\varphi + y\sin\varphi$$

$$y' = r\sin(\theta - \varphi) = r(\sin\theta \cos\varphi - \cos\theta \sin\varphi)$$
$$= (r\sin\theta)\cos\varphi - (r\cos\theta)\sin\varphi$$
$$= y\cos\varphi - x\sin\varphi$$

Thus, the rotation equations are

$$\begin{cases} x' = x\cos\varphi + y\sin\varphi \\ y' = -x\sin\varphi + y\cos\varphi \end{cases} \tag{2}$$

We justify the term "rotation of axes" as follows. From Equations (2) we see that the set of points for which $y' = 0$ has equation $0 = -x\sin\varphi + y\cos\varphi$ (that is, $y = (\tan\varphi)x$, if $\varphi \neq \frac{1}{2}\pi$). This is the line through O with angle of inclination φ. We label it the x'-axis. Similarly, the y'-axis is the set of points for which $x' = 0$; that is, the line $x\cos\varphi + y\sin\varphi = 0$. Note that these lines are perpendicular.

The ray through O and $(\cos\varphi, \sin\varphi)$ is the positive x'-axis, whereas the ray through O and $(-\sin\varphi, \cos\varphi)$ is the positive y'-axis, since these points yield $(x' = 1, y' = 0)$ and $(x' = 0, y' = 1)$ when substituted in Equations (2) (Figure 6-4). As in the case of translation of axes, Figure 6-4 shows two diagrams of R^2, one called the x-y plane, the other the x'-y' plane. The same dot represents both the point $P(x, y)$ and its image $P'(x', y')$ under the rotation. Observe that $|x'|$ and $|y'|$ are precisely the distances from the y' and x' axes to P.

Example 3. Rotate axes through the angle 60° and find the coordinates of the point $(-2, \sqrt{3})$ in the x'-y' plane.

SOLUTION. The equations for the rotation of axes are

$$x' = x \cos 60° + y \sin 60° = \tfrac{1}{2}x + \frac{\sqrt{3}}{2}y$$

$$y' = -x \sin 60° + y \cos 60° = -\frac{\sqrt{3}}{2}x + \tfrac{1}{2}y$$

The coordinates of $(-2, \sqrt{3})$ in the x'-y' plane are

$$x' = \tfrac{1}{2}(-2) + \frac{\sqrt{3}}{2} \cdot \sqrt{3} = \tfrac{1}{2}$$

$$y' = \left(-\frac{\sqrt{3}}{2}\right)(-2) + \tfrac{1}{2}\sqrt{3} = \tfrac{3}{2}\sqrt{3}$$

As with translations, we may compute the transform of the equation $f(x, y) = 0$ under a rotation of axes. We first solve Equation (2) for x and y and substitute in the equation. By the usual process of elimination, we obtain (see Problem 7)

$$\begin{cases} x = x' \cos \varphi - y' \sin \varphi \\ y = x' \sin \varphi + y' \cos \varphi \end{cases} \quad (3)$$

Example 4. Transform the equation $xy = 1$ by a rotation of axes through the angle $\tfrac{1}{4}\pi$.

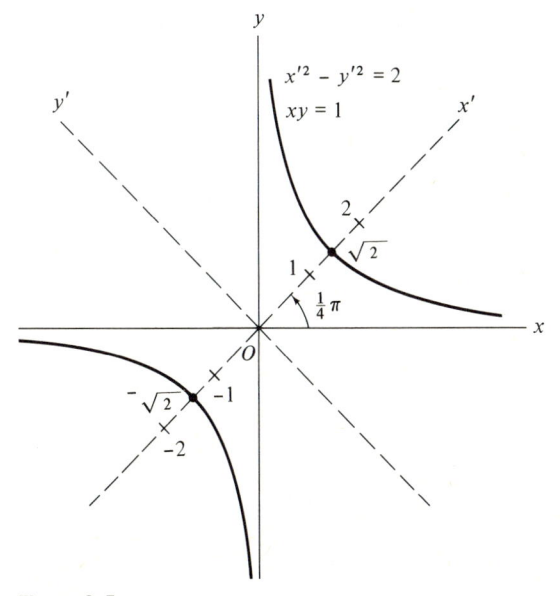

Figure 6-5

SOLUTION. Substituting $\varphi = \frac{1}{4}\pi$ in Equations (3), we have

$$\begin{cases} x = \dfrac{1}{\sqrt{2}}x' - \dfrac{1}{\sqrt{2}}y' \\ y = \dfrac{1}{\sqrt{2}}x' + \dfrac{1}{\sqrt{2}}y' \end{cases}$$

The transform is (see Figure 6–5)

$$\left(\frac{1}{\sqrt{2}}x' - \frac{1}{\sqrt{2}}y'\right)\left(\frac{1}{\sqrt{2}}x' + \frac{1}{\sqrt{2}}y'\right) = 1$$

or

$$x'^2 - y'^2 = 2$$

EXERCISE 1

1. Write the equations for the translation of axes to the point $(-3, 4)$. Find the x', y' coordinates of the following points. Draw a figure.
   (a) $(0, 0)$
   (b) $(-3, 0)$
   (c) $(4, -2)$
   (d) $(1, 2)$
2. Write the equations for the rotation of axes through the angle $\varphi = 120°$. Find the x', y' coordinates of the points of Problem 1.
3. Translate axes to the point $(-2, -3)$ and find the coordinates of $P(-6, 3)$ in the x'-y' plane. Draw a figure.
4. Translate axes so as to remove the linear terms. Do this by two methods. (See Example 2.)
   (a) $4x^2 + y^2 - 16x + 6y + 21 = 0$
   (b) $x^2 + y^2 - 3x + 4y = 5$
   (c) $x^2 - 2y^2 - 5x + 6y = 7$
   (d) $3x^2 - 5x + 6y^2 + 4y - 3 = 0$
   (e) $(x - 4)^3 - 3(y + 5)^3 = 7$
5. Rotate axes through the angle $\varphi = 150°$ and find the coordinates of $(2, 1)$ in the x'-y' plane. Draw a figure.
6. Transform the following equations by a rotation of axes through the indicated angle.
   (a) $x^2 + y^2 = 5$; φ arbitrary
   (b) $y^2 = 4x$; $\varphi = \frac{1}{4}\pi$
   (c) $x^3 = y^2$; $\varphi = \frac{7}{6}\pi$
   (d) $x^2 - y^2 = 4$; $\varphi = \frac{1}{4}\pi$
   (e) $\dfrac{x^2}{4} + \dfrac{y^2}{9} = 1$; $\varphi = \frac{1}{2}\pi$
7. Derive Equations (3).
8. (a) Let (x', y') be the coordinates of (x, y) under the translation of axes to $(3, -4)$.

Let (x'', y'') be the coordinates of (x', y') under the rotation of axes through 120°. Express x'', y'' in terms of x and y.

(b) Let (x', y') be the coordinates of (x, y) under the rotation of axes through 120°. Let (x'', y'') be the coordinates of (x', y') under the translation of axes to $(3, -4)$. Express x'', y'' in terms of x and y.

†9. Let $P'_1(x'_1, y'_1)$ and $P'_2(x'_2, y'_2)$ be the images of $P_1(x_1, y_1)$ and $P_2(x_2, y_2)$, respectively, under

(a) a translation of axes to the point (h, k)
(b) a rotation of axes through the angle φ.

Show that $|\overline{P_1P_2}| = |\overline{P'_1P'_2}|$.

*2. TRANSLATIONS AND ROTATIONS IN MATRIX FORM

In the next chapter, we shall make use of the equations for translations and rotations of axes in matrix form. (See the Appendix for a discussion of matrices.) If we write the coordinates of $P(x, y)$ as a column matrix $\begin{pmatrix} x \\ y \end{pmatrix}$, the equations for the translation of axes to the point (h, k) may be written as a single matrix equation:

$$\begin{pmatrix} x' \\ y' \end{pmatrix} = \begin{pmatrix} x \\ y \end{pmatrix} - \begin{pmatrix} h \\ k \end{pmatrix}$$

The matrix form of the equations for a rotation of axes is

$$\begin{pmatrix} x' \\ y' \end{pmatrix} = \begin{pmatrix} \cos \varphi & \sin \varphi \\ -\sin \varphi & \cos \varphi \end{pmatrix} \begin{pmatrix} x \\ y \end{pmatrix}$$

Letting

$$P = \begin{pmatrix} \cos \varphi & -\sin \varphi \\ \sin \varphi & \cos \varphi \end{pmatrix}$$

this may be written

$$\begin{pmatrix} x' \\ y' \end{pmatrix} = P^t \cdot \begin{pmatrix} x \\ y \end{pmatrix}$$

P (also P^t) is called a *rotation matrix*. Observe that P has determinant $|P| = \cos^2 \varphi + \sin^2 \varphi = 1$, for all φ

Furthermore,

$$PP^t = P^tP = I_2$$

where

$$I_2 = \begin{pmatrix} 1 & 0 \\ 0 & 1 \end{pmatrix}$$

the 2 × 2 identity matrix. Accordingly, $P^t = P^{-1}$, the *inverse* of P.

A matrix B with the property that $B^t = B^{-1}$ is called *orthogonal*. Since $|B^t| = |B|$ and $|B^t B| = |B^t| \cdot |B|$, it follows that for any orthogonal matrix B, $|B|^2 = |I_2| = 1$. Thus, $|B| = \pm 1$. The rotation matrix P above is, therefore, *orthogonal* and $|P| = 1$. It can be shown that any 2 × 2 orthogonal matrix A with $|A| = 1$ is a rotation matrix. (See Problem 6.)

Given the rotation equations in matrix form

$$\begin{pmatrix} x' \\ y' \end{pmatrix} = P^t \cdot \begin{pmatrix} x \\ y \end{pmatrix}$$

where

$$P^t = \begin{pmatrix} \cos \varphi & \sin \varphi \\ -\sin \varphi & \cos \varphi \end{pmatrix}$$

we may solve for x, y in terms of x', y' by multiplying by P. Since $PP^t = I_2$, we obtain

$$\begin{pmatrix} x \\ y \end{pmatrix} = P \cdot \begin{pmatrix} x' \\ y' \end{pmatrix}$$

where

$$P = \begin{pmatrix} \cos \varphi & -\sin \varphi \\ \sin \varphi & \cos \varphi \end{pmatrix}$$

In the next example, we illustrate a composite of a rotation and a translation, that is, a rotation of axes followed by a translation of axes.

Example. Write the matrix equation for the composite of a rotation of axes through 30°, followed by a translation to the point $(3, -4)$. Draw a figure.

SOLUTION. Using (x', y') to denote the coordinates of the point (x, y) after rotation, we have

$$\begin{pmatrix} x' \\ y' \end{pmatrix} = \begin{pmatrix} \cos 30° & \sin 30° \\ -\sin 30° & \cos 30° \end{pmatrix} \begin{pmatrix} x \\ y \end{pmatrix}$$

Letting (x'', y'') denote the coordinates of (x', y') after translation, we have

$$\begin{pmatrix} x'' \\ y'' \end{pmatrix} = \begin{pmatrix} x' \\ y' \end{pmatrix} - \begin{pmatrix} 3 \\ -4 \end{pmatrix}$$

Thus,

$$\begin{pmatrix} x'' \\ y'' \end{pmatrix} = \begin{pmatrix} \cos 30° & \sin 30° \\ -\sin 30° & \cos 30° \end{pmatrix} \begin{pmatrix} x \\ y \end{pmatrix} - \begin{pmatrix} 3 \\ -4 \end{pmatrix}$$

$$= \begin{pmatrix} x \cos 30° + y \sin 30° - 3 \\ -x \sin 30° + y \cos 30° + 4 \end{pmatrix}$$

is the equation for the composite transformation. The three coordinate

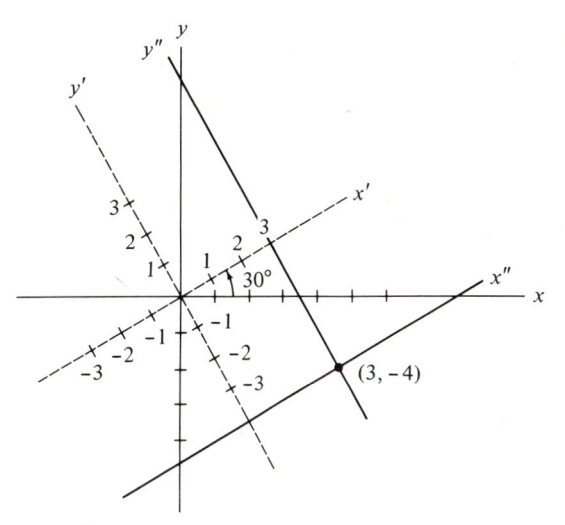

Figure 6-6

systems are shown in Figure 6-6. Note that the x''- and y''-axes are drawn parallel to the x'- and y'-axes, respectively, through the point $(3, -4)$ in the $x'-y'$ plane; that is, the origin in the $x''-y''$ plane is at $x' = 3$ and $y' = -4$ or equivalently, at

$$\begin{cases} x = x' \cos 30° - y' \sin 30° = \dfrac{\sqrt{3}}{2}(3) - \tfrac{1}{2}(-4) \\ \qquad\qquad\qquad\qquad\quad = \dfrac{3\sqrt{3} + 4}{2} \approx 4.6 \\ y = x' \sin 30° + y' \cos 30° = \tfrac{1}{2}(3) + (-4)\dfrac{\sqrt{3}}{2} \\ \qquad\qquad\qquad\qquad\quad = \dfrac{3 - 4\sqrt{3}}{2} \approx -2.0 \end{cases}$$

If in the above example we translate and then rotate, the matrix equation for the composite is as follows:

$$\begin{pmatrix} x' \\ y' \end{pmatrix} = \begin{pmatrix} x \\ y \end{pmatrix} - \begin{pmatrix} 3 \\ -4 \end{pmatrix}$$

$$\begin{pmatrix} x'' \\ y'' \end{pmatrix} = \begin{pmatrix} \cos 30° & \sin 30° \\ -\sin 30° & \cos 30° \end{pmatrix} \begin{pmatrix} x' \\ y' \end{pmatrix}$$

$$= \begin{pmatrix} \cos 30° & \sin 30° \\ -\sin 30° & \cos 30° \end{pmatrix} \left[\begin{pmatrix} x \\ y \end{pmatrix} - \begin{pmatrix} 3 \\ -4 \end{pmatrix} \right]$$

$$= \begin{pmatrix} (x - 3) \cos 30° + (y + 4) \sin 30° \\ -(x - 3) \sin 30° + (y + 4) \cos 30° \end{pmatrix}$$

(Compare with the previous result.)

EXERCISE 2

1. Find the matrix equation for the composite of a rotation of axes through 135° followed by a translation of axes to the point $(-3, 2)$. Then find the composite in the reverse order. Find the coordinates of $(3, 1)$ after each of the above composites. (See the example.)

2. Same as Problem 1 for a rotation through 75° and a translation to the point $(1, -2)$. Find the coordinates of $(4, 0)$ after each composite.

*3. Let A_1, A_2 be 2×2 matrices and let

$$B_1 = \begin{pmatrix} h_1 \\ k_1 \end{pmatrix}, \quad B_2 = \begin{pmatrix} h_2 \\ k_2 \end{pmatrix}$$

Prove that if

$$A_1 \begin{pmatrix} x \\ y \end{pmatrix} + B_1 = A_2 \begin{pmatrix} x \\ y \end{pmatrix} + B_2$$

for all points (x, y), then $A_1 = A_2$ and $B_1 = B_2$. (*Hint:* Substitute special points for (x, y).)

4. Prove that the coefficients of the quadratic terms in the equation

$$Ax^2 + Bxy + Cy^2 + Dx + Ey + F = 0$$

are unchanged after any translation of axes.

5. (a) Show that a translation of axes

$$\begin{pmatrix} x' \\ y' \end{pmatrix} = \begin{pmatrix} x \\ y \end{pmatrix} - \begin{pmatrix} h \\ k \end{pmatrix}$$

may be written as a matrix product

$$\begin{pmatrix} x' \\ y' \\ 1 \end{pmatrix} = \begin{pmatrix} 1 & 0 & -h \\ 0 & 1 & -k \\ 0 & 0 & 1 \end{pmatrix} \begin{pmatrix} x \\ y \\ 1 \end{pmatrix}$$

(b) Show that the rotation of axes through angle φ may be written

$$\begin{pmatrix} x' \\ y' \\ 1 \end{pmatrix} = \begin{pmatrix} \cos\varphi & \sin\varphi & 0 \\ -\sin\varphi & \cos\varphi & 0 \\ 0 & 0 & 1 \end{pmatrix} \begin{pmatrix} x \\ y \\ 1 \end{pmatrix}$$

*6. Prove that an orthogonal 2×2 matrix A with $|A| = +1$ is a rotation matrix. (*Hint:* Let

$$A = \begin{pmatrix} a & b \\ c & d \end{pmatrix}$$

Then using the fact that $AA' = I_2$ and $|A| = 1$, show that

$$A = \begin{pmatrix} a & b \\ -b & a \end{pmatrix}$$

where $a^2 + b^2 = 1$; hence $a = \cos\varphi$ and $b = \sin\varphi$ for some angle φ.)

* 3. ISOMETRIES

The translation and rotation of axes discussed in the preceding sections are examples of mappings called *isometries*. The essential property of such mappings is that they map geometrical figures into *congruent* figures. Clearly, such correspondences must preserve the length of line segments. (See Problem 9, Exercise 1.) Recall that the distance between points P_1 and P_2 is denoted by $|P_1 P_2|$, which is also the length of the segment $\overline{P_1 P_2}$.

Definition. A function F from R^2 onto itself with the property that for all $P_1, P_2 \in R^2$

$$|P_1 P_2| = |\overline{F(P_1)F(P_2)}|$$

is called an *isometry* of R^2. F is also called a *distance preserving* mapping (or transformation) of R^2 onto itself (Figure 6–7).

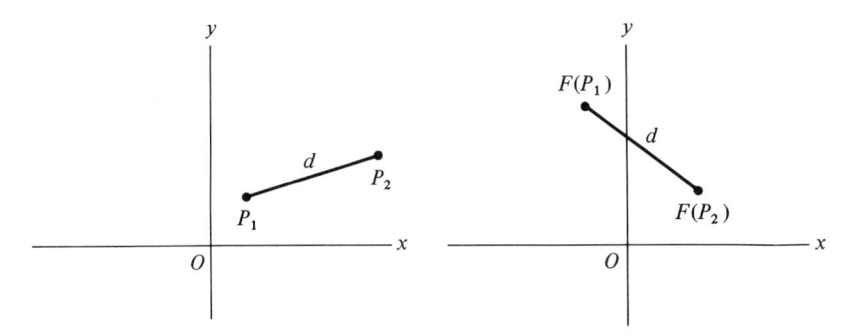

Figure 6-7

Observe that an isometry is automatically one to one: If $P_1 \neq P_2$, then $0 < |P_1 P_2| = |\overline{F(P_1)F(P_2)}|$, so that $F(P_1) \neq F(P_2)$. It is not difficult to show that a composite of two isometries is again an isometry (see Problem 1). There are three basic types of isometries: *translations, rotations,* and *reflections. Every isometry is one of these types or a composite of these.* (We shall not give a proof of this fact.) We have already discussed the first two. A *reflection* is defined as follows.

Definition. The mapping M_x defined by $M_x(x, y) = (x, -y)$ for all points $P(x, y)$ is called a *reflection about the x-axis*. A reflection about the y-axis is defined by $M_y(x, y) = (-x, y)$ (Figure 6–8). The matrix forms for these reflections are, respectively,

$$\begin{pmatrix} x' \\ y' \end{pmatrix} = \begin{pmatrix} 1 & 0 \\ 0 & -1 \end{pmatrix} \begin{pmatrix} x \\ y \end{pmatrix}$$

and

$$\begin{pmatrix} x' \\ y' \end{pmatrix} = \begin{pmatrix} -1 & 0 \\ 0 & 1 \end{pmatrix} \begin{pmatrix} x \\ y \end{pmatrix}$$

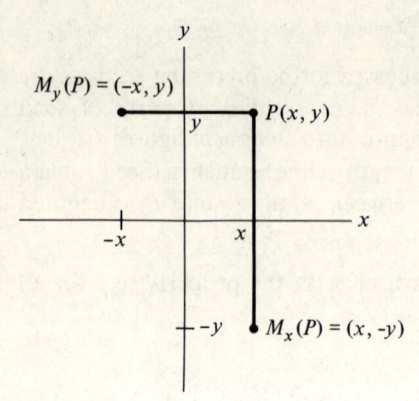

Figure 6-8

Using the law of cosines, it is not difficult to see that an isometry F preserves angles between line segments: Let $\triangle ABC$ be any triangle and let α be the angle at A (Figure 6–9). Let $A' = F(A)$, $B' = F(B)$, $C' = F(C)$, and let α' be the angle at A'. Then

$$a'^2 = b'^2 + c'^2 - 2b'c' \cos \alpha'$$

Therefore,

$$\cos \alpha' = \frac{b'^2 + c'^2 - a'^2}{2b'c'} \qquad (b', c' \neq 0)$$

Since F is an isometry, $a' = a$, $b' = b$, and $c' = c$ (Figure 6–9); so that

$$\cos \alpha = \frac{b^2 + c^2 - a^2}{2bc} = \cos \alpha'$$

Since α and α' are between $0°$ and $180°$, it follows that $\alpha = \alpha'$. We may now give a precise definition of *congruence*: Two point sets S_1 and S_2 are congruent iff there is an isometry which maps S_1 onto S_2.

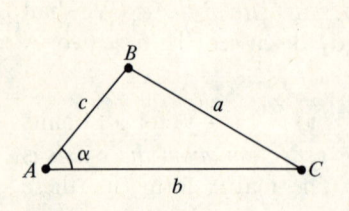

Figure 6-9

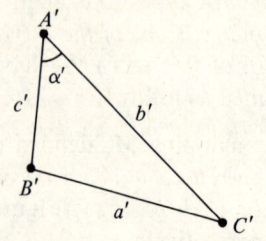

EXERCISE 3

1. Prove that if F and G are isometries, then so is the composite $F(G)$.
2. (a) Prove that the reflections M_x and M_y are isometries.
   (b) Prove that translations and rotations are onto, and hence they are isometries by virtue of Problem 9, Exercise 1.

*3. (a) Since an isometry F is one to one and onto, it has an inverse F^{-1}. Prove that F^{-1} is an isometry.
   (b) Find the equations of the inverse of
   (i) the translation to (h, k)
   (ii) the rotation through angle φ
   (iii) the reflection M_x
   (iv) the reflection M_y.

chapter 7
The Conic Sections

The *conic sections* are a family of plane curves which play an important role in the applications of calculus to geometry and problems of physics and engineering. They are of three types: the *ellipse*, the *parabola*, and the *hyperbola*. The general shape of these curves is shown in Figure 7–1. The name "conic sections" derives from the fact that each is the curve of intersection of a right circular cone with an appropriate plane (Figure 7–2). Although a special definition will ultimately be given for each type, we shall first give a general definition of conic section and consider the three types as special cases.

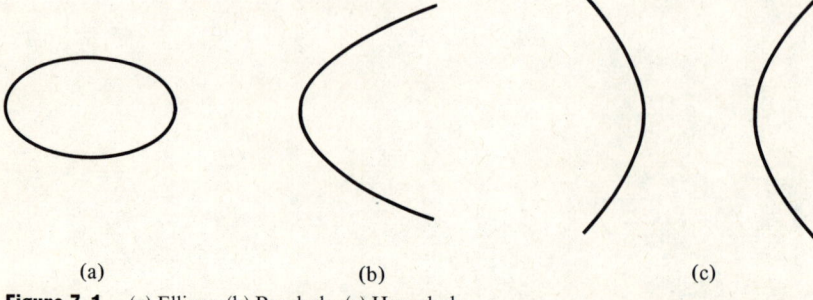

(a)   (b)   (c)

Figure 7-1   (a) Ellipse. (b) Parabola. (c) Hyperbola.

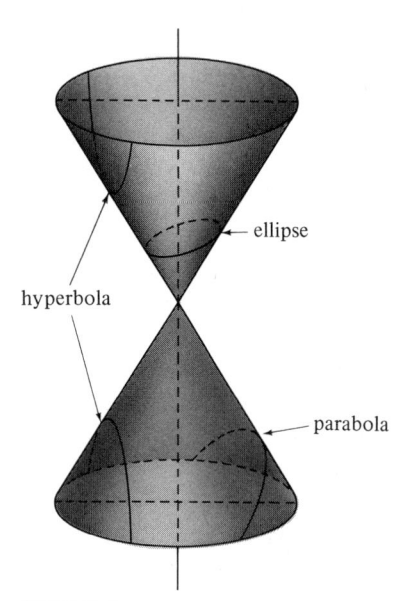

Figure 7-2

1. DEFINITION AND EQUATIONS OF THE CONIC SECTIONS

Definition. Let ℓ be a line, F a point not on ℓ, and e a positive number. The *conic section* (or simply *conic*) with *directrix* ℓ, focus F, and *eccentricity* e is the set of all points P for which the ratio of the distance from P to F to that from P to ℓ is e; that is, if $d_1 = |PF|$ and d_2 is the distance from P to ℓ, then $d_1/d_2 = e$ (Figure 7-3). If $e < 1$, the conic is called an *ellipse*. If $e = 1$, the conic is called a *parabola*. If $e > 1$, the conic is called a *hyperbola*.

REMARK. There are certain other curves (circles, lines, and points) which cannot be obtained from the above definition, but which we shall, nevertheless, call conic sections. The reason for this will become evident when we discuss

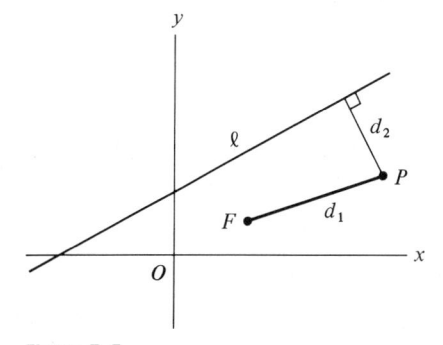

Figure 7-3

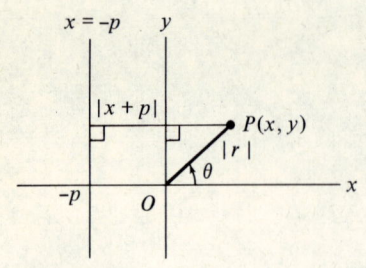

Figure 7-4

the rectangular equations of the parabola, ellipse, and hyperbola.

We first derive the polar equation for a conic section with eccentricity e, focus at the origin (pole), and directrix, the line $x = -p$, where $p > 0$ is the distance from the focus to the directrix. To this end, we observe that if (x, y) is any point with polar coordinates (r, θ), the distance from (x, y) to the origin is $|r|$, and the distance from (x, y) to the line $x = -p$ is $|x + p| = |r \cos \theta + p|$ (Figure 7-4). Thus, (x, y) lies on the conic section iff

$$\frac{|r|}{|r \cos \theta + p|} = e$$

$$r = \pm e(r \cos \theta + p)$$

$$r(1 \mp e \cos \theta) = \pm ep$$

Since $ep \neq 0$, $1 \mp e \cos \theta \neq 0$, and we may write

$$r = \frac{ep}{1 - e \cos \theta} \qquad (1)$$

or

$$r = \frac{-ep}{1 + e \cos \theta} \qquad (2)$$

Now (r, θ) satisfies one of these equations iff $(-r, \theta + \pi)$ satisfies the other, since $\cos(\theta + \pi) = -\cos \theta$. Moreover, (r, θ) and $(-r, \theta + \pi)$ are polar coordinates for the *same point* (x, y). Accordingly, *Equations (1) and (2) have the same graph in the x–y plane.* Either equation is called the *standard polar form* of the equation of a conic section with focus at the origin, eccentricity e, and directrix $x = -p$ $(p > 0)$.

Example 1. Identify the following conic section, giving its eccentricity and the equation of its directrix:

$$r = \frac{5}{2 - 3 \cos \theta}$$

SOLUTION. We divide numerator and denominator of the right member by 2 in order to reduce the equation to standard form. Thus,

$$r = \frac{\frac{5}{2}}{1 - \frac{3}{2}\cos\theta}$$

This is the equation of a hyperbola with $e = \frac{3}{2}$ and $ep = \frac{5}{2}$. Therefore, $p = \frac{5}{3}$, and the directrix is $x = \frac{-5}{3}$.

To find the rectangular equation of the conic section with focus at the origin, eccentricity e, and directrix $x = -p$ ($p > 0$), we use the general definition to obtain

$$\sqrt{x^2 + y^2} = e|x + p|$$

which is equivalent to

$$x^2 + y^2 = e^2(x + p)^2$$

After rearranging terms, we have

$$(1 - e^2)x^2 - 2pe^2 x + y^2 = e^2 p^2 \tag{3}$$

We consider separately the cases $e = 1$ and $e \neq 1$.

CASE 1. $e = 1$ (the *parabola*). Equation (3) reduces to

$$y^2 = 2p\left(x + \frac{p}{2}\right)$$

Translating axes to the point $(-p/2, 0)$, we have $x' = x + p/2$, $y' = y$, and the equation becomes

$$y'^2 = 2px'$$

CASE 2. $e \neq 1$. Completing the square in x in Equation (3), we have

$$(1 - e^2)\left[x^2 - \frac{2pe^2}{1 - e^2}x + \frac{p^2 e^4}{(1 - e^2)^2}\right] + y^2 = e^2 p^2 + \frac{e^4 p^2}{1 - e^2}$$

$$(1 - e^2)\left[x - \frac{pe^2}{1 - e^2}\right]^2 + y^2 = \frac{e^2 p^2}{1 - e^2}$$

Dividing both sides by the right member, we obtain

$$\left(\frac{1 - e^2}{ep}\right)^2 \left[x - \frac{pe^2}{1 - e^2}\right]^2 + \frac{1 - e^2}{(ep)^2} y^2 = 1$$

which may be written

$$\frac{1}{\left(\frac{ep}{1 - e^2}\right)^2}\left(x - e \cdot \frac{ep}{1 - e^2}\right)^2 + \frac{1}{(1 - e^2)\left(\frac{ep}{1 - e^2}\right)^2} y^2 = 1 \tag{4}$$

The equation has been written in this form in order to display the prominence of the term $ep/(1 - e^2)$. To further simplify the equation, we must consider separately the two subcases: (a) $e < 1$, (b) $e > 1$.

(a) $e < 1$ (the *ellipse*). In this case, $1 - e^2 > 0$. Letting $a = ep/(1 - e^2)$, we see that $a > 0$ and that Equation (4) reduces to

$$\frac{(x - ea)^2}{a^2} + \frac{y^2}{(1 - e^2)a^2} = 1$$

We now let $b = \sqrt{1 - e^2} \cdot a > 0$ (so that $b^2 = (1 - e^2)a^2$), and we have

$$\frac{(x - ea)^2}{a^2} + \frac{y^2}{b^2} = 1$$

Translating axes to the point $(ea, 0)$ by means of the translation equations $x' = x - ea$, $y' = y$, we may write the rectangular equation of the ellipse in the form

$$\frac{x'^2}{a^2} + \frac{y'^2}{b^2} = 1$$

(b) $e > 1$ (the *hyperbola*). Since $e^2 > 1$ in this case, we first rewrite Equation (4) in the form

$$\frac{1}{\left(\frac{ep}{e^2 - 1}\right)^2}\left(x + e \cdot \frac{ep}{e^2 - 1}\right)^2 - \frac{1}{(e^2 - 1)\left(\frac{ep}{e^2 - 1}\right)^2} y^2 = 1$$

Letting $a = ep/(e^2 - 1)$ and $b = \sqrt{e^2 - 1} \cdot a$, we see that $a > 0$, $b > 0$, and that the equation reduces to

$$\frac{(x + ea)^2}{a^2} - \frac{y^2}{b^2} = 1$$

After a translation of axes to $(-ea, 0)$ by means of the translation equations $x' = x + ea$, $y' = y$, the rectangular equation of the hyperbola assumes the form

$$\frac{x'^2}{a^2} - \frac{y'^2}{b^2} = 1$$

Note: There is no need, for the present, to memorize the relationships among the quantities a, b, e, and p or the equations of the various conics.

Example 2. Using the general definition, find the rectangular equation of the conic section with focus at $(1, -2)$, directrix $x = 3$, and eccentricity $e = \frac{1}{2}$. (Note that the conic is an ellipse.)

SOLUTION. A point (x, y) is on the graph iff

$$\frac{\sqrt{(x - 1)^2 + (y + 2)^2}}{|x - 3|} = \frac{1}{2}$$

Clearing of fractions and squaring both sides, we have

$$4[(x - 1)^2 + (y + 2)^2] = (x - 3)^2$$

which reduces to

$$3x^2 - 2x + 4y^2 + 16y + 11 = 0$$

EXERCISE 1

1. Using the definition, write the rectangular equation of the parabola with the given focus and directrix.
   - (a) $(0, 3); x = -2$
   - (b) $(-1, 4); y = 0$
   - (c) $(1, 3); y = x$
   - (d) $(0, 5); y = 10$
   - (e) $(0, 0); 3x + 4y - 12 = 0$
2. Using the definition, write the rectangular equation of the conic section with the given focus, directrix, and eccentricity.
   - (a) $(0, 0); x = -3; \frac{2}{3}$
   - (b) $(1, 2); y = -4; 2$
   - (c) $(-2, 4); x = 2; \frac{1}{4}$
   - (d) $(1, 1); y = 5; 3$
   - (e) $(-1, 2); x - y = 3; \frac{4}{3}$
3. Show that the standard polar form of the conic section with focus at the origin, eccentricity e, and
   - (a) directrix $x = p(p > 0)$ is

   $$r = \frac{-ep}{1 - e \cos \theta}$$

   or

   $$r = \frac{ep}{1 + e \cos \theta}$$

   (b) directrix $y = -p(p > 0)$ is

   $$r = \frac{ep}{1 - e \sin \theta}$$

   or

   $$r = \frac{-ep}{1 + e \sin \theta}$$

   (c) directrix $y = p(p > 0)$ is

   $$r = \frac{-ep}{1 - e \sin \theta}$$

   or

   $$r = \frac{ep}{1 + e \sin \theta}$$

4. Identify each of the following conics, giving the eccentricity and the equation of the directrix. (See Problem 3 and Example 1.)

(a) $r = \dfrac{3}{1 - \cos\theta}$

(b) $r = \dfrac{6}{2 + \sin\theta}$

(c) $r = \dfrac{-9}{3 + 4\sin\theta}$

(d) $3r - r\sin\theta + 6 = 0$

(e) $r - 5r\cos\theta + 7 = 0$

(f) $2r + 2r\cos\theta + 9 = 0$

5. Find the coordinates of the focus and the equation of the directrix in the x'–y' plane (translated as in the text) for the cases $e = 1$, $e < 1$, and $e > 1$.

†*6. (a) For the case of the ellipse ($e < 1$) show that the point $(2ea, 0)$ and the line $x = 2ea + p$ ($p > 0$) are also a focus and directrix; that is, show that the equation (in rectangular coordinates) of the ellipse with eccentricity e, focus $(2ea, 0)$ and directrix $x = 2ea + p$ is also

$$\frac{(x - ea)^2}{a^2} + \frac{y^2}{(1 - e^2)a^2} = 1$$

Thus, each ellipse has two foci and two directrices.
(*Hint:* Use the general definition of conic section.)

(b) For the case of the hyperbola ($e > 1$), show that the point $(-2ea, 0)$ and the line $x = -2ea + p$ ($p > 0$) are also a focus and directrix; that is, show that the equation (in rectangular coordinates) of the hyperbola with eccentricity e, focus $(-2ea, 0)$, and directrix $x = -2ea + p$ is also

$$\frac{(x + ea)^2}{a^2} - \frac{y^2}{(e^2 - 1)a^2} = 1$$

Thus, each hyperbola has two foci and two directrices.
(*Hint:* Use the general definition of conic section.)

2. THE PARABOLA ($e = 1$)

We first note that, since $e = 1$, the parabola may be defined as *the set of all points equidistant from a fixed line (the directrix) and a fixed point (the focus) not on the line.* In Section 1 we found that with an appropriate translation of axes, the equation becomes (after dropping the primes)

$$y^2 = 2px \tag{1}$$

Relative to the new axes, the focus is $F(p/2, 0)$ and the directrix is $\ell: x = -p/2$ ($p > 0$). p is the distance from the focus to the directrix (Figure 7–5).

The graph is symmetric with respect to the x-axis and consists of the graphs

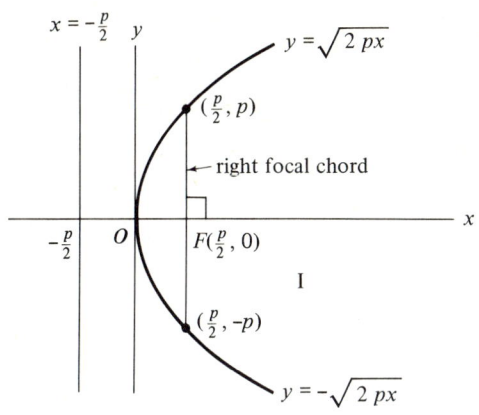

Figure 7-5 The parabola: $y^2 = 2px$ $(p > 0)$.

of the two functions $y = \sqrt{2px}$ and $y = -\sqrt{2px}$. The domain of each function is $x \geq 0$; so the graph lies to the right of the y-axis. Since $0 \leq x_1 < x_2$ implies $\sqrt{x_1} < \sqrt{x_2}$, the graph of $y = \sqrt{2px}$ is *increasing*, and so the graph of $y = -\sqrt{2px}$ is *decreasing*. (See Problem 12, Exercise 3, Chapter 2.) The only intercept is the origin $(0, 0)$, which is called the *vertex* of the parabola. The x-axis is called the *axis (axis of symmetry)* of the parabola. When $x = p/2$, $y = \pm p$. The segment $\overline{(p/2, p)(p/2, -p)}$ is called the *right focal chord* and has length $2p$. The endpoints of the right focal chord are useful in plotting the graph.

An interesting property of the parabola (and one which justifies our drawing its graph as shown in Figure 7-5) is that of *convexity*. (See Chapter 4, Section 7.) Let $I = \{(x, y) | x \geq 0 \text{ and } y^2 < 2px\}$. Now a point (x, y) lies in I iff $|y| < \sqrt{2px}$; that is,

$$-\sqrt{2px} < y < \sqrt{2px}$$

Hence (x, y) lies below the graph of $y = \sqrt{2px}$ and above that of $y = -\sqrt{2px}$, and so must lie in the region labeled I in Figure 7-5. I is called the *interior* of the parabola.

Theorem. *The interior I of the parabola is a convex set.*

PROOF. Let (x_1, y_1) and (x_2, y_2) be distinct points in I. We must show that the segment $\overline{(x_1, y_1)(x_2, y_2)}$ lies in I. The parametric equations of the segment are

$$\begin{cases} x = (1 - t)x_1 + tx_2 \\ y = (1 - t)y_1 + ty_2, \quad 0 \leq t \leq 1 \end{cases} \qquad (2)$$

Let (x, y) be any point on the segment other than the endpoints, so that $0 < t < 1$. We must show that (x, y) belongs to I; that is, $|y| < \sqrt{2px}$. Now

$$|y| = |(1-t)y_1 + ty_2| \le (1-t)|y_1| + t|y_2| < (1-t)\sqrt{2px_1} + t\sqrt{2px_2} \quad (3)$$

since (x_1, y_1) and (x_2, y_2) belong to I. For the inequality $|y| < \sqrt{2px}$ to hold, it is sufficient to have (using Equations (2) and (3))

$$(1-t)\sqrt{2px_1} + t\sqrt{2px_2} < \sqrt{2p[(1-t)x_1 + tx_2]}$$

Since all terms are positive, we may square both sides. After canceling $2p > 0$, we obtain

$$(1-t)^2 x_1 + 2t(1-t)\sqrt{x_1 x_2} + t^2 x_2 < (1-t)x_1 + tx_2$$

Transposing all terms to one side, we see that this is equivalent to

$$t(1-t)x_1 - 2t(1-t)\sqrt{x_1 x_2} + t(1-t)x_2 > 0$$

that is,

$$t(1-t)(\sqrt{x_1} - \sqrt{x_2})^2 > 0$$

But this inequality holds, provided $x_1 \ne x_2$, since $0 < t < 1$. In case $x_1 = x_2$, we have $x = x_1 = x_2$ for all t (vertical segment), and so (from Equation (3)),

$$|y| < (1-t)\sqrt{2px} + t\sqrt{2px} = \sqrt{2px}$$

Therefore,

$$|y| < \sqrt{2px}$$

Thus, in either case, the inequality $|y| < \sqrt{2px}$ holds, and (x, y) belongs to I. Therefore, I is convex.

Other choices for directrix and focus are shown in Figure 7–6. The corresponding equations are (a) $x^2 = 2py$, (b) $y^2 = -2px$, and (c) $x^2 = -2py$. (See Problem 4.) These are called the *standard forms* of the parabola with vertex at the origin and axis along a coordinate axis.

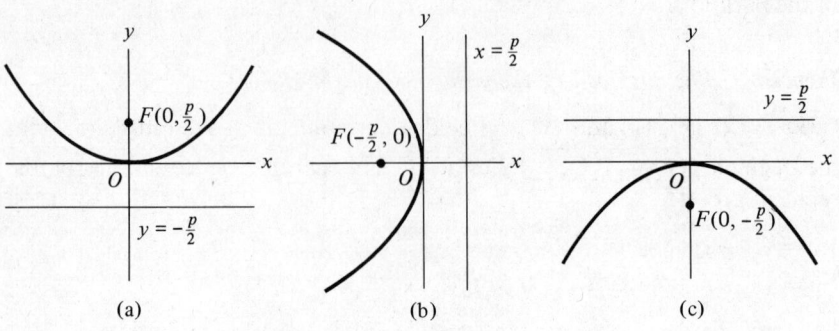

Figure 7-6 (a) $x^2 = 2py$ $(p>0)$. (b) $y^2 = -2px$ $(p>0)$. (c) $x^2 = -2py$ $(p>0)$.

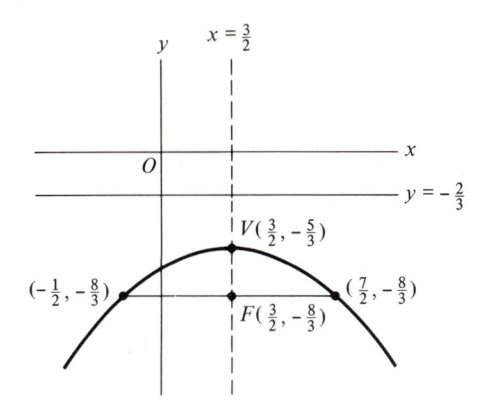

Figure 7-8 $(x-\tfrac{3}{2})^2 = -4(y+\tfrac{5}{3})$.

EXERCISE 2

1. Sketch the graphs of the following parabolas, giving the vertex, focus, axis of symmetry, directrix, and the ends of the right focal chord.
   (a) $x^2 + 4y = 0$
   (b) $2x^2 - 5y = 0$
   (c) $3y^2 + 8x = 0$
   (d) $y^2 - 6x = 0$.

2. Same as Problem 1.
   (a) $(y + 2)^2 = 4(x - 3)$
   (b) $x^2 = 3(y + 1)$
   (c) $x + \tfrac{1}{2} = -\tfrac{1}{2}(y + 3)^2$
   (d) $(x - 3)^2 = -3y + 6$

3. Reduce the following to standard form and sketch their graphs, giving vertex, focus, and so on, wherever appropriate.
   (a) $y^2 + 6y + 2x + 7 = 0$
   (b) $9x^2 - 12x - 27y - 23 = 0$
   (c) $4y^2 - 12y + 16x + 41 = 0$
   (d) $x^2 + 3y + 6 = 0$
   (e) $y^2 - 3y + 2 = 0$
   (f) $3x^2 + 7x - 6 = 0$
   (g) $y^2 - 6 = 3x$
   (h) $5x^2 - 10x + 4y - 3 = 0$
   (i) $y^2 + 2y + 8x = 31$

†4. Derive the standard forms of the parabola shown in Figure 7–6. Use the definition.

5. Find the equation of the parabola in standard position satisfying the following conditions. Discuss and sketch its graph.
   (a) vertex at $(0, 0)$; focus at $(4, 0)$
   (b) vertex at $(2, 3)$; directrix $x = 5$
   (c) focus at $(-1, 2)$; directrix $y = -2$

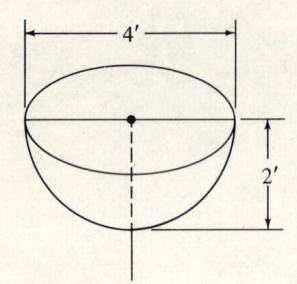

Figure 7-9

  (d) vertex at $(3, -2)$; passing through $(1, -4)$ (two solutions)
  (e) focus at $(-2, 3)$; one end of right focal chord at $(-2, 5)$ (two solutions)
  (f) vertex at $(\frac{3}{2}, 4)$; axis vertical; length of right focal chord 6 (two solutions)
  (g) axis $y = 2$; passing through $(0, 4)$ and $(3, -2)$
  (h) axis $x = -2$; passing through $(0, \frac{11}{3})$ and $(-3, \frac{19}{6})$
6. A church window has the shape of a parabola with vertex at the top and axis vertical. It is 6 feet wide at the base and 10 feet high. How wide is it half way up from the base?
7. (a) Find the equation of the set of all points equidistant from the point $(2, -1)$ and the line $y = 3$. Sketch the graph.
  (b) Find the equation of the set of all points equidistant from the point $(-5, 1)$ and the y-axis.
8. A parabolic reflector is formed by revolving a parabola about its axis. If a light source is placed at the focus of the parabola, the rays are reflected in lines parallel to the axis. Find the point where the light source should be placed in the reflector shown in Figure 7-9.

3. THE ELLIPSE ($e < 1$)

In analyzing the graph of the ellipse, we shall use the equation obtained in Section 1 by a translation of axes, which (after dropping the primes) is

$$\frac{x^2}{a^2} + \frac{y^2}{b^2} = 1$$

From Problem 6, Exercise 1, the ellipse has two foci and two directrices. The foci are $F_1(-ea, 0)$ and $F_2(ea, 0)$. For convenience, let

$$c = ea > 0 \tag{2}$$

Then from Section 1, $b^2 = (1 - e^2)a^2 = a^2 - c^2$. Thus,

$$c^2 = a^2 - b^2 \tag{3}$$

Observe that $c < a$ and $b < a$.

 From Equation (1) we draw the following conclusions:
(1) The ellipse is symmetric with respect to both the x- and y-axes and hence with respect to the origin. For this reason, the origin is called the *center of symmetry*, or simply the *center*, of the ellipse.

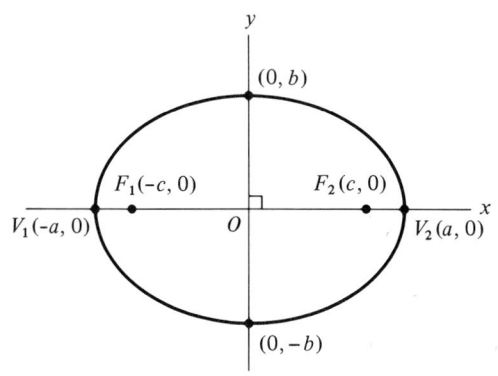

Figure 7-10 $\dfrac{x^2}{a^2} + \dfrac{y^2}{b^2} = 1; c^2 = a^2 - b^2; e = \dfrac{c}{a} < 1.$

(2) The intercepts are $(\pm a, 0)$ and $(0, \pm b)$.
(3) The extent in x is $-a \leq x \leq a$ and the extent in y is $-b \leq y \leq b$.
(4) Solving for y in terms of x, we obtain

$$y = \pm \frac{b}{a}\sqrt{a^2 - x^2}$$

Thus, the ellipse consists of the graphs of the two functions

$$y = (b/a)\sqrt{a^2 - x^2} \quad \text{and} \quad y = -(b/a)\sqrt{a^2 - x^2}$$

The former is decreasing in the interval $0 \leq x \leq a$. (See Problem 12, Exercise 3, Chapter 2.) Now let

$$I = \left\{(x, y) \,\middle|\, |y| < \frac{b}{a}\sqrt{a^2 - x^2}\right\}$$

I is called the *interior* or "inside" of the ellipse (Figure 7–10). A point (x, y) lies in I iff it lies between the two functions given in Part 4 above. One can prove that the interior of the ellipse is convex in essentially the same way that the corresponding proof for the parabola was carried out. The above properties provide the motivation for drawing the ellipse as in Figure 7–10.

The segment $\overline{(-a, 0)(a, 0)}$ is the *major axis*, and $\overline{(0, -b)(0, b)}$ is the *minor axis*. The ends of the major axis are called *vertices*. a is the *length of the semimajor axis*, and b the *length of the semiminor axis*. (We shall use the term "major axis" and "length of major axis" interchangeably; the same is true for minor axis.) Except for the endpoints, the major and minor axes are contained in the interior I.

REMARK. For an ellipse with eccentricity close to 0, a and b are approximately the same, since $b = \sqrt{1 - e^2} \cdot a$. Such an ellipse is approximately the circle $(x^2/a^2) + (y^2/a^2) = 1$. For this reason, we shall regard the circle as an "ellipse with eccentricity zero." Accordingly, when identifying a curve as an ellipse, we shall allow the possibility that it may indeed be a circle. On the other hand, if e is close to 1, b is small relative to a, and the ellipse appears "flat."

The equation

$$\frac{y^2}{a^2} + \frac{x^2}{b^2} = 1 \qquad (4)$$

is an ellipse with major axis of length $2a$ along the y-axis, minor axis of length $2b$ along the x-axis, and foci $F_1(0, -c)$ and $F_2(0, c)$, where $c^2 = a^2 - b^2$. As before, the eccentricity is $e = c/a$. Remember that the major axis is always the longer one and that the foci and vertices are always on the major axis.

Example 1. Discuss the equation

$$9x^2 + 4y^2 = 36$$

and sketch its graph.

DISCUSSION. Dividing the equation by 36 reduces it to the form

$$\frac{x^2}{4} + \frac{y^2}{9} = 1$$

This is the equation of an ellipse with major axis along the y-axis.

$a^2 = 9; \qquad a = 3$
$b^2 = 4; \qquad b = 2$

Therefore,

$$c^2 = 9 - 4 = 5; \qquad c = \sqrt{5}$$

The intercepts are $(0, -3)$, $(0, 3)$, $(-2, 0)$, and $(2, 0)$. The foci are $F_1(0, -\sqrt{5})$ and $F_2(0, \sqrt{5})$. The eccentricity is $e = \sqrt{5}/3 < 1$ (Figure 7–11).

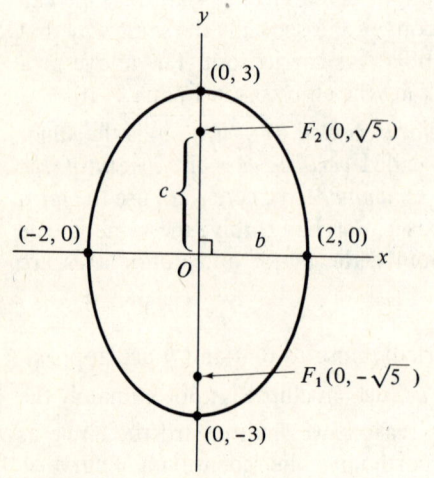

Figure 7-11 $x^2/4 + y^2/9 = 1$.

The following theorem asserts that an ellipse may be defined as *the set of all points (x, y), the sum of whose distances from two fixed points* (the foci) *is a constant.*

Theorem. *A point P(x, y) lies on the ellipse*
$(x^2/a^2) + (y^2/b^2) = 1$ *iff* $|\overline{PF_1}| + |\overline{PF_2}| = 2a$
where $F_1(-c, 0)$ *and* $F_2(c, 0)$ $(c^2 = a^2 - b^2)$ *are the foci. (Note that the constant sum is the length of the major axis.)*

PROOF. (See Problem 5.)

By a translation of axes to the point (h, k) we see that the equation

$$\frac{(x-h)^2}{a^2} + \frac{(y-k)^2}{b^2} = 1 \tag{5}$$

is the equation of the ellipse with center at (h, k), major axis (of length $2a$) parallel to the x-axis, and vertices and foci as shown in Figure 7–12. The formulas relating a, b, e, and c are the same as before. Similarly, the equation

$$\frac{(y-k)^2}{a^2} + \frac{(x-h)^2}{b^2} = 1 \tag{6}$$

represents the ellipse with center (h, k) and major axis (of length $2a$) parallel to the y-axis. Equations (5) and (6) are called *standard forms* of the equation of the ellipse. An ellipse with axes parallel to the coordinate axes is said to be in *standard position*. Note that a^2 is the larger of the denominators in Equations (5) and (6). If a^2 is in the term involving x, the major axis is parallel to the x-axis; similarly for the y-axis.

After squaring the binomials in Equations (5) and (6) and rearranging the terms, we obtain in both cases an equation of the form

$$Ax^2 + By^2 + Cx + Dy + E = 0 \tag{7}$$

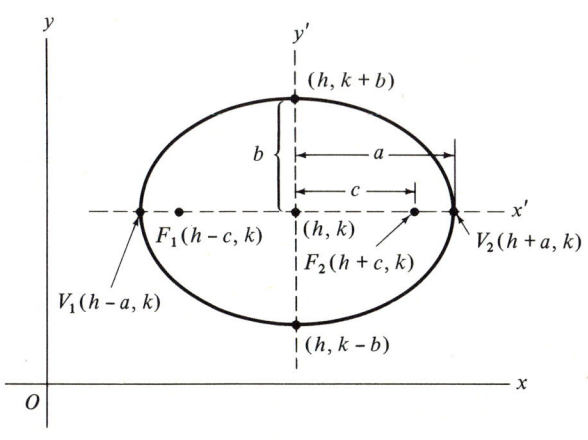

Figure 7–12 $(x-h)^2/a^2 + (y-k)^2/b^2 = 1$.

where *A* and *B* are *nonzero* and have the *same sign*. Moreover, by completing the squares in x and y in Equation (7), one sees that every such equation represents an ellipse, a *degenerate* ellipse (circle or single point), or no graph (see Problem 6).

Example 2.   Discuss the graph of the following equation:

$$9x^2 + 16y^2 + 54x - 64y + 1 = 0$$

DISCUSSION.   We first reduce the equation to standard form by completing the squares:

$$9x^2 + 54x \quad\quad + 16y^2 - 64y \quad\quad = -1$$
$$9(x^2 + 6x + 9) + 16(y^2 - 4y + 4) = -1 + 81 + 64$$
$$9(x + 3)^2 \quad\quad + 16(y - 2)^2 \quad\quad = 144$$
$$\frac{(x + 3)^2}{16} + \frac{(y - 2)^2}{9} = 1$$

This is the equation of an ellipse with center at $(-3, 2)$, and major axis parallel to the x-axis. $a = 4$ and $b = 3$. Therefore, $c^2 = 16 - 9 = 7$; so $c = \sqrt{7} \approx 2.6$ and $e = \sqrt{7}/4 < 1$. The vertices are $V_1(-7, 2)$ and $V_2(1, 2)$. The ends of the minor axes are $(-3, 5)$ and $(-3, -1)$. The foci are $F_1(-3 - \sqrt{7}, 2)$ and $F_2(-3 + \sqrt{7}, 2)$.

EXERCISE 3

1. For the following ellipses find the center, the major and minor axes, foci, vertices, and eccentricity. Sketch the graph.
   (a) $4x^2 + y^2 = 4$
   (b) $4x^2 + 9y^2 - 36 = 0$
   (c) $16x^2 + 25y^2 = 400$
   (d) $25y^2 + 49x^2 - 1225 = 0$
   (e) $3x^2 + 4y^2 = 12$
   (f) $36y^2 + 100x^2 = 3600$
2. Discuss the following equations and sketch their graphs (if they exist). For those that represent ellipses give the information called for in Problem 1.
   (a) $x^2 + 4y^2 + 2x - 8y + 1 = 0$
   (b) $-4x^2 + 16x - 9y^2 - 18y + 11 = 0$
   (c) $4y^2 + 16x^2 - 12y - 96x + 89 = 0$
   (d) $100y^2 + 36x^2 - 100y + 96x - 811 = 0$
   (e) $-2x^2 - 3y^2 + 8x + 6y - 11 = 0$
   (f) $4x^2 + y^2 - 16x + 6y + 29 = 0$
3. Find the equations of the following ellipses in standard position. Sketch their graphs.
   (a) major axis 18, parallel to x-axis; minor axis $2\sqrt{5}$; center at the origin

(b) foci $(\pm 3, 0)$; major axis 8
(c) eccentricity $\sqrt{33}/7$; ends of minor axis $(0, \pm 4)$
(d) vertices $V_1(0, 3)$, $V_2(0, -3)$; passing through $(\sqrt{2}, \frac{3}{2}\sqrt{2})$
(e) foci $F_1(-2 - \sqrt{8}, -3)$, $F_2(-2 + \sqrt{8}, -3)$; one end of minor axis at $(-2, -2)$

4. A right focal chord of an ellipse is a chord through a focus, perpendicular to the major axis. Show that the length of a right focal chord is $2b^2/a$.
*5. Prove the theorem of this section.
*6. Prove that Equation (7) represents an ellipse, a single point, or no graph.
7. (a) Find the equation of the set of all points $P(x, y)$, the sum of whose distances from $(4, 0)$ and $(-4, 0)$ is 12. Sketch the graph.
   (b) Find the equation satisfied by all points $P(x, y)$, the sum of whose distances from $(-2, 1)$ and $(-2, 5)$ is 10. Sketch the graph.

4. THE HYPERBOLA ($e > 1$)

Referring again to Section 1, we find that after a translation of axes, the equation of the hyperbola is (dropping the primes)

$$\frac{x^2}{a^2} - \frac{y^2}{b^2} = 1 \tag{1}$$

From Problem 6 Exercise 1, the hyperbola has two foci and two directrices. The foci are $F_1(-ea, 0)$ and $F_2(ea, 0)$. Again, we let

$$c = ea > 0 \tag{2}$$

Then from Section 1, $b^2 = (e^2 - 1)a^2 = c^2 - a^2$. Thus,

$$c^2 = a^2 + b^2 \tag{3}$$

Note that $c > a$, while no fixed relationship between a and b exists.

From Equation (1), we obtain the following:

(1) The hyperbola is symmetric with respect to the x- and y-axes and hence to the origin. The origin is called the *center of symmetry* or, simply, the *center* of the hyperbola.
(2) The x-intercepts are $(\pm a, 0)$. There are no y-intercepts.
(3) Solving for y in terms of x, we have

$$y = \pm \frac{b}{a}\sqrt{x^2 - a^2}$$

Thus, the extent in x is $|x| \geq a$; that is, $x \geq a$ or $x \leq -a$. The extent in y is the entire y-axis. The hyperbola is the union of the graphs of the two functions $y = (b/a)\sqrt{x^2 - a^2}$ and $y = -(b/a)\sqrt{x^2 - a^2}$. The former function is increasing in the interval $[a, \infty)$. (See Problem 9.)

Let

$$I^+ = \left\{(x, y) \,\Big|\, |y| < \frac{b}{a}\sqrt{x^2 - a^2}, x > a\right\}$$

and

$$I^- = \left\{(x, y) \,\Big|\, |y| < \frac{b}{a}\sqrt{x^2 - a^2}, x < -a\right\}$$

Then I^+ and I^- are convex sets. The proof is analogous to that for the interior of the parabola. The set $I^+ \cup I^-$ is called the *interior* of the hyperbola. A point (x, y) lies in the interior iff it lies between the graphs of the two functions given above. Observe that the foci $F_1(-c, 0)$ and $F_2(c, 0)$ belong to the interior, whereas the center does not (Figure 7-13).

The points $V_1(-a, 0)$ and $V_2(a, 0)$ are called *vertices*, the segment $\overline{(-a, 0)(a, 0)}$ the *transverse* axis, the segment $\overline{(0, -b)(0, b)}$ the *conjugate* axis, and a, b the length of the semitransverse and semiconjugate axes, respectively. The above considerations suggest that the hyperbola be drawn as in Figure 7-13.

The lines $y = (b/a)x$ and $y = -(b/a)x$ shown in Figure 7-13 are asymptotes of the hyperbola in the following sense: If we rotate axes through the angle $\theta = \tan^{-1} b/a$, the line $y = (b/a)x$ becomes the x'-axis in the new coordinate system, since $\tan \theta = b/a$ is the slope of this line. Moreover, the x'-axis is a horizontal asymptote in the x'-y' plane, according to the definition of Section 4 in Chapter 2. In the same sense, the line $y = -(b/a)x$ is an asymptote (see Problem 10). As x "becomes very large," the difference between $(b/a)x$ and $(b/a)\sqrt{x^2 - a^2}$ "becomes arbitrarily small." Thus, the graph of the hyperbola approaches ever nearer to the asymptotes. The easiest way to sketch the asymptotes is first to draw the *central rectangle* with center at the center of the hyperbola and sides of length $2a$ and $2b$ parallel to the transverse and conjugate axes, respectively. The asymptotes are the lines

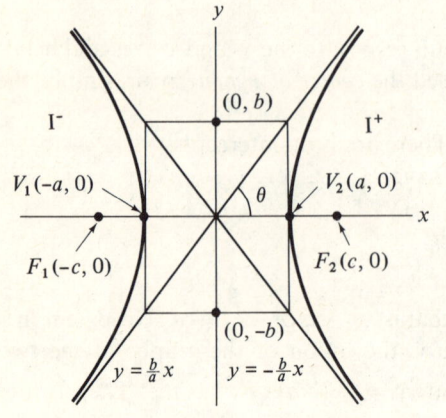

Figure 7-13 $x^2/a^2 - y^2/b^2 = 1$; $c^2 = a^2 + b^2$; $e = c/a > 1$.

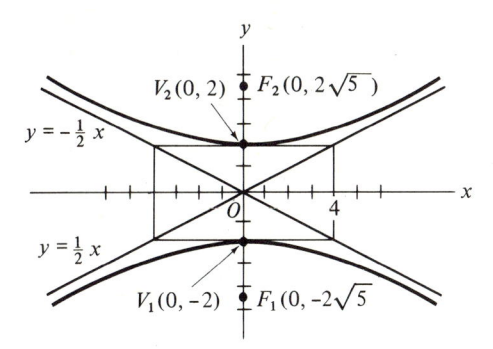

Figure 7-14 $y^2/4 - x^2/16 = 1$.

containing the diagonals of this rectangle.
The equation

$$\frac{y^2}{a^2} - \frac{x^2}{b^2} = 1 \tag{4}$$

is the equation of the hyperbola with center at the origin, transverse axis (of length $2a$) along the y-axis, and foci $F_1(0, -c)$ and $F_2(0, c)$. (The transverse axis always contains the vertices and foci.)

Example 1. Discuss the equation $4x^2 - 16y^2 + 64 = 0$ and sketch its graph.

DISCUSSION. The equation may be reduced to the form

$$\frac{y^2}{4} - \frac{x^2}{16} = 1$$

This is a hyperbola with transverse axis along the y-axis (since the term in y has a positive coefficient). $a = 2$, $b = 4$, and $c^2 = 4 + 16 = 20$; hence $c = 2\sqrt{5}$, and $e = 2\sqrt{5}/2 = \sqrt{5} > 1$. To sketch the hyperbola, we first draw the central rectangle and the asymptotes as in Figure 7–14. The equations of the asymptotes are $y = \frac{1}{2}x$ and $y = -\frac{1}{2}x$ (that is, $y = \pm(a/b)x$). We then draw the hyperbola as shown. The foci are $F_1(0, -2\sqrt{5})$ and $F_2(0, 2\sqrt{5})$. The vertices are $V_1(0, -2)$ and $V_2(0, 2)$.

The hyperbola may be defined as *the set of all points (x, y), the difference of whose distances from two fixed points (the foci) is numerically equal to a constant.*

Theorem. *A point $P(x, y)$ lies on the hyperbola $(x^2/a^2) - (y^2/b^2) = 1$ iff $||\overline{PF_1}| - |\overline{PF_2}|| = 2a$ where $F_1(-c, 0)$ and $F_2(c, 0)$ $(c^2 = a^2 + b^2)$ are the foci.* (Note that *the constant difference is the length of the transverse axis.*)

PROOF. (See Problem 8.)

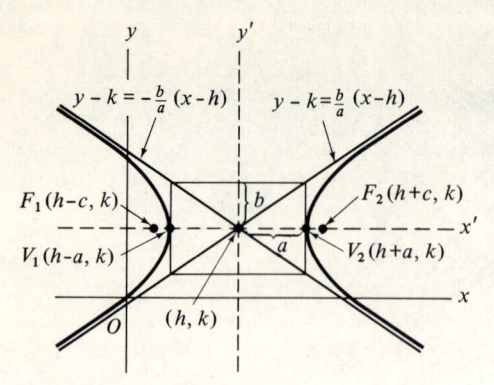

Figure 7-15 $(x-h)^2/a^2 - (y-k)^2/b^2 = 1$.

The equation

$$\frac{(x-h)^2}{a^2} - \frac{(y-k)^2}{b^2} = 1 \tag{5}$$

is the equation of a hyperbola with center (h, k), transverse axis (of length $2a$) parallel to the x-axis, and vertices, foci and asymptotes as shown in Figure 7-15. Similarly, the equation

$$\frac{(y-k)^2}{a^2} - \frac{(x-h)^2}{b^2} = 1 \tag{6}$$

represents the hyperbola with transverse axis (of length $2a$) parallel to the y-axis. Equations (5) and (6) are called *standard forms*. A hyperbola with axes parallel to the coordinate axis is said to be in *standard position*. Note that the transverse axis is always parallel to that coordinate axis whose variable occurs in the term with the positive coefficient.

After squaring the binomials in Equations (5) and (6) and rearranging terms, we obtain in both cases an equation of the form

$$Ax^2 + By^2 + Cx + Dy + E = 0 \tag{7}$$

where A and B are nonzero and have *opposite sign*. It is left as an exercise to show that Equation (7) always represents a hyperbola or two intersecting lines (a degenerate hyperbola). (See Problem 7.)

Example 2. Discuss the following equation and sketch its graph:

$$9x^2 - 4y^2 - 54x - 16y + 29 = 0$$

DISCUSSION. We first reduce to standard form by completing the squares:

$$9(x^2 - 6x + 9) - 4(y^2 + 4y + 4) = -29 + 81 - 16$$
$$9(x - 3)^2 - 4(y + 2)^2 = 36$$
$$\frac{(x-3)^2}{4} - \frac{(y+2)^2}{9} = 1$$

This is the equation of the hyperbola with center $(3, -2)$ and transverse axis parallel to the x-axis. $a = 2$, $b = 3$, and $c^2 = 4 + 9 = 13$; hence $c = \sqrt{13}$ and $e = \sqrt{13}/2 > 1$. The graph is similar to Figure 7–15.

Example 3. Discuss the equation

$$2x^2 - 3y^2 + 4x + 12y - 10 = 0$$

DISCUSSION. Completing the squares, we obtain

$$\frac{(x+1)^2}{3} - \frac{(y-2)^2}{2} = 0$$

that is,

$$\left(\frac{x+1}{\sqrt{3}} - \frac{y-2}{\sqrt{2}}\right)\left(\frac{x+1}{\sqrt{3}} + \frac{y-2}{\sqrt{2}}\right) = 0$$

This is the equation of two lines

$$y - 2 = \frac{\sqrt{2}}{\sqrt{3}}(x+1)$$

$$y - 2 = -\frac{\sqrt{2}}{\sqrt{3}}(x+1)$$

intersecting at $(-1, 2)$. Thus, the given equation represents a degenerate hyperbola.

EXERCISE 4

1. Give a complete discussion of the following hyperbolas. Sketch their graphs, showing vertices, foci, and asymptotes.
   (a) $4x^2 - y^2 = 4$
   (b) $9y^2 - 4x^2 - 36 = 0$
   (c) $16x^2 - 25y^2 + 400 = 0$
   (d) $25y^2 - 49x^2 = -1225$
   (e) $3x^2 - 4y^2 = 12$
   (f) $5x^2 - 2y^2 + 10 = 0$

2. Give a complete discussion of the following equations. Sketch their graphs, showing all important aspects.
   (a) $4x^2 - 9y^2 + 8x + 36y - 68 = 0$
   (b) $x^2 - 4y^2 - 4x + 12y - 1 = 0$
   (c) $81x^2 - 4y^2 + 486x + 8y + 725 = 0$
   (d) $16y^2 - 16x^2 - 16y - 48x - 96 = 0$
   (e) $9y^2 - x^2 + 18y + 4x + 5 = 0$
   (f) $16x^2 - 9y^2 - 160x - 54y + 175 = 0$

3. Find the equations of the following hyperbolas in standard position. Sketch their graphs.
   (a) foci $(\pm 5, 0)$; vertices $(\pm 3, 0)$
   (b) asymptotes $y = \pm\frac{3}{2}x$; a vertex at $(4, 0)$
   (c) vertices $(\pm 4, 0)$; eccentricity $\frac{3}{2}$
   (d) foci $(2, -1), (2, 7)$; eccentricity 2
   (e) vertices $(0, \pm 3)$; passing through $(2, 3\sqrt{2})$
   (f) axes along the coordinate axes and passing through $(2, 2\sqrt{3}), (-3, -4\sqrt{2})$
4. Identify the following curves. Discuss and sketch their graphs.
   (a) $27x^2 - 216x + 9y^2 - 12y + 355 = 0$
   (b) $4x^2 + 28y - 7y^2 + 4x - 55 = 0$
   (c) $-9y^2 - 54y - 36x^2 + 48x - 61 = 0$
   (d) $4x^2 - 24y + 4y^2 - 12x + 29 = 0$
   (e) $25y^2 - 40y = 75x - 166$.
   (f) $9y^2 - 12y - 27x^2 - 216x = 455$
5. Find the equation of the set of all points, the difference of whose distances from
   (a) $(4, 0)$ and $(-4, 0)$ is numerically equal to 4
   (b) $(-2, 1)$ and $(-2, 5)$ is numerically equal to 2
   (c) $(-1, -1)$ and $(1, 1)$ is numerically equal to 2
6. A *right focal chord* of a hyperbola is a chord passing through a focus, perpendicular to the transverse axis. Show that the length of a right focal chord is $2b^2/a$.
*7. Prove that Equation (7) represents a hyperbola or two intersecting lines.
*8. Prove the theorem of this section.
*9. Prove that the function $y = (b/a)\sqrt{x^2 - a^2}$ is increasing in the interval $[a, \infty)$.
*10. Prove that the line $y = (b/a)x$ is an asymptote of the hyperbola

$$(x^2/a^2) - (y^2/b^2) = 1$$

in the sense described in the text. *Hint:* Show that after a rotation of axes through the angle $\varphi = \tan^{-1} b/a$ (so that $\cos \varphi = a/\sqrt{a^2 + b^2}$ and $\sin \varphi = b/\sqrt{a^2 + b^2}$), the line $y = (b/a)x$ becomes the x'-axis (that is, the line $y' = 0$). Then show that the equation of the hyperbola in the x', y' system is $(b^2 - a^2)y'^2 - 2abx'y' = a^2b^2$ and that the line $y' = 0$ is an asymptote.

5. THE GENERAL QUADRATIC EQUATION IN TWO VARIABLES—FIRST METHOD

In the preceding sections, we saw that every conic section is represented by a second-degree (that is, quadratic) equation in x and y. It is the purpose of this section to show that the *general quadratic equation*

$$ax^2 + bxy + cy^2 + dx + ey + f = 0 \tag{1}$$

represents a conic section, a degenerate conic, or no graph. (We assume $b \neq 0$, since the case $b = 0$ has already been discussed.) If we rotate axes through an arbitrary angle φ, the transform of Equation (1) is of the form

$$a'x'^2 + b'x'y' + c'y'^2 + d'x' + e'y' + f' = 0 \tag{2}$$

The coefficients $a', b', c', \ldots$ depend upon the coefficients $a, b, c, \ldots$ and φ. We shall show that if $b \neq 0$, it is possible to choose φ so that $b' = 0$; that is, so that Equation (2) has no *product term* $x'y'$. Since Equations (1) and (2) represent the same set of points in the x–y plane, it follows that Equation (1) represents a conic. (It cannot represent a circle; for the equation of a circle has no product term.)

We first observe that the second-degree terms in Equation (2) arise solely from the second-degree terms in Equation (1). Consequently, we need consider only the *quadratic form*

$$ax^2 + bxy + cy^2. \tag{3}$$

Recall that the equations for a rotation of axes are

$$\begin{cases} x = x' \cos \varphi - y' \sin \varphi \\ y = x' \sin \varphi + y' \cos \varphi \end{cases} \tag{4}$$

Substitution of Equations (4) in (3) yields what we shall call the *transform* of (3)

$$a'x'^2 + b'x'y' + c'y'^2, \tag{5}$$

where

$$\begin{cases} a' = a \cos^2 \varphi + b \sin \varphi \cos \varphi + c \sin^2 \varphi \\ b' = 2(c - a) \sin \varphi \cos \varphi + b(\cos^2 \varphi - \sin^2 \varphi) \\ c' = a \sin^2 \varphi - b \sin \varphi \cos \varphi + c \cos^2 \varphi \end{cases} \tag{6}$$

(See Problem 2.) We now determine φ so that $b' = 0$. Using the double-angle formulas $\sin 2\varphi = 2 \sin \varphi \cos \varphi$ and $\cos 2\varphi = \cos^2 \varphi - \sin^2 \varphi$, we have

$$b' = (c - a) \sin 2\varphi + b \cos 2\varphi$$

Setting $b' = 0$ and solving for φ, we obtain

$$\cot 2\varphi = \frac{a - c}{b} \quad (b \neq 0) \tag{7}$$

Since the range of values of $\cot 2\varphi$ is the entire real line, it follows that Equation (7) always has a solution. Thus, if we choose φ so as to satisfy Equation (7), b' will be zero in (5).

In practice, we do not need to compute φ. All that is needed are $\sin \varphi$ and $\cos \varphi$ to be used in Equations (4). From Equation (7) we obtain $\cot 2\varphi$, where $0 < 2\varphi < \pi$; that is, $0 < \varphi < \frac{1}{2}\pi$. We then compute $\cos 2\varphi$, using trigonometric relations. Finally $\sin \varphi$ and $\cos \varphi$ are obtained from the half-angle formulas

$$\sin \varphi = \sqrt{\frac{1 - \cos 2\varphi}{2}}; \quad \cos \varphi = \sqrt{\frac{1 + \cos 2\varphi}{2}} \tag{8}$$

Example 1. Rotate axes so that the transform of the equation $2x^2 + 3xy - 2y^2 = 25$ has no product term. Sketch the graph in the x'–y' plane.

SOLUTION. From Equation (7), with $a = 2, b = 3$, and $c = -2$,

$$\cot 2\varphi = \frac{2 - (-2)}{3} = \frac{4}{3}$$

Choosing 2φ in the first quadrant (since $\cot 2\varphi > 0$), we have $\cos 2\varphi = \frac{4}{5}$. From Equations (8),

$$\sin \varphi = \frac{1}{\sqrt{10}} \quad \text{and} \quad \cos \varphi = \frac{3}{\sqrt{10}}$$

The rotation equations (4) are

$$\begin{cases} x = \frac{3}{\sqrt{10}}x' - \frac{1}{\sqrt{10}}y' \\ y = \frac{1}{\sqrt{10}}x' - \frac{3}{\sqrt{10}}y' \end{cases}$$

Substituting in the given equation and simplifying yields

$$\frac{x'^2}{10} - \frac{y'^2}{10} = 1$$

This is the equation of a hyperbola with center at the origin and transverse axis along the x'-axis. In the x'-y' plane, the vertices are $(-\sqrt{10}, 0), (\sqrt{10}, 0)$ and the asymptotes are $y' = \pm x'$. To sketch the graph, we first draw the x'-, y'-axes as shown in Figure 7–16. The x'-axis makes the angle φ with the x-axis. Since $\tan \varphi = \sin \varphi / \cos \varphi = \frac{1}{3}$, the x'-, y'-axes can readily be drawn without computing φ (Figure 7–16).

An important property of the quadratic form (3) is that the quantity $b^2 - 4ac$, called the *discriminant* of the quadratic form (or of the quadratic equation (1)), is *invariant under an arbitrary rotation of axes*. This means that

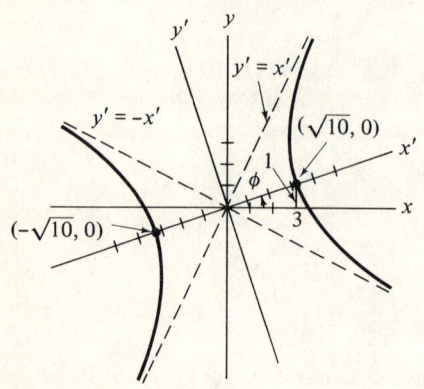

Figure 7–16

if we rotate axes through any angle φ, so that $ax^2 + bxy + cy^2$ is transformed into $a'x'^2 + b'x'y' + c'y'^2$, then

$$b'^2 - 4a'c' = b^2 - 4ac \tag{9}$$

that is, *a quadratic form and its transform under a rotation of axes have the same discriminant*. This can be verified from Equations (6) by a direct (though tedious) calculation, making use of elementary trigonometric relations (see Problem 3). A simple proof, using matrices, is given in Section 6. (It is clear that the discriminant is invariant also under a translation of axes, since the coefficients of the quadratic terms are unchanged.)

For the special case in which φ is chosen so that $b' = 0$, we have

$$b^2 - 4ac = -4a'c' \tag{10}$$

Formula (10) provides a simple means for identifying a conic section directly from the general equation (1). Using the conclusions of Sections 2, 3, 4, regarding quadratic equations with no product term, we may state the following rules (we include the possibility that the conic may be degenerate or that there is no graph):

(1) $b^2 - 4ac = 0$: parabola (since $a' = 0$ or $c' = 0$)
(2) $b^2 - 4ac < 0$: ellipse   (since a' and c' have the same sign)
(3) $b^2 - 4ac > 0$: hyperbola (since a' and c' have opposite signs)

Example 2. Identify the following conic and sketch the graph:

$$x^2 + 3xy + 5y^2 + 2x - 8y + 1 = 0$$

SOLUTION. $b^2 - 4ac = 9 - 20 < 0$. Hence the graph is an ellipse.

In the case of the ellipse or the hyperbola, the computations are usually simpler if we first eliminate the linear terms by a translation of axes and then rotate axes to eliminate the product term. (For the parabola, it is sometimes not possible to eliminate the linear terms.)

We first determine a translation of axes $x = x' + h$, $y = y' + k$, so as to eliminate the first-degree terms. Substituting in the given equation, we obtain

$$x'^2 + 3x'y' + 5y'^2 + (2h + 3k + 2)x' + (3h + 10k - 8)y'$$
$$+ (h^2 + 3hk + 5k^2 + 2h - 8k + 1) = 0$$

Setting the coefficients of x' and y' equal to zero, we have

$$\begin{cases} 2h + 3k + 2 = 0 \\ 3h + 10k - 8 = 0 \end{cases}$$

The solution is $h = -4$, $k = 2$. The transform is

$$x'^2 + 3x'y' + 5y'^2 - 11 = 0$$

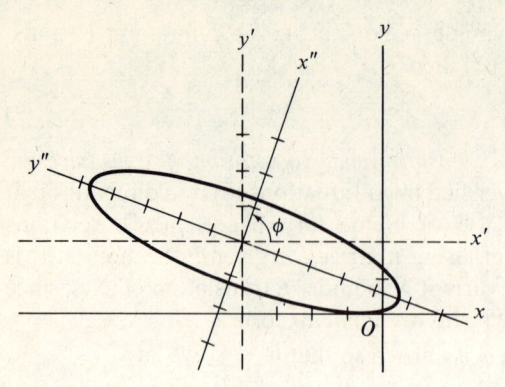

Figure 7-17 $x''^2/2 + y''^2/22 = 1$.

We next rotate axes so as to eliminate the product term $x'y'$:

$$\cot 2\varphi = \frac{1-5}{3} = -\frac{4}{3}$$

$$\cos 2\varphi = -\tfrac{4}{5}; \qquad \cos \varphi = \frac{1}{\sqrt{10}}; \qquad \sin \varphi = \frac{3}{\sqrt{10}}$$

$$x' = \frac{1}{\sqrt{10}} x'' - \frac{3}{\sqrt{10}} y''$$

$$y' = \frac{3}{\sqrt{10}} x'' + \frac{1}{\sqrt{10}} y''$$

Under this rotation, the transform of the above equation in x' and y' is (see Figure 7–17 for the graph)

$$\frac{x''^2}{2} + \frac{y''^2}{22} = 1$$

EXERCISE 5

1. Identify the following conics by means of the discriminant. Simplify by translations and rotations of axes. Sketch the graphs.
   (a) $xy - 7x + 7y + 5 = 0$
   (b) $x^2 - xy + y^2 + 2x - y - 5 = 0$
   (c) $x^2 + 2xy + y^2 - 4x - 4y + 4 = 0$
   (d) $x^2 - xy + 3x + 2 = 0$
   (e) $4x^2 + 4xy + y^2 - 24x + 38y - 139 = 0$
   (f) $3x^2 + 2\sqrt{3}xy + y^2 + 4x - 4\sqrt{3}y + 16 = 0$ (*Hint:* Rotate axes first.)
2. Derive Equation (6) by substituting Equations (4) in (3).
*3. Prove Equation (9) $b'^2 - 4a'c' = b^2 - 4ac$ from Equation (6).

*6. THE GENERAL QUADRATIC EQUATION IN TWO VARIABLES—SECOND METHOD

Recall from Chapter 6 that the rotation equations, written in matrix form, are

$$X' = P^t X \tag{1}$$
$$X = PX' \tag{2}$$

where

$$X = \begin{pmatrix} x \\ y \end{pmatrix}, \quad X' = \begin{pmatrix} x' \\ y' \end{pmatrix}$$

$$P = \begin{pmatrix} \cos \varphi & -\sin \varphi \\ \sin \varphi & \cos \varphi \end{pmatrix}$$

$$P^t = \begin{pmatrix} \cos \varphi & \sin \varphi \\ -\sin \varphi & \cos \varphi \end{pmatrix} \quad \text{(the transpose of } P\text{)}$$

P is an orthogonal matrix; that is, $P^t = P^{-1}$, so that

$$PP^t = P^t P = I_2 = \begin{pmatrix} 1 & 0 \\ 0 & 1 \end{pmatrix}$$

Also $|P| = 1$; P is called a *rotation matrix*.

The quadratic form $ax^2 + bxy + cy^2$ may be written in matrix form as follows:

$$(x \; y) \begin{pmatrix} a & \frac{b}{2} \\ \frac{b}{2} & c \end{pmatrix} \begin{pmatrix} x \\ y \end{pmatrix} = (x \; y) \begin{pmatrix} ax + \frac{b}{2}y \\ \frac{b}{2}x + cy \end{pmatrix}$$

$$= x\left(ax + \frac{b}{2}y\right) + y\left(\frac{b}{2}x + cy\right)$$

$$= ax^2 + bxy + cy^2$$

Now let

$$Q = \begin{pmatrix} a & \frac{b}{2} \\ \frac{b}{2} & c \end{pmatrix}$$

Q is *symmetric* (that is, $Q^t = Q$), and we have

$$X^t Q X = ax^2 + bxy + cy^2 \tag{3}$$

(*Note:* We identify the one-by-one matrix (a_{11}) with the number a_{11}.) It is easy to show that Q is the only two-by-two symmetric matrix satisfying Equation (3). (See Problem 7.) Q is called *the symmetric* matrix for the quadratic form.

To compute the transform of Equation (3), we substitute Equation (2) in it to obtain

$$(PX)^t Q(PX') = a'x'^2 + b'x'y' + c'y'^2$$
$$X'^t(P^tQP)X' = a'x'^2 + b'x'y' + c'y'^2$$

Since the matrix P^tQP is symmetric (see Problem 8), it follows that P^tQP is *the* symmetric matrix Q' for the quadratic form $a'x'^2 + b'x'y' + c'y'^2$, where

$$Q' = \begin{pmatrix} a' & \dfrac{b'}{2} \\ \dfrac{b'}{2} & c' \end{pmatrix}$$

It is now a simple matter to establish the invariance of the discriminant of a quadratic form (Equation (9), Section 5). From

$$Q' = P^tQP$$

we have (by Theorem 11, Appendix),

$$|Q'| = |P^tQP| = |P^t| \cdot |Q| \cdot |P| = |Q|$$

since

$$|P^t| = |P| = 1$$

Thus,

$$a'c' - \frac{b'^2}{4} = ac - \frac{b^2}{4}$$

that is,

$$b'^2 - 4a'c' = b^2 - 4ac$$

To determine the angle of rotation φ such that $b' = 0$, it is necessary and sufficient that

$$Q' = P^tQP = \begin{pmatrix} a' & 0 \\ 0 & c' \end{pmatrix}$$

The matrix on the right is said to be in *diagonal form* (zero's off the main diagonal). Thus, the problem of eliminating the product term by a rotation of axes is equivalent to the following: Given a *symmetric matrix Q*, find an orthogonal matrix P (with $|P| = 1$) which satisfies

$$P^tQP = D \tag{4}$$

where D is a diagonal matrix; that is

$$D = \begin{pmatrix} \lambda_1 & 0 \\ 0 & \lambda_2 \end{pmatrix}$$

for some numbers λ_1 and λ_2. The transform of the quadratic form under the rotation P is

$$\lambda_1 x'^2 + \lambda_2 y'^2$$

The "second method" consists of using Equation (4) and the orthogonality of P to determine P. We present this method not because it is more efficient than the first, but for the reason that it readily extends to quadratic equations in three variables.

Let the unknown matrix P be written

$$P = \begin{pmatrix} p_{11} & p_{12} \\ p_{21} & p_{22} \end{pmatrix}$$

and let

$$X_1 = \begin{pmatrix} p_{11} \\ p_{21} \end{pmatrix}, \quad X_2 = \begin{pmatrix} p_{12} \\ p_{22} \end{pmatrix}$$

X_1 and X_2 are the columns of P. We call them *column matrices*, or *column vectors*, or simply *vectors*. So we may write

$$P = (X_1 \ X_2)$$

Since P is to be orthogonal ($P^t = P^{-1}$), Equation (4) takes the form

$$P^{-1}QP = D$$

that is,

$$QP = PD$$

Thus,

$$Q \cdot (X_1 \ X_2) = (X_1 \ X_2) \cdot \begin{pmatrix} \lambda_1 & 0 \\ 0 & \lambda_2 \end{pmatrix}$$

From the definition of matrix multiplication, this last equation is equivalent to

$$(QX_1 \ QX_2) = (\lambda_1 X_1 \ \lambda_2 X_2)$$

Accordingly, the matrix P satisfies Equation (4) iff there exist numbers λ_1 and λ_2 such that

$$QX_1 = \lambda_1 X_1 \quad \text{and} \quad QX_2 = \lambda_2 X_2$$

Thus, our problem is to find numbers λ and nonzero vectors X for which

$$QX = \lambda X$$

that is,

$$QX - \lambda X = (0)$$

where

$$(0) = \begin{pmatrix} 0 \\ 0 \end{pmatrix}$$

The last equation takes the form

$$(Q - \lambda I_2)X = (0) \tag{5}$$

Definition. Let Q be any square matrix. If λ is a number and X is a nonzero column vector for which $QX = \lambda X$ (equivalently, $(Q - \lambda I_2)X = (0)$), then λ is called an *eigenvalue* or *characteristic value* of Q, and X is called an *eigenvector* or *characteristic vector* of Q corresponding to λ. We call $Q - \lambda I_2$ the *characteristic matrix* of Q.

REMARK. It is important to note that if X is an eigenvector of Q corresponding to λ, then for every number $c \neq 0$, cX is also an eigenvector corresponding to λ. Thus, to each λ correspond infinitely many eigenvectors.

Since the columns of P must satisfy Equation (5), they are necessarily eigenvectors corresponding to certain eigenvalues of Q. Furthermore, since P is to be orthogonal, $P^t P = I_2$. Now

$$P^t = \begin{pmatrix} p_{11} & p_{21} \\ p_{12} & p_{22} \end{pmatrix} = \begin{pmatrix} X_1^t \\ X_2^t \end{pmatrix}$$

that is, the *rows* of P^t are X_1^t and X_2^t. Hence

$$P^t P = \begin{pmatrix} X_1^t \\ X_2^t \end{pmatrix} (X_1 \ X_2) = \begin{pmatrix} X_1^t X_1 & X_1^t X_2 \\ X_2^t X_1 & X_2^t X_2 \end{pmatrix}$$

If we now identify a column matrix $X = \begin{pmatrix} a \\ b \end{pmatrix}$ with the vector $a\mathbf{i} + b\mathbf{j}$, the inner product (dot product) of two vectors X and Y may be written $X \circ Y = X^t Y$. Note that if $Y = \begin{pmatrix} c \\ d \end{pmatrix}$, we have $X \circ Y = ac + bd = X^t Y$. Accordingly,

$$P^t P = \begin{pmatrix} X_1 \circ X_1 & X_1 \circ X_2 \\ X_2 \circ X_1 & X_2 \circ X_2 \end{pmatrix}$$

(Recall that $X_1 \circ X_2 = X_2 \circ X_1$.)
Since

$$P^t P = I_2 = \begin{pmatrix} 1 & 0 \\ 0 & 1 \end{pmatrix}$$

we must have $X_1 \circ X_1 = X_2 \circ X_2 = 1$, and $X_1 \circ X_2 = X_2 \circ X_1 = 0$. Now $X_1 \circ X_1 = |X_1|^2 = 1$, so that X_1 and X_2 have length 1. In other words, *the columns of P must be orthogonal unit vectors.* (This accounts for the name "orthogonal matrix.") We shall have solved our problem, then, if we find two orthogonal unit eigenvectors of Q. We first find the eigenvalues. As before, let

$$Q = \begin{pmatrix} a & \frac{b}{2} \\ \frac{b}{2} & c \end{pmatrix}$$

be the symmetric matrix of the quadratic form $ax^2 + bxy + cy^2$. If, for the moment, we write the unknown column vector X as

$$X = \begin{pmatrix} p_1 \\ p_2 \end{pmatrix}$$

the matrix equation (5) may be written

$$\left[\begin{pmatrix} a & \frac{b}{2} \\ \frac{b}{2} & c \end{pmatrix} - \begin{pmatrix} \lambda & 0 \\ 0 & \lambda \end{pmatrix} \right] \begin{pmatrix} p_1 \\ p_2 \end{pmatrix} = \begin{pmatrix} 0 \\ 0 \end{pmatrix}$$

that is,

$$\begin{pmatrix} a - \lambda & \frac{b}{2} \\ \frac{b}{2} & c - \lambda \end{pmatrix} \begin{pmatrix} p_1 \\ p_2 \end{pmatrix} = \begin{pmatrix} 0 \\ 0 \end{pmatrix}$$

This matrix equation is equivalent to the following system of linear homogeneous equations:

$$\begin{cases} (a - \lambda)p_1 + \frac{b}{2}p_2 = 0 \\ \frac{b}{2}p_1 + (c - \lambda)p_2 = 0 \end{cases} \quad (6)$$

From elementary algebra, we know that the system (6) has a nonzero solution p_1, p_2 (that is, $p_1 \neq 0$ or $p_2 \neq 0$) iff the determinant of the coefficient matrix is zero; that is,

$$\begin{vmatrix} a - \lambda & \frac{b}{2} \\ \frac{b}{2} & c - \lambda \end{vmatrix} = 0 \quad (7)$$

Thus, Equation (5) has a nonzero solution X iff

$$|Q - \lambda I_2| = 0 \quad (8)$$

Equation (8) (or Equation (7)) is called the *characteristic equation* of Q. Expanding the determinant in Equation (7), we see that the characteristic equation is a quadratic equation in λ, namely,

$$\lambda^2 - (a + c)\lambda + ac - \frac{b^2}{4} = 0 \tag{9}$$

The discriminant of Equation (9) is

$$(a + c)^2 - 4\left(ac - \frac{b^2}{4}\right) = (a - c)^2 + b^2$$

If $b \neq 0$, the discriminant is positive, and the characteristic equation has two (distinct) real roots λ_1 and λ_2. To find P, then, we must find eigenvectors corresponding to λ_1 and λ_2, respectively.

Let X_1^0 and X_2^0 denote eigenvectors corresponding to λ_1 and λ_2, respectively. Then, by the remark following the definition of eigenvector, the vectors

$$X_1 = \frac{X_1^0}{|X_1^0|}, \quad X_2 = \frac{X_2^0}{|X_2^0|}$$

are also eigenvectors corresponding to λ_1 and λ_2. Since $|X_1| = |X_2| = 1$, X_1 and X_2 are *unit* vectors. (When a vector is divided by its length to produce a unit vector, we say the vector has been "normalized.") The problem remaining is the orthogonality of X_1 and X_2, or, equivalently, of X_1^0 and X_2^0. This is automatic, according to the following theorem.

Theorem. *If Q is a symmetric 2×2 matrix with distinct eigenvalues λ_1, λ_2 and corresponding eigenvectors X_1, X_2, then $X_1 \circ X_2 = 0$; that is, X_1 and X_2 are orthogonal.*

PROOF. We have

$$QX_1 = \lambda_1 X_1 \quad \text{and} \quad QX_2 = \lambda_2 X_2$$

Multiplying the second equation on the right by X_1^t yields

$$X_1^t Q X_2 = \lambda_2 X_1^t X_2 \tag{10}$$

Taking the transpose of both members of the first equation, we obtain

$$X_1^t Q^t = \lambda_1 X_1^t$$

Multiplying this equation by X_2, we have

$$X_1^t Q^t X_2 = \lambda_1 X_1^t X_2$$

Comparing this with Equation (10) and using the hypothesis that $Q = Q^t$, we conclude that

$$\lambda_1 X_1^t X_2 = \lambda_2 X_1^t X_2$$

that is,

$$(\lambda_1 - \lambda_2) X_1 \circ X_2 = 0$$

Since $\lambda_1 - \lambda_2 \neq 0$, by hypothesis, it follows that $X_1 \circ X_2 = 0$. Thus, X_1 and X_2 are orthogonal.

The General Quadratic Equation in Two Variables — Second Method

We conclude with the observation that from the rotation equations

$$\begin{pmatrix} x \\ y \end{pmatrix} = P \begin{pmatrix} x' \\ y' \end{pmatrix}$$

a sketch of the x'-, y'-axes may be made in the x–y plane by plotting the points $(x' = 1, y' = 0)$ and $(x' = 0, y' = 1)$ in the x–y plane. Since $P = (X_1 \ X_2)$, the coordinates of $(x' = 1, y' = 0)$ are given by

$$(X_1 \ X_2) \begin{pmatrix} 1 \\ 0 \end{pmatrix} = X_1$$

and those of $(x' = 0, y' = 1)$ by

$$(X_1 \ X_2) \begin{pmatrix} 0 \\ 1 \end{pmatrix} = X_2$$

Thus, *the columns of P are the unit vectors along the positive x'-, y'-axes, respectively.* We illustrate the second method with an example.

Example. Rotate axes so that the transform of the equation

$$2x^2 - \sqrt{3}xy + y^2 = 1$$

has no product term. Sketch the graph.

SOLUTION. First we observe that since $b^2 - 4ac = 3 - 4 \cdot 2 \cdot 1 = -5 < 0$, the conic is an ellipse.

$$Q = \begin{pmatrix} 2 & -\frac{\sqrt{3}}{2} \\ -\frac{\sqrt{3}}{2} & 1 \end{pmatrix} \qquad X = \begin{pmatrix} x \\ y \end{pmatrix}$$

The given equation in matrix form is

$$X'QX = 1$$

To find the eigenvalues of Q, we solve the characteristic equation

$$|Q - \lambda I_2| = \begin{vmatrix} 2 - \lambda & -\frac{\sqrt{3}}{2} \\ -\frac{\sqrt{3}}{2} & 1 - \lambda \end{vmatrix} = \lambda^2 - 3\lambda + \tfrac{5}{4} = 0$$

that is, $4\lambda^2 - 12\lambda + 5 = (2\lambda - 5)(2\lambda - 1) = 0$. Therefore, $\lambda_1 = \tfrac{1}{2}$ and $\lambda_2 = \tfrac{5}{2}$. To find an eigenvector for $\lambda_1 = \tfrac{1}{2}$, we solve $(Q - \tfrac{1}{2}I_2)X_1^0 = (0)$ for X_1^0;

that is,

$$\begin{pmatrix} \frac{3}{2} & -\frac{\sqrt{3}}{2} \\ -\frac{\sqrt{3}}{2} & \frac{1}{2} \end{pmatrix} \begin{pmatrix} x_1^0 \\ y_1^0 \end{pmatrix} = \begin{pmatrix} 0 \\ 0 \end{pmatrix}$$

The equivalent system of linear equations is

$$\begin{cases} \frac{3}{2}x_1^0 - \frac{\sqrt{3}}{2}y_1^0 = 0 \\ -\frac{\sqrt{3}}{2}x_1^0 + \frac{1}{2}y_1^0 = 0 \end{cases}$$

Since both equations are equivalent to $y_1^0 = \sqrt{3}x_1^0$, there are infinitely many solutions (as was remarked earlier). Thus, we may choose $x_1^0 \neq 0$ arbitrarily and solve for y_1^0. For instance, we may take $x_1^0 = 1$, so that $y_1^0 = \sqrt{3}$, and

$$X_1^0 = \begin{pmatrix} 1 \\ \sqrt{3} \end{pmatrix}$$

is an eigenvector corresponding to $\lambda_1 = \frac{1}{2}$. A *unit* eigenvector is

$$X_1 = \frac{X_1^0}{|X_1^0|} = \frac{1}{2} \cdot \begin{pmatrix} 1 \\ \sqrt{3} \end{pmatrix} = \begin{pmatrix} \frac{1}{2} \\ \frac{\sqrt{3}}{2} \end{pmatrix}$$

Repeating the above process for $\lambda_2 = \frac{5}{2}$, we solve

$$\left(Q - \frac{5}{2}I_2\right)X_2^0 = \begin{pmatrix} -\frac{1}{2} & -\frac{\sqrt{3}}{2} \\ -\frac{\sqrt{3}}{2} & -\frac{3}{2} \end{pmatrix} \begin{pmatrix} x_2^0 \\ y_2^0 \end{pmatrix} = \begin{pmatrix} 0 \\ 0 \end{pmatrix}$$

and obtain

$$X_2^0 = \begin{pmatrix} 3 \\ -\sqrt{3} \end{pmatrix}$$

which when normalized gives

$$X_2 = \begin{pmatrix} \frac{\sqrt{3}}{2} \\ -\frac{1}{2} \end{pmatrix}$$

The General Quadratic Equation in Two Variables — Second Method

Let

$$P_1 = \begin{pmatrix} \frac{1}{2} & \frac{\sqrt{3}}{2} \\ \frac{\sqrt{3}}{2} & -\frac{1}{2} \end{pmatrix}$$

Although P_1 is an orthogonal matrix, it is not a rotation matrix, since $|P_1| = -1$ (rather that $+1$). However, by multiplying a column by -1 (say the second), we obtain a rotation matrix

$$P = \begin{pmatrix} \frac{1}{2} & -\frac{\sqrt{3}}{2} \\ \frac{\sqrt{3}}{2} & \frac{1}{2} \end{pmatrix}$$

A direct computation yields

$$Q' = P^t Q P = \begin{pmatrix} \frac{1}{2} & 0 \\ 0 & \frac{5}{2} \end{pmatrix}$$

(Note the eigenvalues.) The transform of the given equation is, therefore,

$$\tfrac{1}{2}x'^2 + \tfrac{5}{2}y'^2 = 1$$

To sketch the graph, we first plot $(1/2, \sqrt{3}/2)$ and $(-\sqrt{3}/2, 1/2)$, the columns of P, these being points on the positive x'- and y'-axes, respectively. We then sketch the ellipse in standard position in the x'-y' plane (Figure 7–18).

Observe that if λ_1 and λ_2 are the eigenvalues of Q, we know, without computing P, that the transform of the quadratic part of the given equation is either

$$\lambda_1 x'^2 + \lambda_2 y'^2 \quad \text{or} \quad \lambda_2 x'^2 + \lambda_1 y'^2$$

depending upon the rotation matrix P used to eliminate the product term. (P is not unique.)

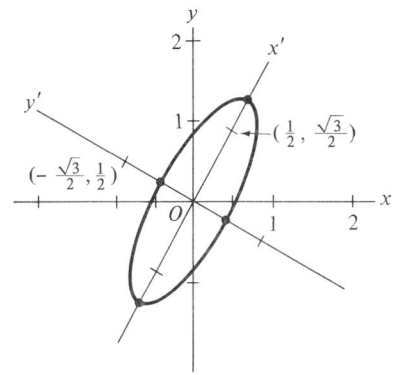

Figure 7-18 $2x^2 - \sqrt{3}xy + y^2 = 1;\ x'^2/2 + y'^2/\tfrac{2}{5} = 1$.

EXERCISE 6

In Problems 1 to 6, identify the conic section. Use the method of this section (the eigenvector method) to obtain a rotation matrix P that eliminates the product term. Obtain the transform of the equation. Draw the x'-, y'-axes in the x–y plane and sketch the graph in the x'–y' plane.

1. $x^2 - xy + y^2 + 2x - y - 5 = 0$
2. $xy = 4$
3. $3x^2 + 2\sqrt{3}xy + y^2 + 4x - 4\sqrt{3}y + 16 = 0$
4. $5x^2 - 6xy + 5y^2 = 0$
5. $2x^2 + 4\sqrt{3}xy - 2y^2 = 0$
6. $7x^2 - 4xy + 4y^2 = 240$

*7. (a) Write three matrices Q_1, Q_2, and Q_3 such that

$$X^t Q_i X = 2x^2 + 4xy + 3y^2 \ (i = 1, 2, 3)$$

(b) Write the symmetric matrix Q for the quadratic form $2x^2 + 4xy + 3y^2$.

(c) Let

$$\begin{pmatrix} a & b \\ b & d \end{pmatrix}$$

be any symmetric matrix such that

$$X^t \begin{pmatrix} a & b \\ b & d \end{pmatrix} X = 2x^2 + 4xy + 3y^2$$

for all x, y. Solve for a, b, and d, thereby showing that the symmetric matrix for the quadratic form is unique.

(d) Prove in general that the symmetric matrix for the quadratic form $ax^2 + bxy + cy^2$ is unique.

8. Prove that if Q is a symmetric 2×2 matrix and A is any 2×2 matrix, then $A^t Q A$ is symmetric.

9. If Q is the symmetric matrix for the quadratic form $ax^2 + bxy + cy^2$ ($b \neq 0$), and λ_1, λ_2 are the eigenvalues of Q, show that the corresponding conic section is a parabola, ellipse, or hyperbola according as $\lambda_1 \cdot \lambda_2 = 0$, $\lambda_1 \cdot \lambda_2 > 0$, or $\lambda_1 \cdot \lambda_2 < 0$, respectively.

chapter 8
Three-Dimensional Vectors

The development of vectors in three-dimensional space R^3 closely parallels that of vectors in the plane R^2 as presented in Chapter 3. We begin by defining a *point* as an ordered triple of real numbers. Then, as in R^2, we define an *arrow* to be an ordered pair of points. Finally, a *vector* is defined as a class of equivalent arrows.

1. THREE-DIMENSIONAL SPACE R^3

Definition. Three-dimensional space R^3 (read "R three") is the set

$$R^3 = \{(x, y, z) | x, y, z \in R\}$$

The ordered triple (x, y, z) is called a *point* in R^3 and is written $P(x, y, z)$ or $P = (x, y, z)$. The numbers x, y, z are called, respectively, the x-, y-, z-coordinates of P.

The following subsets of R^3 are called *coordinate axes* (see Figure 8-1):

$\ell_x = \{(x, y, z) | y = z = 0\}$ (x-axis)
$\ell_y = \{(x, y, z) | x = z = 0\}$ (y-axis)

[Figure 8-1]

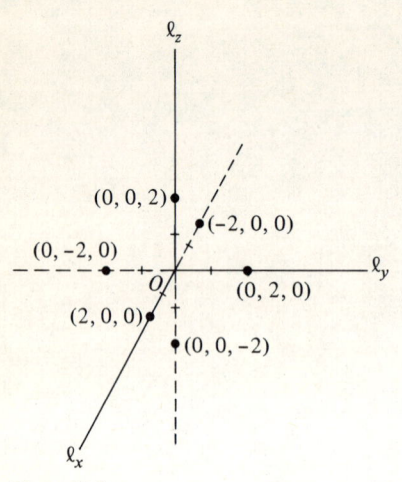

Figure 8-1

$$\ell_z = \{(x, y, z) | x = y = 0\} \quad (z\text{-axis})$$

The only point common to all three axes is the point $O = (0, 0, 0)$, called the *origin*. Each axis has the property that two of the coordinates are fixed at zero, while the third varies over the entire set R. Thus, we represent each axis as a line (as we did in R^2). We picture the axes as mutually perpendicular lines intersecting at O. The *coordinate planes* in R^3 are the sets (see Figure 8-2)

$$\pi_{xy} = \{(x, y, z) | z = 0\} \quad (x\text{-}y \text{ plane})$$
$$\pi_{yz} = \{(x, y, z) | x = 0\} \quad (y\text{-}z \text{ plane})$$
$$\pi_{xz} = \{(x, y, z) | y = 0\} \quad (x\text{-}z \text{ plane})$$

In addition to the coordinate planes, there are eight mutually exclusive

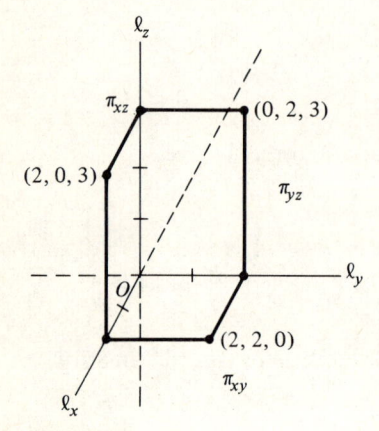

Figure 8-2

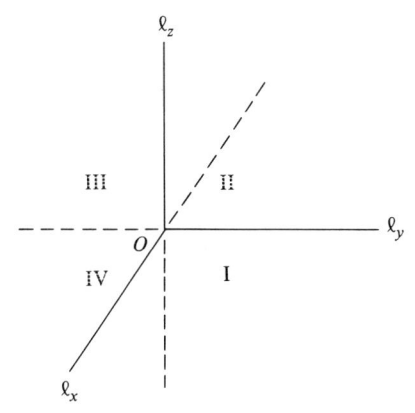

Figure 8-3

subsets called *octants* defined as follows (see Figure 8–3):

I: $\{(x, y, z) | x > 0, y > 0, z > 0\}$
II: $\{(x, y, z) | x < 0, y > 0, z > 0\}$
III: $\{(x, y, z) | x < 0, y < 0, z > 0\}$
IV: $\{(x, y, z) | x > 0, y < 0, z > 0\}$

Octants V, VI, VII, and VIII are similar to I, II, III, and IV, respectively, except that $z < 0$. The first four octants are said to be "above" the x–y plane, while the others are "below" the x–y plane. Every point in R^3 lies in a coordinate plane or in one of the octants.

A point $P(x, y, z)$ in R^3 will be represented as in Figure 8-4. The point illustrated lies in Octant I. The axes will be labeled x, y, z rather than ℓ_x, ℓ_y, ℓ_z.

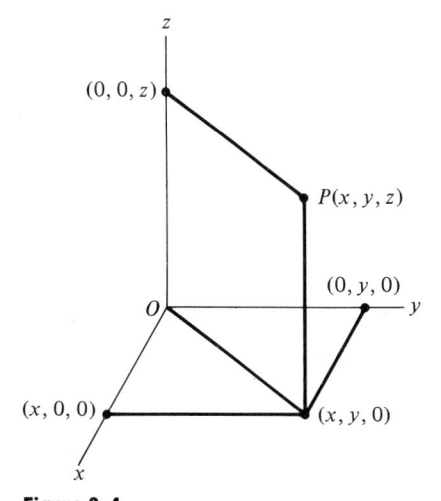

Figure 8-4

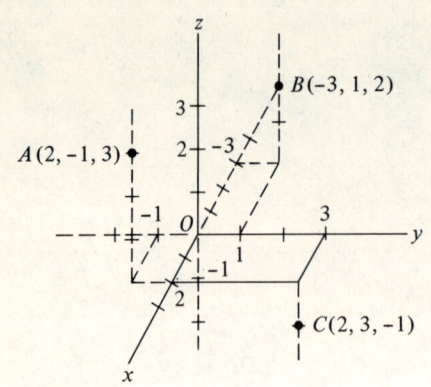

Figure 8-5

Example 1. Plot the points $A(2, -1, 3)$, $B(-3, 1, 2)$, and $C(2, 3, -1)$ (Figure 8-5).

Definition. Let $P_1(x_1, y_1, z_1)$ and $P_2(x_2, y_2, z_2)$ be (equal or distinct) points in R^3. The *arrow* $\overrightarrow{P_1P_2}$ with the *initial point* P_1 and *terminal point* P_2 is the ordered pair (P_1, P_2). The components of $\overrightarrow{P_1P_2}$ are the numbers $x_2 - x_1$, $y_2 - y_1$, $z_2 - z_1$ called the x-, y-, z-components, respectively.

If $P_1 \neq P_2$, $\overrightarrow{P_1P_2}$ will be drawn as in Figure 8-6 with the arrowhead at the terminal point P_2. Note that if $P_1 = (0, 0, 0)$, then the components of $\overrightarrow{P_1P_2}$ are precisely the coordinates of P_2. We shall find it convenient to refer to $\overrightarrow{P_1P_2}$ as "an arrow in R^3."

Example 2. Let $P_1 = (3, -2, 1)$ and $P_2 = (-3, 4, 2)$. Plot the arrow $\overrightarrow{P_1P_2}$ and find its components.

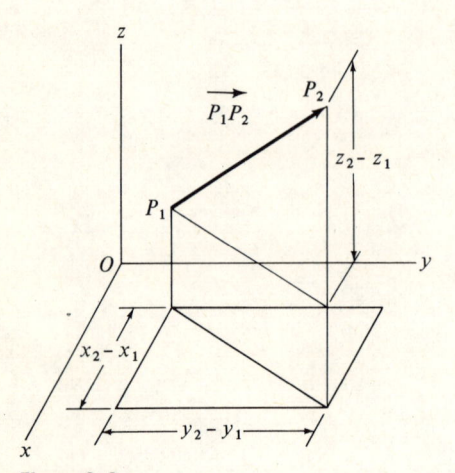

Figure 8-6

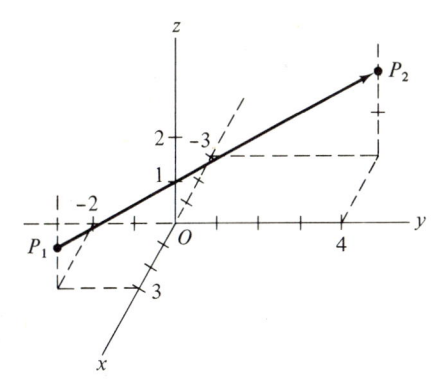

Figure 8-7

SOLUTION. (See Figure 8-7.)

x-component: $-3 - 3 = -6$
y-component: $4 - (-2) = 6$
z-component: $2 - 1 = 1$

Definition. Let $P_1 = (x_1, y_1, z_1)$, $P_2 = (x_2, y_2, z_2)$, $Q_1 = (u_1, v_1, w_1)$, and $Q_2 = (u_2, v_2, w_2)$. $\overrightarrow{P_1P_2}$ is equivalent to $\overrightarrow{Q_1Q_2}$, written $\overrightarrow{P_1P_2} \sim \overrightarrow{Q_1Q_2}$, iff $x_2 - x_1 = u_2 - u_1$, $y_2 - y_1 = v_2 - v_1$, and $z_2 - z_1 = w_2 - w_1$.

Thus, $\overrightarrow{P_1P_2}$ and $\overrightarrow{Q_1Q_2}$ are equivalent iff corresponding components are equal. Equivalent arrows are called *translates* of one another. In Section 3, we shall see that equivalent arrows have the same *length*, and in Section 4 that they have the same *direction*. As in R^2, the relation of *equivalence* between arrows is an equivalence relation. (See Chapter 3, Section 2.) The definition of a vector in R^3 as a class of equivalent arrows is the same as in R^2.

Definition. The *vector* **v** in R^3 with *components a, b, c in that order* is the class of all (equivalent) *arrows* whose x-, y-, z-components are a, b, c, respectively. We write

$$\mathbf{v} = (a, b, c)$$

and call a, b, c the x-, y-, z-components, respectively, of **v**.

In the terminology of Section 2, Chapter 3, an arrow $\overrightarrow{P_1P_2}$ representing **v** will often be identified with **v**. We write $\overrightarrow{P_1P_2} = \mathbf{v}$, and call P_1 the initial point and P_2 the terminal point of **v**. Thus, if $P_1 = (2, -1, 3)$ and $P_2 = (-3, 0, 5)$, we may write

$$\overrightarrow{P_1P_2} = (-3 - 2, 0 - (-1), 5 - 3) = (-5, 1, 2)$$

Example 3. Let $\mathbf{v} = (1, 2, 3)$. Find and plot the arrow representing **v** with (a) initial point $P_1(0, 0, 0)$, (b) initial point $P_1(1, 1, -1)$, and (c) terminal point $P_2(2, 1, 4)$.

(a)   (b)   (c)

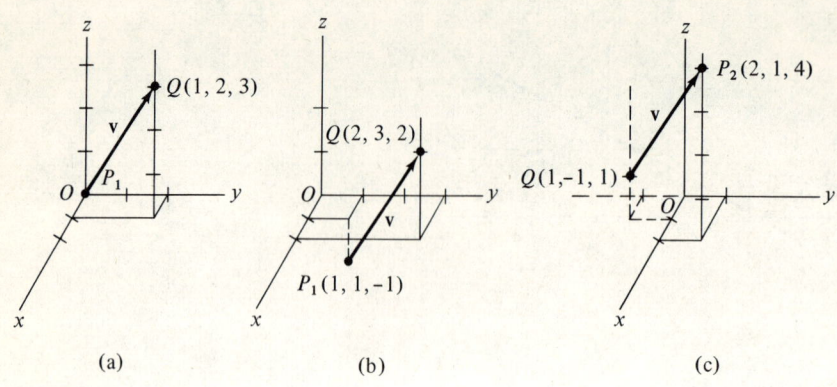

Figure 8-8

SOLUTION

(a) Let $Q(x, y, z)$ be the terminal point of the arrow. Then $\overrightarrow{P_1Q} = \mathbf{v}$, and so $x - 0 = 1$, $y - 0 = 2$, $z - 0 = 3$. Therefore, $Q = (1, 2, 3)$ (Figure 8-8(a)).

(b) We have (as in Part (a)) $\overrightarrow{P_1Q} = \mathbf{v}$; so $Q = (2, 3, 2)$ (Figure 8-8(b)).

(c) Let $Q(x, y, z)$ be the initial point. Then $\overrightarrow{QP_2} = \mathbf{v}$; so $2 - x = 1$, $1 - y = 2$, $4 - z = 3$. Therefore, $Q = (1, -1, 1)$ (Figure 8-8(c)).

EXERCISE 1

1. Plot the following pairs of arrows and determine whether the arrows of a given pair are equivalent.
   (a) $\overrightarrow{P_1P_2}$ and $\overrightarrow{Q_1Q_2}$, where
   $P_1 = (2, -1, 3)$ $P_2 = (4, 0, -1)$, $Q_1 = (3, -5, -2)$, and $Q_2 = (5, -6, -6)$.
   (b) $\overrightarrow{P_1P_2}$ and $\overrightarrow{Q_2Q_1}$, where
   $P_1 = (1, 1, 1), P_2 = (2, -1, 2), Q_1 = (-3, 0, 2),$ and $Q_2 = (-4, 2, 1)$.
   (c) $\overrightarrow{P_2P_1}$ and $\overrightarrow{Q_1Q_2}$, where
   $P_1 = (1, 2, 3), P_2 = (-2, 1, -3), Q_1 = (0, 4, 3),$ and $Q_2 = (3, 5, 9)$.
   (d) $\overrightarrow{P_1P_2}$ and $\overrightarrow{Q_1Q_2}$, where
   $P_1 = (2, -4, 3), P_2 = (2, 0, -1), Q_1 = (1, 4, -2),$ and $Q_2 = (1, 8, 2)$.

2. Let $\mathbf{v} = (2, -1, 3)$. Plot $\mathbf{v}$ with (a) initial point $P_1(0, 0, 0)$, (b) initial point $P_1(1, 1, 1)$, (c) initial point $P_1(-2, 1, -3)$, (d) terminal point $P_2(0, 1, -2)$, and (e) terminal point $P_2(-3, 2, 0)$.

2. VECTOR OPERATIONS

The two basic vector operations are *addition* and *scalar multiplication*.

Definition. Let $\mathbf{v}_1 = (a_1, b_1, c_1)$ and $\mathbf{v}_2 = (a_2, b_2, c_2)$. The sum of $\mathbf{v}_1$ and $\mathbf{v}_2$ (in that order) is the vector

$$\mathbf{v}_1 + \mathbf{v}_2 = (a_1 + a_2, b_1 + b_2, c_1 + c_2)$$

Figure 8-9 (a) $\overrightarrow{P_1Q} = a\overrightarrow{P_1P_2}$ $(a>0)$. (b) $\overrightarrow{P_1Q} = a\overrightarrow{P_1P_2}$ $(a<0)$.

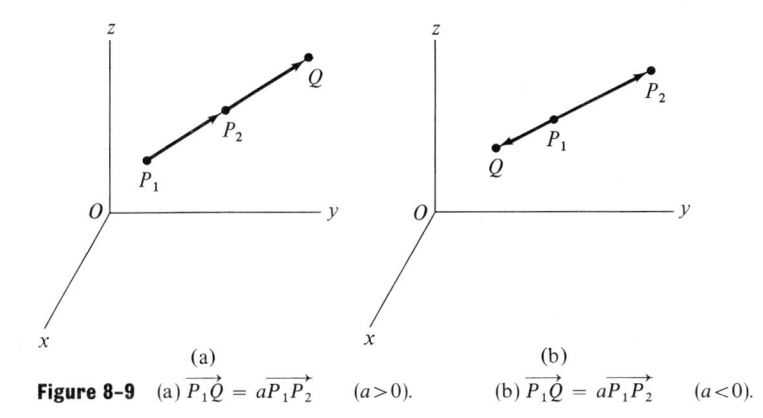

If $\overrightarrow{P_1P_2}$ and $\overrightarrow{P_2P_3}$ are representatives of v_1 and v_2, respectively, then $v_1 + v_2$ is represented by $\overrightarrow{P_1P_3}$. (See Chapter 3, Section 2.)

Definition. Let $P_1 = (x_1, y_1, z_1)$, $P_2 = (x_2, y_2, z_2)$ and a be a real number (called a *scalar*). The *scalar product* of $\overrightarrow{P_1P_2}$ by a, denoted by $a\overrightarrow{P_1P_2}$, is the arrow

$$\overrightarrow{P_1Q} = a\overrightarrow{P_1P_2} = (a(x_2 - x_1), a(y_2 - y_1), a(z_2 - z_1))$$

Thus, if $Q = (x, y, z)$, then

$$x = x_1 + a(x_2 - x_1)$$
$$y = y_1 + a(y_2 - y_1)$$
$$z = z_1 + a(z_2 - z_1)$$

We draw $a\overrightarrow{P_1P_2}$ as in Figure 8-9. If $a = 0$, $a\overrightarrow{P_1P_2} = \overrightarrow{P_1P_1}$, which is represented by the point P_1.

We extend the definition of scalar multiplication to vectors as follows.

Definition. Let $v = (a, b, c)$ and let k be a scalar. The *scalar product* of v by k is the vector $kv = (ka, kb, kc)$.

We call kv a *scalar multiple* (or simply *multiple*) of v. If $v \neq (0, 0, 0)$ and $k > 0$, we say kv has the "same direction" as v. If $k < 0$, we say kv has the "opposite direction" to v. If $\overrightarrow{P_1P_2}$ represents v and $\overrightarrow{P_1Q}$ represents kv, then $\overrightarrow{P_1Q} = k\overrightarrow{P_1P_2}$ according to the previous definition.

We collect the basic properties of addition and scalar multiplication of vectors in the following theorem. The proof is merely an extension of the proof of Theorem 1, Section 2, Chapter 3.

Theorem 1. *For all vectors v, v_1, v_2, v_3 and scalars a, b the following laws hold:*
(1) (a) $(v_1 + v_2) + v_3 = v_1 + (v_2 + v_3)$ *(associative law)*
    (b) $v_1 + v_2 = v_2 + v_1$ *(commutative law)*

(c) $\mathbf{v} + \mathbf{0} = \mathbf{v}$, where $\mathbf{0} = (0, 0, 0)$ (*identity law*)
(d) $\mathbf{v} + (-\mathbf{v}) = \mathbf{0}$, where $-\mathbf{v} = (-1)\mathbf{v}$ (*inverse law*)
($\mathbf{0}$ is called the *zero vector*, and $-\mathbf{v}$ the *negative* of $\mathbf{v}$.)

(2) (a) $a(b\mathbf{v}) = (ab)\mathbf{v}$
    (b) $a(\mathbf{v}_1 + \mathbf{v}_2) = a\mathbf{v}_1 + a\mathbf{v}_2$
    (c) $(a + b)\mathbf{v} = a\mathbf{v} + b\mathbf{v}$
    (d) $1\mathbf{v} = \mathbf{v}$

PROOF. The proof is similar to that of Theorem 1, Section 2, Chapter 3. (See Problem 4.)

Definition. Let $\mathbf{v}_1 = (a_1, b_1, c_1)$ and $\mathbf{v}_2 = (a_2, b_2, c_2)$. The *difference* $\mathbf{v}_1 - \mathbf{v}_2$ is the vector $\mathbf{v}_1 + (-\mathbf{v}_2)$.

It follows that

$$\mathbf{v}_1 - \mathbf{v}_2 = (a_1 - a_2, b_1 - b_2, c_1 - c_2)$$

and

$$\mathbf{v}_1 - \mathbf{v}_2 = \mathbf{v}_3 \text{ iff } \mathbf{v}_1 = \mathbf{v}_2 + \mathbf{v}_3$$

If $\mathbf{v}_1 = \overrightarrow{OP_1}$ and $\mathbf{v}_2 = \overrightarrow{OP_2}$, then $\mathbf{v}_1 - \mathbf{v}_2 = \overrightarrow{P_2P_1}$ (Figure 8–10).

Theorem 2. *For all vectors* $\mathbf{v}, \mathbf{v}_1, \mathbf{v}_2, \mathbf{v}_3$ *and all scalars* c, c_1, c_2, *the following laws hold:*
(1) $0\mathbf{v} = \mathbf{0}$
(2) $c\mathbf{0} = \mathbf{0}$
(3) $c\mathbf{v} = \mathbf{0}$ iff $c = 0$ or $\mathbf{v} = \mathbf{0}$
(4) $c\mathbf{v}_1 = c\mathbf{v}_2$ and $c \neq 0 \Rightarrow \mathbf{v}_1 = \mathbf{v}_2$
(5) $c_1\mathbf{v} = c_2\mathbf{v}$ and $\mathbf{v} \neq \mathbf{0} \Rightarrow c_1 = c_2$ (*cancellation laws*)
(6) $\mathbf{v}_1 + \mathbf{v}_2 = \mathbf{v}_1 + \mathbf{v}_3 \Rightarrow \mathbf{v}_2 = \mathbf{v}_3$
(7) $-(\mathbf{v}_1 + \mathbf{v}_2) = -\mathbf{v}_1 - \mathbf{v}_2$

PROOF. The proof is similar to that of Theorem 2, Section 2, Chapter 3. (See Problem 5.)

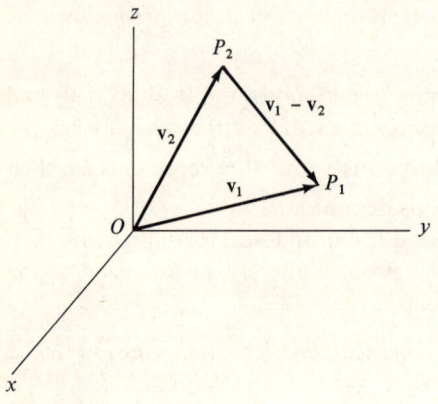

Figure 8–10

Example 1. Given $v_1 - 2v_2 = (0, 2, 4)$ and $3v_1 + v_2 = (-1, 2, 5)$. Find v_1 and v_2.

SOLUTION. Multiplying the first equation by 3 and subtracting the result from the second equation, we obtain

$$7v_2 = (-1, -4, -7)$$

Therefore,

$$v_2 = (-\tfrac{1}{7}, -\tfrac{4}{7}, -1)$$

Substituting this in the first equation yields

$$v_1 = (-\tfrac{2}{7}, -\tfrac{6}{7}, 2)$$

Definition. Let v_1 and v_2 be nonzero vectors. v_1 is *parallel* to v_2, written $v_1 \parallel v_2$, iff there exists $t \in R$ such that $v_1 = tv_2$ (that is, v_1 is a multiple of v_2). If $\overrightarrow{P_1P_2}$ represents v_1 and $\overrightarrow{Q_1Q_2}$ represents v_2, then $\overrightarrow{P_1P_2} \parallel \overrightarrow{Q_1Q_2}$ iff $v_1 \parallel v_2$.

Note that $t \neq 0$ in the above definition. It follows immediately from the definition that parallelism of vectors is an equivalence relation. (See Problem 7.) Two vectors v_1 and v_2 are said to be *nonparallel* ($v_1 \not\parallel v_2$) iff v_1 and v_2 are nonzero and not parallel. Thus, the terms *parallel* and *nonparallel* apply only to *nonzero* vectors. The following theorem provides a useful criterion for determining whether two (nonzero) vectors are parallel. Recall that

$$\begin{vmatrix} a & b \\ c & d \end{vmatrix}$$

is a determinant of order two, and

$$\begin{vmatrix} a & b \\ c & d \end{vmatrix} = ad - bc$$

Theorem 3. Let $v_1 = (a_1, b_1, c_1)$ and $v_2 = (a_2, b_2, c_2)$ be nonzero vectors. Then v_1 and v_2 are parallel iff

$$\begin{vmatrix} a_1 & b_1 \\ a_2 & b_2 \end{vmatrix} = \begin{vmatrix} b_1 & c_1 \\ b_2 & c_2 \end{vmatrix} = \begin{vmatrix} a_1 & c_1 \\ a_2 & c_2 \end{vmatrix} = 0$$

PROOF. We prove that if the determinants are zero, then $v_1 \parallel v_2$. The proof of the converse is left as an exercise. (See Problem 8.)

Suppose all three determinants are zero. Without loss of generality we may assume $c_2 \neq 0$. By hypothesis $a_1b_2 = a_2b_1$, $b_1c_2 = b_2c_1$, and $a_1c_2 = a_2c_1$. Since $c_2 \neq 0$, we may let $t = c_1/c_2$. Then $a_1 = (c_1/c_2)a_2 = ta_2$, $b_1 = (c_1/c_2)b_2 = tb_2$, and $c_1 = tc_2$. Thus, $v_1 = tv_2$. Moreover $t \neq 0$, since $v_1 \neq 0$, and so $v_1 \parallel v_2$.

Corollary. If v_2 has all nonzero components, then $v_1 \parallel v_2$ iff

$$\frac{a_1}{a_2} = \frac{b_1}{b_2} = \frac{c_1}{c_2}$$

(In which case v_1 also has all nonzero components.)

REMARK. The criterion for parallelism given in Theorem 3 will be used in Section 5. In practice it is simpler to use the corollary or the definition.

Example 2. Determine whether the following pairs of vectors are parallel.
(a) $v_1 = (4, -3, 0)$, $v_2 = (\frac{8}{3}, -2, 0)$
(b) $v_1 = (\frac{1}{2}, 6, -4)$, $v_2 = (\frac{1}{3}, 4, -2)$

SOLUTION

(a) $\dfrac{4}{\frac{8}{3}} = \dfrac{-3}{-2} = \dfrac{3}{2} = t$ and $v_1 = \frac{3}{2} v_2$

Therefore, v_1 and v_2 are parallel.

(b) $\dfrac{\frac{1}{2}}{\frac{1}{3}} = \dfrac{6}{4} \neq \dfrac{-4}{-2}$

Therefore, $v_1 \not\parallel v_2$.

EXERCISE 2

1. Let $v_1 = (1, 1, 3)$ and $v_2 = (1, 5, 2)$.
   (a) Plot v_1 and v_2 with initial point at the origin.
   (b) Find and plot $v_1 + v_2$, $v_1 - v_2$, and $v_2 - v_1$.
   (c) Find and plot $4v_1$ with initial point $P_1(-2, 1, -6)$.
   (d) Find and plot $-3v_2$ with terminal point $P_2(1, -1, 2)$.
2. (a) Given $v_1 - v_2 = (4, -6, 2)$ and $v_1 = (-2, -1, 3)$. Find v_2.
   (b) Given $(v_1 + v_2)/3 = (0, 1, -2)$ and $(v_1 - v_2)/2 = (3, -4, 5)$. Find v_1 and v_2. Check your answer.
3. Determine whether the following pairs of vectors are parallel.
   (a) $v_1 = (3, -6, 4)$, $v_2 = (-1, 2, -\frac{4}{3})$
   (b) $v_1 = (4, 0, \frac{1}{2})$, $v_2 = (3, 2, -1)$
   (c) $v_1 = (1, 1, 0)$, $v_2 = (1, 0, 1)$
4. Prove Theorem 1.
5. Prove Theorem 2.
6. (a) Show that if $v_1 + v_2 \parallel v_1$, then $v_2 \parallel v_1$ or $v_2 = 0$.
   (b) Given that $v_1 + v_2 \parallel v_1 - v_2$, $v_1 \neq 0$, and $v_2 \neq 0$, what can you say about v_1 and v_2?
7. Prove that prallelism of vectors is an equivalence relation.
8. Complete the proof of Theorem 3.

3. LENGTH; DOT PRODUCT

Definition. The *length* of the vector $\mathbf{v} = (a, b, c)$ is the (nonnegative) real number

$$|\mathbf{v}| = (a^2 + b^2 + c^2)^{1/2}$$

If $\overrightarrow{P_1P_2}$ is a representative of $\mathbf{v}$, the length of $\overrightarrow{P_1P_2}$ is defined by $|\overrightarrow{P_1P_2}| = |\mathbf{v}|$.

Thus, if $P_1 = (x_1, y_1, z_1)$ and $P_2 = (x_2, y_2, z_2)$, then

$$|\overrightarrow{P_1P_2}| = [(x_2 - x_1)^2 + (y_2 - y_1)^2 + (z_2 - z_1)^2]^{1/2}$$

This is also called the *distance* between P_1 and P_2. Clearly, this definition is motivated by the Pythagorean theorem of Euclidean geometry (Figure 8–6). (Observe that equivalent arrows have the same length.)

Example 1. Let $P_1 = (2, -3, 1)$ and $P_2 = (-4, -2, 6)$. Then

$$|\overrightarrow{P_1P_2}| = [(-4 - 2)^2 + (-2 - (-3))^2 + (6 - 1)^2]^{1/2} = \sqrt{62}$$

The important properties of *length* are contained in Theorem 1. We state the properties for both vectors and arrows.

Theorem 1

(1) $|\mathbf{v}| \geq 0; |\mathbf{v}| = 0 \Leftrightarrow \mathbf{v} = \mathbf{0}$
(2) $|-\mathbf{v}| = |\mathbf{v}|$
(3) $|k\mathbf{v}| = |k||\mathbf{v}|$
(4) $|\mathbf{v}_1 + \mathbf{v}_2| \leq |\mathbf{v}_1| + |\mathbf{v}_2|$
(5) $|\mathbf{v}_1 - \mathbf{v}_2| \geq ||\mathbf{v}_1| - |\mathbf{v}_2||$
(1′) $|\overrightarrow{P_1P_2}| \geq 0; |\overrightarrow{P_1P_2}| = 0 \Leftrightarrow P_1 = P_2$
(2′) $|\overrightarrow{P_1P_2}| = |\overrightarrow{P_2P_1}|$
(3′) $|k\overrightarrow{P_1P_2}| = |k||\overrightarrow{P_1P_2}|$
(4′) $|\overrightarrow{P_1P_3}| \leq |\overrightarrow{P_1P_2}| + |\overrightarrow{P_2P_3}|$
(5′) $|\overrightarrow{P_1P_2} - \overrightarrow{P_1P_3}| \geq ||\overrightarrow{P_1P_2}| - |\overrightarrow{P_1P_3}||$

Part 4 is called the *triangle inequality*. Equality holds in Part 4 iff one of the vectors is a nonnegative scalar times the other.

We shall prove Part 4 only, leaving the proof of the remaining parts as an exercise (see Problem 5). Before presenting the proof of the triangle inequality, we state and prove a very important lemma. The triangle inequality will then follow immediately from this.

Lemma *(Cauchy's inequality).* Let $\mathbf{v}_1 = (a_1, b_1, c_1)$ and $\mathbf{v}_2 = (a_2, b_2, c_2)$. Then

$$|a_1a_2 + b_1b_2 + c_1c_2| \leq (a_1^2 + b_1^2 + c_1^2)^{1/2} \cdot (a_2^2 + b_2^2 + c_2^2)^{1/2}$$

with equality holding iff one of the vectors is **0**, or one is a nonzero scalar times the other.

PROOF. The inequality is equivalent to
$$(a_1a_2 + b_1b_2 + c_1c_2)^2 \leq (a_1^2 + b_1^2 + c_1^2)(a_2^2 + b_2^2 + c_2^2)$$
Performing the indicated multiplications, simplifying, and transposing all terms to one side, we obtain the equivalent inequality
$$(a_1^2b_2^2 - 2a_1a_2b_1b_2 + b_1^2a_2^2) + (a_1^2c_2^2 - 2a_1a_2c_1c_2 + c_1^2a_2^2) \\ + (b_1^2c_2^2 - 2b_1b_2c_1c_2 + c_1^2b_2^2) \geq 0$$
that is,
$$(a_1b_2 - b_1a_2)^2 + (a_1c_2 - c_1a_2)^2 + (b_1c_2 - c_1b_2)^2 \geq 0$$
Since a sum of squares is never negative, the last inequality holds for all real numbers, and so Cauchy's inequality holds.

Now assuming that equality holds, we proceed as above and obtain
$$(a_1b_2 - b_1a_2)^2 + (a_1c_2 - c_1a_2)^2 + (b_1c_2 - c_1b_2)^2 = 0$$
Hence $a_1b_2 - b_1a_2 = 0$, $a_1c_2 - c_1a_2 = 0$, and $b_1c_2 - c_1b_2 = 0$. Therefore,
$$\begin{vmatrix} a_1 & b_1 \\ a_2 & b_2 \end{vmatrix} = \begin{vmatrix} b_1 & c_1 \\ b_2 & c_2 \end{vmatrix} = \begin{vmatrix} a_1 & c_1 \\ a_2 & c_2 \end{vmatrix} = 0$$
By Theorem 3, Section 2, if $\mathbf{v}_1$ and $\mathbf{v}_2$ are nonzero, then $\mathbf{v}_1 \parallel \mathbf{v}_2$, so that $\mathbf{v}_1 = t\mathbf{v}_2$ ($t \neq 0$).

If $\mathbf{v}_1 = \mathbf{0}$ or $\mathbf{v}_2 = \mathbf{0}$, then both members of the inequality equal zero, and so equality holds. If $\mathbf{v}_1 = t\mathbf{v}_2$ for some $t \neq 0$, then both members reduce to $|t| \cdot (a_2^2 + b_2^2 + c_2^2)$.

Corollary. For all vectors $\mathbf{v}_1 = (a_1, b_1, c_1)$ and $\mathbf{v}_2 = (a_2, b_2, c_2)$,
$$a_1a_2 + b_1b_2 + c_1c_2 \leq (a_1^2 + b_1^2 + c_1^2)^{1/2}(a_2^2 + b_2^2 + c_2^2)^{1/2}$$
Moreover, equality holds iff one of the vectors is **0**, or one is a positive scalar times the other.

PROOF. (See Problem 8.)

The proof of Part 4 of Theorem 1 now proceeds as follows:
$$|\mathbf{v}_1 + \mathbf{v}_2| \leq |\mathbf{v}_1| + |\mathbf{v}_2|$$
iff
$$|\mathbf{v}_1 + \mathbf{v}_2|^2 \leq (|\mathbf{v}_1| + |\mathbf{v}_2|)^2$$
iff
$$(a_1 + a_2)^2 + (b_1 + b_2)^2 + (c_1 + c_2)^2 \\ \leq a_1^2 + b_1^2 + c_1^2 + a_2^2 + b_2^2 + c_2^2 + 2|\mathbf{v}_1| \cdot |\mathbf{v}_2|$$

iff
$$a_1a_2 + b_1b_2 + c_1c_2 \leq (a_1^2 + b_1^2 + c_1^2)^{1/2} \cdot (a_2^2 + b_2^2 + c_2^2)^{1/2}$$

From the corollary, the last inequality holds for all vectors $\mathbf{v}_1$ and $\mathbf{v}_2$; hence so does the triangle inequality. Moreover, equality holds iff one of the vectors is a nonnegative scalar times the other.

Example 2. Verify Cauchy's inequality and the triangle inequality for the vectors $\mathbf{v}_1 = (1, 2, -1)$ and $\mathbf{v}_2 = (2, 3, -2)$.

SOLUTION

$$a_1a_2 + b_1b_2 + c_1c_2 = 1 \cdot 2 + 2 \cdot 3 + (-1)(-2) = 10$$
$$|\mathbf{v}_1| = \sqrt{6}, \quad |\mathbf{v}_2| = \sqrt{17}$$

and $10 < \sqrt{6} \cdot \sqrt{17} = \sqrt{102}$. Also, $\mathbf{v}_1 + \mathbf{v}_2 = (3, 5, -3)$, $|\mathbf{v}_1 + \mathbf{v}_2| = \sqrt{43}$, and $\sqrt{43} < \sqrt{6} + \sqrt{17}$.

Definition. The *dot product* of $\mathbf{v}_1 = (a_1, b_1, c_1)$ and $\mathbf{v}_2 = (a_2, b_2, c_2)$ (in that order) written $\mathbf{v}_1 \circ \mathbf{v}_2$ is the number

$$\mathbf{v}_1 \circ \mathbf{v}_2 = a_1a_2 + b_1b_2 + c_1c_2$$

Cauchy's inequality may now be written

$$|\mathbf{v}_1 \circ \mathbf{v}_2| \leq |\mathbf{v}_1| |\mathbf{v}_2|$$

The basic properties of the *dot product* are summarized in Theorem 2.

Theorem 2. *For all vectors* $\mathbf{v}, \mathbf{v}_1, \mathbf{v}_2, \mathbf{v}_3$ *and scalars* c,
(1) $\mathbf{v}_1 \circ \mathbf{v}_2 = \mathbf{v}_2 \circ \mathbf{v}_1$ (*commutative law*)
(2) $\mathbf{v}_1 \circ (\mathbf{v}_2 + \mathbf{v}_3) = \mathbf{v}_1 \circ \mathbf{v}_2 + \mathbf{v}_1 \circ \mathbf{v}_3$
(3) $\mathbf{v}_1 \circ (\mathbf{v}_2 - \mathbf{v}_3) = \mathbf{v}_1 \circ \mathbf{v}_2 - \mathbf{v}_1 \circ \mathbf{v}_3$ (*distributive laws*)
(4) $\mathbf{v} \circ \mathbf{v} = |\mathbf{v}|^2$
(5) $(\mathbf{v}_1 + \mathbf{v}_2) \circ (\mathbf{v}_1 - \mathbf{v}_2) = |\mathbf{v}_1|^2 - |\mathbf{v}_2|^2$
$\qquad\qquad\qquad\qquad\quad = (|\mathbf{v}_1| + |\mathbf{v}_2|)(|\mathbf{v}_1| - |\mathbf{v}_2|)$
(6) $c(\mathbf{v}_1 \circ \mathbf{v}_2) = (c\mathbf{v}_1) \circ \mathbf{v}_2 = \mathbf{v}_1 \circ (c\mathbf{v}_2)$
(7) $\mathbf{0} \circ \mathbf{v} = 0$
(8) $\mathbf{v} \circ \mathbf{v} = 0$ *iff* $\mathbf{v} = \mathbf{0}$
(9) $|\mathbf{v}_1 \circ \mathbf{v}_2| \leq |\mathbf{v}_1| \cdot |\mathbf{v}_2|$ (*Cauchy's inequality*)
(10) $|\mathbf{v}_1 + \mathbf{v}_2|^2 = |\mathbf{v}_1|^2 + |\mathbf{v}_2|^2$ *iff* $\mathbf{v}_1 \circ \mathbf{v}_2 = 0$

PROOF. We prove Part 5 and leave the proofs of the others as exercises (see Problem 6).

$(\mathbf{v}_1 + \mathbf{v}_2) \circ (\mathbf{v}_1 - \mathbf{v}_2) = (\mathbf{v}_1 + \mathbf{v}_2) \circ \mathbf{v}_1 - (\mathbf{v}_1 + \mathbf{v}_2) \circ \mathbf{v}_2$ (by Part 3)
$\qquad\qquad\qquad\quad = \mathbf{v}_1 \circ \mathbf{v}_1 + \mathbf{v}_2 \circ \mathbf{v}_1 - \mathbf{v}_1 \circ \mathbf{v}_2 - \mathbf{v}_2 \circ \mathbf{v}_2$
$\qquad\qquad\qquad$ (by Parts 1 and 2)

$$= |v_1|^2 - |v_2|^2 + v_1 \circ v_2 - v_1 \circ v_2$$
(by Parts 1 and 4)
$$= |v_1|^2 - |v_2|^2.$$

EXERCISE 3

1. Find the distance between the following pairs of points.
    (a) $P_1(2, 3, -1)$, $P_2(-1, 2, -4)$
    (b) $P_1(0, 2, -3)$, $P_2(5, 0, -4)$
    (c) $P_1(2, -1, 3)$, $P_2(-2, 3, 5)$
    (d) $P_1(1, 0, 0)$, $P_2(0, 1, 0)$
2. Let $v_1 = (-1, 2, 3)$ and $v_2 = (1, 2, -1)$. Show that $|v_1 + v_2|^2 = |v_1|^2 + |v_2|^2$.
3. Verify the triangle inequality and Cauchy's inequality for each of the following pairs of vectors.
    (a) $v_1 = (2, -1, 3)$, $v_2 = (-1, 4, 2)$
    (b) $v_1 = (0, 4, 3)$, $v_2 = (0, -1, 2)$
4. Verify the triangle inequality (Part 4 of Theorem 1) for the following triples of points.
    (a) $P_1(2, 3, -1)$, $P_2(-1, 2, -4)$, $P_3(2, 1, 1)$
    (b) $P_1(1, 0, -2)$, $P_2(-1, 1, -1)$, $P_3(0, 4, -2)$
5. Prove Parts 1, 2, 3, and 5 of Theorem 1.
6. (a) Prove Parts 1 to 4 of Theorem 2.
    (b) Prove Parts 6, 7, 8, and 10 of Theorem 2.
7. Let a, b, and c be arbitrary real numbers. Prove that
    (a) $|a| + |b| + |c| \le \sqrt{3}\sqrt{a^2 + b^2 + c^2}$
    (*Hint*: Consider $(|a|, |b|, |c|) \circ (1, 1, 1)$.)
    (b) $\sqrt{a^2 + b^2 + c^2} \le |a| + |b| + |c|$
    (*Hint*: $(a, b, c) = (a, 0, 0) + (0, b, 0) + (0, 0, c)$.)
8. Prove the corollary to the lemma.

4. THE ANGLE BETWEEN TWO VECTORS; DIRECTION OF A VECTOR

In this section, we shall define the angle θ between two nonzero vectors v_1 and v_2. We motivate this definition as follows: Let

$$\overrightarrow{P_0P_1} = v_1 = (a_1, b_1, c_1) \text{ and } \overrightarrow{P_0P_2} = v_2 = (a_2, b_2, c_2)$$

In the next chapter, we define a triangle in R^3 and show that the law of cosines holds as in R^2. Thus,

$$|\overrightarrow{P_1P_2}|^2 = |\overrightarrow{P_0P_1}|^2 + |\overrightarrow{P_0P_2}|^2 - 2|\overrightarrow{P_0P_1}| \cdot |\overrightarrow{P_0P_2}| \cos \theta$$

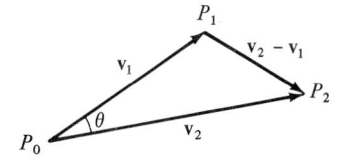

The Angle Between Two Vectors; Direction of a Vector

(Figure 8–11). Equivalently,

$$|v_2 - v_1|^2 = |v_1|^2 + |v_2|^2 - 2|v_1|\cdot|v_2|\cos\theta$$

Also,

$$\begin{aligned}|v_2 - v_1|^2 &= (v_2 - v_1)\circ(v_2 - v_1)\\ &= v_2\circ v_2 - 2v_1\circ v_2 + v_1\circ v_1\\ &= |v_2|^2 - 2v_1\circ v_2 + |v_1|^2\end{aligned}$$

Hence

$$|v_1|\cdot|v_2|\cos\theta = v_1\circ v_2$$

Since v_1 and v_2 are nonzero,

$$\cos\theta = \frac{v_1\circ v_2}{|v_1|\cdot|v_2|}$$

In component form, we have

$$\cos\theta = \frac{a_1a_2 + b_1b_2 + c_1c_2}{(a_1^2 + b_1^2 + c_1^2)^{1/2}(a_2^2 + b_2^2 + c_2^2)^{1/2}}$$

Thus, we see that $\cos\theta$ can be expressed entirely in terms of the vectors v_1 and v_2. This suggests the following definition.

Definition. Let v_1 and v_2 be nonzero vectors. The angle θ between v_1 and v_2 is the angle defined by

$$\cos\theta = \frac{v_1\circ v_2}{|v_1|\cdot|v_2|}, \qquad 0 \le \theta \le \pi$$

REMARK. From Cauchy's inequality, it follows that $\theta = 0$ iff $v_1 = tv_2$, $t > 0$; that is v_1 and v_2 are parallel and in the same direction. Similarly, $\theta = \pi$ iff $v_1 = tv_2$, $t < 0$; that is v_1 and v_2 are parallel and in opposite directions.

Observe that for nonzero vectors, $\theta = \tfrac{1}{2}\pi$ iff $v_1\circ v_2 = 0$. This motivates the following definition.

Definition. v_1 and v_2 are called orthogonal (perpendicular) iff $v_1\circ v_2 = 0$. We write $v_1 \perp v_2$. Note that $\mathbf{0}$ is orthogonal to every vector (see Problem 12). Moreover, θ is acute $(0 \le \theta < \tfrac{1}{2}\pi)$ iff $v_1\circ v_2 > 0$, and θ is obtuse $(\tfrac{1}{2}\pi < \theta \le \pi)$ iff $v_1\circ v_2 < 0$.

Example 1. Find the angle θ between the following pairs of vectors.
(a) $\mathbf{v}_1 = (1, -1, 2)$, $\mathbf{v}_2 = (-2, 4, 3)$
(b) $\mathbf{v}_1 = (1, -2, 3)$, $\mathbf{v}_2 = (-\frac{1}{2}, 1, -\frac{3}{2})$
(c) $\mathbf{v}_1 = (1, -2, 1)$, $\mathbf{v}_2 = (1, 1, 0)$

SOLUTION

(a) $\mathbf{v}_1 \circ \mathbf{v}_2 = 1 \cdot (-2) + (-1) \cdot 4 + 2 \cdot 3 = -2 - 4 + 6 = 0$

Therefore, $\mathbf{v}_1 \perp \mathbf{v}_2$ and $\theta = \dfrac{\pi}{2}$.

(b) $\mathbf{v}_1 \circ \mathbf{v}_2 = 1 \cdot (-\frac{1}{2}) + (-2) \cdot 1 + 3(-\frac{3}{2})$
$= -\frac{1}{2} - 2 - \frac{9}{2} = -7$
$|\mathbf{v}_1| = \sqrt{1 + 4 + 9} = \sqrt{14}$
$|\mathbf{v}_2| = \sqrt{\frac{1}{4} + 1 + \frac{9}{4}} = \sqrt{\frac{14}{4}}$

$\cos \theta = \dfrac{-7}{\sqrt{14} \cdot \frac{\sqrt{14}}{2}} = -1$

Therefore, $\theta = \pi$ (note that $\mathbf{v}_2 = -\frac{1}{2} \cdot \mathbf{v}_1$).

(c) $\mathbf{v}_1 \circ \mathbf{v}_2 = 1 \cdot 1 + (-2) \cdot 1 + 1 \cdot 0 = -1$
$|\mathbf{v}_1| = \sqrt{1 + 4 + 1} = \sqrt{6}$
$|\mathbf{v}_2| = \sqrt{1 + 1} = \sqrt{2}$

Therefore, $\cos \theta = \dfrac{-1}{\sqrt{6}\sqrt{2}} = -\dfrac{1}{2\sqrt{3}} = -\dfrac{\sqrt{3}}{6}$.

Therefore, $\theta = \arccos\left(-\dfrac{\sqrt{3}}{6}\right)$.

Let $\mathbf{v}_1$ and $\mathbf{v}_2$ be nonzero vectors and let $\mathbf{u} = \mathbf{v}_1/|\mathbf{v}_1|$. Then $\mathbf{u}$ is a vector of length 1 in the same direction as $\mathbf{v}_1$. $\mathbf{u}$ is called a *unit vector*. The *projection* of $\mathbf{v}_2$ on $\mathbf{v}_1$ is defined to be the vector (see Figure 8–12)

$(\text{proj } \mathbf{v}_2)_{\mathbf{v}_1} = (|\mathbf{v}_2| \cos \theta)\mathbf{u}$
$= \left(\dfrac{\mathbf{v}_1 \circ \mathbf{v}_2}{|\mathbf{v}_1|^2}\right) \mathbf{v}_1$

| $\mathbf{v}_2$ || cos θ |

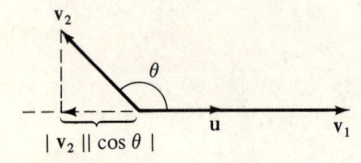

Figure 8–12

The Angle Between Two Vectors; Direction of a Vector

Example 2. For the vectors $\mathbf{v}_1 = (2, -1, 3)$ and $\mathbf{v}_2 = (4, 0, -2)$, find $(\text{proj } \mathbf{v}_1)_{\mathbf{v}_2}$.

SOLUTION

$$(\text{proj } \mathbf{v}_1)_{\mathbf{v}_2} = \left(\frac{\mathbf{v}_1 \circ \mathbf{v}_2}{|\mathbf{v}_2|^2}\right)\mathbf{v}_2$$

$$\mathbf{v}_1 \circ \mathbf{v}_2 = 2 \cdot 4 + (-1)(0) + 3(-2) = 2$$

$$|\mathbf{v}_2|^2 = 16 + 0 + 4 = 20$$

$$(\text{proj } \mathbf{v}_1)_{\mathbf{v}_2} = \tfrac{2}{20}(4, 0, -2) = (\tfrac{2}{5}, 0, -\tfrac{1}{5})$$

Note that

$$(\text{proj } \mathbf{v}_2)_{\mathbf{v}_1} = \tfrac{2}{14}(2, -1, 3) = (\tfrac{2}{7}, -\tfrac{1}{7}, \tfrac{3}{7})$$

Let $\mathbf{v} = (a, b, c)$ be an arbitrary vector in R^3. We may write

$$\mathbf{v} = (a, 0, 0) + (0, b, 0) + (0, 0, c)$$
$$= a(1, 0, 0) + b(0, 1, 0) + c(0, 0, 1)$$

Letting $\mathbf{i} = (1, 0, 0)$, $\mathbf{j} = (0, 1, 0)$, and $\mathbf{k} = (0, 0, 1)$, we have

$$\mathbf{v} = a\mathbf{i} + b\mathbf{j} + c\mathbf{k}$$

Such a sum is called a *linear combination* of $\mathbf{i}$, $\mathbf{j}$, and $\mathbf{k}$. Furthermore, the representation of $\mathbf{v}$ as a linear combination of $\mathbf{i}$, $\mathbf{j}$, and $\mathbf{k}$ is *unique*. For, suppose $\mathbf{v} = a\mathbf{i} + b\mathbf{j} + c\mathbf{k} = a_1\mathbf{i} + b_1\mathbf{j} + c_1\mathbf{k}$. It follows that $(a, b, c) = (a_1, b_1, c_1)$, and so $a = a_1$, $b = b_1$, and $c = c_1$. Observe that

$$|\mathbf{i}| = |\mathbf{j}| = |\mathbf{k}| = 1$$

and that $\mathbf{i} \circ \mathbf{j} = \mathbf{j} \circ \mathbf{k} = \mathbf{k} \circ \mathbf{i} = 0$. For this reason, $\mathbf{i}$, $\mathbf{j}$, and $\mathbf{k}$ are called *mutually orthogonal unit vectors*. We also say they form an *orthonormal set* of vectors (Figure 8-13).

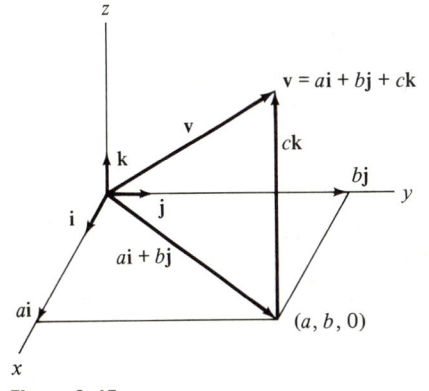

Figure 8-13

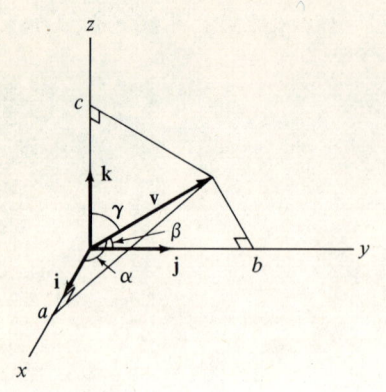

Figure 8-14

Example 3

$$v_1 = (3, -4, 5) = 3i - 4j + 5k$$
$$v_2 = (4, 0, -2) = 4i - 2k$$
$$v_1 + v_2 = (7, -4, 3) = 7i - 4j + 3k$$
$$v_1 - v_2 = (-1, -4, 7) = -i - 4j + 7k$$

Definition. Let $v = ai + bj + ck$ be a nonzero vector. The *direction angles* α, β, γ of v are the angles between v and the basis vectors i, j, k, respectively. The ordered triple (α, β, γ) is called the *direction of* v (Figure 8-14). (We do not define direction for the vector 0.)

Thus,

$$\cos \alpha = \frac{v \circ i}{|v| \cdot |i|}$$

$$\cos \beta = \frac{v \circ j}{|v| \cdot |j|}$$

$$\cos \gamma = \frac{v \circ k}{|v| \cdot |k|}$$

where $0 \leq \alpha, \beta, \gamma \leq \pi$.

Since $v \circ i = a$, $v \circ j = b$, $v \circ k = c$, and i, j, k are unit vectors, the direction angles are given by $\cos \alpha = a/|v|$, $\cos \beta = b/|v|$, and $\cos \gamma = c/|v|$. These are called the *direction cosines* of v. Note that

$$\cos^2 \alpha + \cos^2 \beta + \cos^2 \gamma = 1$$

The sum of the squares of the direction cosines is 1. Furthermore, the direction (α, β, γ) is completely determined by the triple $(\cos \alpha, \cos \beta, \cos \gamma)$.

Since $a = |v| \cos \alpha$, $b = |v| \cos \beta$, $c = |v| \cos \gamma$,

$$v = |v| \cos \alpha i + |v| \cos \beta j + |v| \cos \gamma k$$

so that the vector v is completely determined by its *magnitude* (length) and

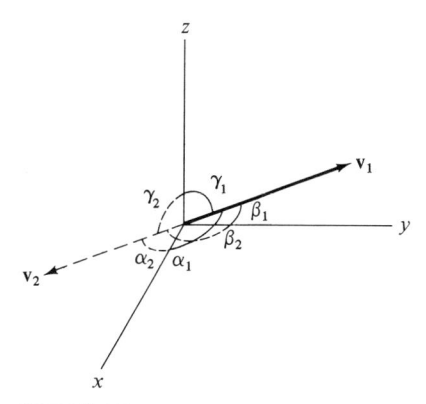

The Angle Between Two Vectors; Direction of a Vector

Figure 8-15

its *direction*. Hence *two vectors are equal iff they have the same length and direction*.

Theorem. *Let v_1 have direction $(\alpha_1, \beta_1, \gamma_1)$, and v_2 have direction $(\alpha_2, \beta_2, \gamma_2)$. Then*
(1) v_1 *and* v_2 *are parallel and in the same direction iff*

$$(\alpha_1, \beta_1, \gamma_1) = (\alpha_2, \beta_2, \gamma_2)$$

(2) v_1 *and* v_2 *are parallel and in opposite directions iff*

$$(\alpha_1, \beta_1, \gamma_1) = (\pi - \alpha_2, \pi - \beta_2, \pi - \gamma_2) \quad (Figure\ 8\text{-}15)$$

PROOF. We shall prove Part 2 of the above theorem. The proof of Part 1 is similar (see Problem 15). Let $v_1 = a_1 i + b_1 j + c_1 k$ and $v_2 = a_2 i + b_2 j + c_2 k$. Suppose v_1 and v_2 are in opposite directions. Then $v_1 = tv_2$ with $t < 0$, and consequently

$$\cos \alpha_1 = \frac{a_1}{|v_1|} = \frac{ta_2}{|t||v_2|} = -\frac{a_2}{|v_2|}$$

since $|t| = -t$. So $\cos \alpha_1 = -\cos \alpha_2 = \cos(\pi - \alpha_2)$. Therefore, $\alpha_1 = \pi - \alpha_2$, since $0 \leq \alpha_1, \pi - \alpha_2 \leq \pi$. In exactly the same way, we show that $\beta_1 = \pi - \beta_2$, and $\gamma_1 = \pi - \gamma_2$. Accordingly, $(\alpha_1, \beta_1, \gamma_1) = (\pi - \alpha_2, \pi - \beta_2, \pi - \gamma_2)$.

Conversely, suppose that $(\alpha_1, \beta_1, \gamma_1) = (\pi - \alpha_2, \pi - \beta_2, \pi - \gamma_2)$. Then $\cos \alpha_1 = -\cos \alpha_2$, $\cos \beta_1 = -\cos \beta_2$, $\cos \gamma_1 = -\cos \gamma_2$, and so

$$\begin{aligned} v_1 &= |v_1|(\cos \alpha_1 i + \cos \beta_1 j + \cos \gamma_1 k) \\ &= |v_1|(-\cos \alpha_2 i - \cos \beta_2 j - \cos \gamma_2 k) \\ &= \frac{-|v_1|}{|v_2|}[|v_2| \cdot (\cos \alpha_2 i + \cos \beta_2 j + \cos \gamma_2 k)] \\ &= -\frac{|v_1|}{|v_2|} v_2 \end{aligned}$$

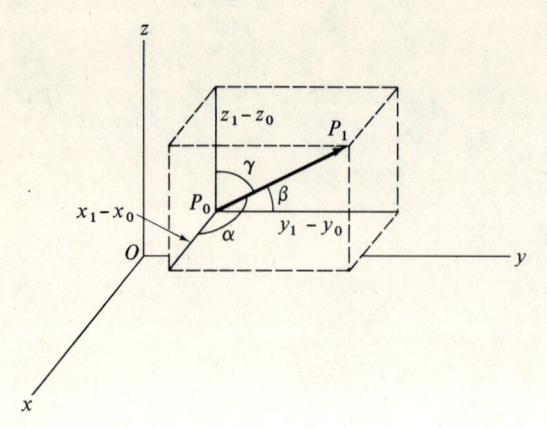

Figure 8-16

Thus,

$$\mathbf{v}_1 = t\mathbf{v}_2$$

where

$$t = -\frac{|\mathbf{v}_1|}{|\mathbf{v}_2|} < 0$$

REMARKS

(1) If $\mathbf{v} = a\mathbf{i} + b\mathbf{j} + c\mathbf{k}$ has direction (α, β, γ), then the unit vector $\mathbf{u} = \mathbf{v}/|\mathbf{v}| = \cos \alpha \mathbf{i} + \cos \beta \mathbf{j} + \cos \gamma \mathbf{k}$.

(2) When $P_0 \neq P_1$, the *direction of the arrow* $\overrightarrow{P_0 P_1}$ is defined to be the direction of the vector $\mathbf{v}$ which $\overrightarrow{P_0 P_1}$ represents, namely,

$$\mathbf{v} = (x_1 - x_0, y_1 - y_0, z_1 - z_0)$$

where $P_0 = (x_0, y_0, z_0)$ and $P_1 = (x_1, y_1, z_1)$. The direction angles of $\overrightarrow{P_0 P_1}$ are usually pictured as in Figure 8–16. Two arrows are equivalent iff they have the same direction and magnitude.

(3) If $\mathbf{u}_1$ and $\mathbf{u}_2$ are unit vectors in the direction of $\mathbf{v}_1$ and $\mathbf{v}_2$, respectively, so that

$$\mathbf{u}_1 = \cos \alpha_1 \mathbf{i} + \cos \beta_1 \mathbf{j} + \cos \gamma_1 \mathbf{k}$$
$$\mathbf{u}_2 = \cos \alpha_2 \mathbf{i} + \cos \beta_2 \mathbf{j} + \cos \gamma_2 \mathbf{k}$$

then the angle θ between $\mathbf{v}_1$ and $\mathbf{v}_2$ is given by

$$\cos \theta = \mathbf{u}_1 \circ \mathbf{u}_2 = \cos \alpha_1 \cos \alpha_2 + \cos \beta_1 \cos \beta_2 + \cos \gamma_1 \cos \gamma_2$$

Example 4. Find a unit vector having equal direction angles.

SOLUTION. The vector $\mathbf{v}$ is of the form

$$\mathbf{v} = \cos \alpha \mathbf{i} + \cos \beta \mathbf{j} + \cos \gamma \mathbf{k}$$

The Angle Between Two Vectors; Direction of a Vector

where

$\cos \alpha = \cos \beta = \cos \gamma$

Hence

$\cos^2 \alpha + \cos^2 \beta + \cos^2 \gamma = 3 \cos^2 \alpha = 1$

Therefore,

$\cos^2 \alpha = \frac{1}{3}$

$\cos \alpha = \pm \frac{1}{\sqrt{3}}$

There are two solutions:

$\mathbf{v}_1 = \frac{1}{\sqrt{3}} \mathbf{i} + \frac{1}{\sqrt{3}} \mathbf{j} + \frac{1}{\sqrt{3}} \mathbf{k}$

$\mathbf{v}_2 = -\frac{1}{\sqrt{3}} \mathbf{i} - \frac{1}{\sqrt{3}} \mathbf{j} - \frac{1}{\sqrt{3}} \mathbf{k} = -\mathbf{v}_1$

Example 5. Let $\mathbf{v}$ have direction $(45°, 45°, \gamma)$. Find γ.

SOLUTION

$\cos^2 45° + \cos^2 45° + \cos^2 \gamma = 1$

$\frac{1}{2} + \frac{1}{2} + \cos^2 \gamma = 1$

$\cos^2 \gamma = 0$

$\gamma = \frac{\pi}{2} = 90°$

Thus, $\mathbf{v}$ lies in the x–y plane.

Example 6. Find a vector $\mathbf{v}$ for which $\alpha = 30°$, $\beta = 135°$.

SOLUTION

$\cos^2 30° + \cos^2 135° + \cos^2 \gamma = 1$

$\left(\frac{\sqrt{3}}{2}\right)^2 + \left(-\frac{\sqrt{2}}{2}\right)^2 + \cos^2 \gamma = 1$

$\cos^2 \gamma = -\frac{1}{4}$

Since there is no solution to this equation, no such vector exists.

Example 7. Let $\mathbf{v}_1$ and $\mathbf{v}_2$ have direction cosines $(\frac{1}{3}, -\frac{2}{3}, \frac{2}{3})$ and $(a, b, \frac{1}{3})$, respectively. Find a and b such that $\mathbf{v}_1 \perp \mathbf{v}_2$.

SOLUTION. a and b must satisfy the equations

$\begin{cases} a^2 + b^2 + (\frac{1}{3})^2 = 1 \\ \frac{1}{3}a - \frac{2}{3}b + \frac{2}{3} \cdot \frac{1}{3} = 0 \end{cases}$

that is,

$$\begin{cases} a^2 + b^2 = \frac{8}{9} \\ a - 2b = -\frac{2}{3} \end{cases}$$

Therefore, $a = 2b - \frac{2}{3}$.

Substituting in the first equation, we obtain

$45b^2 - 24b - 4 = 0$

$(15b + 2)(3b - 2) = 0$

$$b = -\frac{2}{15}, \frac{2}{3}$$

When $b = -\frac{2}{15}$, $a = -\frac{14}{15}$. When $b = \frac{2}{3}$, $a = \frac{2}{3}$. Thus, there are two solutions for the direction cosines of v_2:

$(-\frac{14}{15}, -\frac{2}{15}, \frac{1}{3})$ and $(\frac{2}{3}, \frac{2}{3}, \frac{1}{3})$

EXERCISE 4

1. Find the angle θ between v_1 and v_2.
   (a) $v_1 = (4, -2, 3)$, $v_2 = (\frac{2}{3}, -\frac{1}{3}, \frac{1}{2})$
   (b) $v_1 = (3, 1, -2)$, $v_2 = (-\frac{1}{2}, -\frac{1}{6}, \frac{1}{3})$
   (c) $v_1 = (2, -4, 5)$, $v_2 = (-1, 1, \frac{6}{5})$
   (d) $v_1 = (1, 2, -1)$, $v_2 = (-1, -1, 0)$
   (e) $v_1 = (2, 4, -2)$, $v_2 = (7, 7, 12)$

2. Find the direction cosines of the following vectors.
   (a) $v = 2i - j + 4k$
   (b) $v = -2j + k$
   (c) $v = i + j + k$
   (d) $v = \frac{1}{2}i + \frac{1}{3}j - \frac{\sqrt{23}}{6}k$

3. Write the following arrows in the form $ai + bj + ck$ and find the direction cosines of each.
   (a) $\overrightarrow{P_0P_1}$, where $P_0 = (1, -2, 3)$, $P_1 = (-4, 2, -3)$
   (b) $\overrightarrow{P_1P_0}$, where $P_0 = (-2, 3, 4)$, $P_1 = (0, -1, 5)$
   (c) $\overrightarrow{P_0P_1}$, where $P_0 = (0, 0, 0)$, $P_1 = (-1, -2, -3)$
   (d) $\overrightarrow{P_1P_0}$, where $P_0 = (2, 0, -1)$, $P_1 = (-2, 3, -1)$

4. Determine whether the following pairs of vectors are parallel, perpendicular, or neither. If neither, state whether the angle between them is acute or obtuse. If parallel, state whether they are in the same direction or opposite directions.
   (a) $v_1 = 5i - 2j + 6k$, $v_2 = -\frac{5}{2}i + j - 3k$
   (b) $v_1 = 7i - 3k$, $v_2 = -3j + 4k$
   (c) $v_1 = -i + 4j + 7k$, $v_2 = 6i - 2j + 2k$
   (d) $v_1 = 2i - j + k$, $v_2 = 3i + 4j - 2k$

(e) $\mathbf{v}_1 = \frac{1}{2}\mathbf{i} + \frac{7}{3}\mathbf{j} - 7\mathbf{k}$, $\mathbf{v}_2 = \frac{2}{7}\mathbf{i} + \frac{4}{3}\mathbf{j} - 4\mathbf{k}$

(f) $\mathbf{v}_1 = -2\mathbf{i} + \frac{1}{2}\mathbf{j} + \frac{2}{3}\mathbf{k}$, $\mathbf{v}_2 = \mathbf{i} + \frac{4}{5}\mathbf{j} + \frac{12}{5}\mathbf{k}$

5. For the vectors of Problem 4, find $(\text{proj } \mathbf{v}_1)_{\mathbf{v}_2}$.

6. (a) Let $\mathbf{v}$ have direction $(60°, 150°, \gamma)$. Find γ.
   (b) Let $\mathbf{v}$ have direction $(\alpha, 30°, 45°)$. Find α.
   (c) Let $\mathbf{v} = (a, b, c)$ have direction $(150°, 75°, \gamma)$ and let $c < 0$. Find the direction cosines of $\mathbf{v}$.

7. Prove that $(\text{proj }(\mathbf{v}_1 + \mathbf{v}_2))_{\mathbf{v}_3} = (\text{proj } \mathbf{v}_1)_{\mathbf{v}_3} + (\text{proj } \mathbf{v}_2)_{\mathbf{v}_3}$.

8. (a) Prove that if θ is the angle between $\mathbf{v}_1$ and $\mathbf{v}_2$, then
$$|\mathbf{v}_1 + \mathbf{v}_2|^2 = |\mathbf{v}_1|^2 + |\mathbf{v}_2|^2 + 2|\mathbf{v}_1| \cdot |\mathbf{v}_2| \cos \theta$$
   (b) Prove that $\mathbf{v}_1 \perp \mathbf{v}_2$ iff $|\mathbf{v}_1 + \mathbf{v}_2|^2 = |\mathbf{v}_1|^2 + |\mathbf{v}_2|^2$ (Pythagorean theorem).

9. Let $\mathbf{v}_1$ and $\mathbf{v}_2$ be nonzero vectors and a, b be nonzero scalars. Let θ_1 be the angle between $\mathbf{v}_1$ and $\mathbf{v}_2$, and let θ_2 be the angle between $a\mathbf{v}_1$ and $b\mathbf{v}_2$. Show that
   (a) $\theta_2 = \theta_1$ iff a and b have the same sign;
   (b) $\theta_2 = \pi - \theta_1$ iff a and b have opposite sign.

10. (a) Prove that if $\mathbf{v} \perp \mathbf{v}_1$ and $\mathbf{v} \perp \mathbf{v}_2$, then $\mathbf{v} \perp (a\mathbf{v}_1 + b\mathbf{v}_2)$ for all scalars a and b.
    (b) Prove that if $\mathbf{v}$ is perpendicular to each of n vectors $\mathbf{v}_1, \mathbf{v}_2, \ldots, \mathbf{v}_n$, then $\mathbf{v} \perp (a_1\mathbf{v}_1 + a_2\mathbf{v}_2 + \ldots + a_n\mathbf{v}_n)$ for all scalars $a_1, a_2, \ldots, a_n$. (Use mathematical induction.)

*11. Let $\{\mathbf{v}_1, \mathbf{v}_2, \mathbf{v}_3\}$ be a set of three mutually orthogonal nonzero vectors (that is, any two are orthogonal).
    (a) Prove that if $a_1\mathbf{v}_1 + a_2\mathbf{v}_2 + a_3\mathbf{v}_3 = \mathbf{0}$, then $a_1 = a_2 = a_3 = 0$.
    (b) Prove that if $a_1\mathbf{v}_1 + a_2\mathbf{v}_2 + a_3\mathbf{v}_3 = b_1\mathbf{v}_1 + b_2\mathbf{v}_2 + b_3\mathbf{v}_3$, then $a_1 = b_1$, $a_2 = b_2$, and $a_3 = b_3$.

*12. (a) Prove that $\mathbf{0}$ is the *only* vector which is orthogonal to *every* vector.
    (b) Prove that if $\mathbf{v} = c_1\mathbf{v}_1 + c_2\mathbf{v}_2 + c_3\mathbf{v}_3$ is orthogonal to each of the three mutually orthogonal nonzero vectors $\mathbf{v}_1, \mathbf{v}_2,$ and $\mathbf{v}_3$, then $\mathbf{v} = \mathbf{0}$.

*13. Let $\mathbf{v}_1 = (1, -2, -1)$. Find nonzero vectors $\mathbf{v}_2$ and $\mathbf{v}_3$ so that $\{\mathbf{v}_1, \mathbf{v}_2, \mathbf{v}_3\}$ is a mutually orthogonal set.

*14. Let $\{\mathbf{v}_1, \mathbf{v}_2, \mathbf{v}_3\}$ be a set of mutually orthogonal nonzero vectors. Let $\mathbf{v} = c_1\mathbf{v}_1 + c_2\mathbf{v}_2 + c_3\mathbf{v}_3$. Prove that
$$\mathbf{v} = (\text{proj } \mathbf{v})_{\mathbf{v}_1} + (\text{proj } \mathbf{v})_{\mathbf{v}_2} + (\text{proj } \mathbf{v})_{\mathbf{v}_3}$$

*15. Prove Part 1 of the theorem of this section.

5. THE CROSS PRODUCT

Another important operation with vectors is the *cross product* (or *vector product*) of two vectors. In this section, we denote vectors by **a**, **b**, and **c**, while the components of a vector are denoted by the same letter with subscripts 1, 2, and 3. Thus, we write

$\mathbf{a} = a_1\mathbf{i} + a_2\mathbf{j} + a_3\mathbf{k}$, $\mathbf{b} = b_1\mathbf{i} + b_2\mathbf{j} + b_3\mathbf{k}$

Definition. Let $\mathbf{a} = a_1\mathbf{i} + a_2\mathbf{j} + a_3\mathbf{k}$ and $\mathbf{b} = b_1\mathbf{i} + b_2\mathbf{j} + b_3\mathbf{k}$. The *cross product (vector product)* of $\mathbf{a}$ and $\mathbf{b}$ (*in that order*) is the *vector*

$$\mathbf{a} \times \mathbf{b} = \begin{vmatrix} a_2 & a_3 \\ b_2 & b_3 \end{vmatrix}\mathbf{i} - \begin{vmatrix} a_1 & a_3 \\ b_1 & b_3 \end{vmatrix}\mathbf{j} + \begin{vmatrix} a_1 & a_2 \\ b_1 & b_2 \end{vmatrix}\mathbf{k}$$
$$= (a_2 b_3 - a_3 b_2)\mathbf{i} - (a_1 b_3 - a_3 b_1)\mathbf{j} + (a_1 b_2 - a_2 b_1)\mathbf{k}$$

Note that $\mathbf{a} \times \mathbf{b}$ may be obtained by formally expanding the determinant

$$\begin{vmatrix} \mathbf{i} & \mathbf{j} & \mathbf{k} \\ a_1 & a_2 & a_3 \\ b_1 & b_2 & b_3 \end{vmatrix}$$

by minors along the first row. (See the Appendix.) Thus, we may write

$$\mathbf{a} \times \mathbf{b} = \begin{vmatrix} \mathbf{i} & \mathbf{j} & \mathbf{k} \\ a_1 & a_2 & a_3 \\ b_1 & b_2 & b_3 \end{vmatrix}$$

Example 1. Let $\mathbf{a} = 2\mathbf{i} - 3\mathbf{k}$ and $\mathbf{b} = 4\mathbf{i} + 3\mathbf{j} - 2\mathbf{k}$. Find $\mathbf{a} \times \mathbf{b}$ and $\mathbf{b} \times \mathbf{a}$.

SOLUTION

$$\mathbf{a} \times \mathbf{b} = \begin{vmatrix} \mathbf{i} & \mathbf{j} & \mathbf{k} \\ 2 & 0 & -3 \\ 4 & 3 & -2 \end{vmatrix} = (0(-2) - (-3)3)\mathbf{i} - (2(-2) - (-3)4)\mathbf{j} + (2 \cdot 3 - 0 \cdot 4)\mathbf{k}$$
$$= 9\mathbf{i} - 8\mathbf{j} + 6\mathbf{k}$$

$$\mathbf{b} \times \mathbf{a} = \begin{vmatrix} \mathbf{i} & \mathbf{j} & \mathbf{k} \\ 4 & 3 & -2 \\ 2 & 0 & -3 \end{vmatrix} = (3(-3) - (-2)0)\mathbf{i} - (4(-3) - (-2)2)\mathbf{j} + (4 \cdot 0 - 3 \cdot 2)\mathbf{k}$$
$$= -9\mathbf{i} + 8\mathbf{j} - 6\mathbf{k}$$

Observe that $\mathbf{b} \times \mathbf{a} = -(\mathbf{a} \times \mathbf{b})$. This is to be expected since the second determinant is obtained from the first by interchanging two rows. The cross product may be used to determine parallelism of vectors. From Theorem 3, Section 2, the nonzero vectors $\mathbf{a}$ and $\mathbf{b}$ are parallel iff

$$\begin{vmatrix} a_2 & a_3 \\ b_2 & b_3 \end{vmatrix} = \begin{vmatrix} a_1 & a_3 \\ b_1 & b_3 \end{vmatrix} = \begin{vmatrix} a_1 & a_2 \\ b_1 & b_2 \end{vmatrix} = 0$$

Thus, $\mathbf{a} \parallel \mathbf{b}$ *iff* $\mathbf{a} \times \mathbf{b} = \mathbf{0}$.

We collect the basic properties of the cross product in the following theorem.

Theorem

(1) $\mathbf{0} \times \mathbf{a} = \mathbf{a} \times \mathbf{0} = \mathbf{0}$
(2) $\mathbf{b} \times \mathbf{a} = -(\mathbf{a} \times \mathbf{b})$ (*anticommutative law*)
(3) $\mathbf{a} \times \mathbf{b} = \mathbf{0}$ *iff* $\mathbf{a}$ and $\mathbf{b}$ are parallel ($\mathbf{a}, \mathbf{b} \neq \mathbf{0}$)

(4) $(c\mathbf{a}) \times \mathbf{b} = \mathbf{a} \times (c\mathbf{b}) = c(\mathbf{a} \times \mathbf{b})$ *for all scalars c*
(5) $\mathbf{a} \perp \mathbf{a} \times \mathbf{b}$ *and* $\mathbf{b} \perp \mathbf{a} \times \mathbf{b}$
(6) $\mathbf{a} \times (\mathbf{b} + \mathbf{c}) = \mathbf{a} \times \mathbf{b} + \mathbf{a} \times \mathbf{c}$ *(distributive law)*
(7) $|\mathbf{a} \times \mathbf{b}| = |\mathbf{a}| \cdot |\mathbf{b}| \sin \theta$, *where θ is the angle between* $\mathbf{a}$ *and* $\mathbf{b}$

PROOF. Properties 1 and 2 are immediate from the definition of $\mathbf{a} \times \mathbf{b}$. Property 3 has already been discussed. We leave the proof of the remaining parts as an exercise. (See Problem 4.)

REMARKS. From Property 2, we see that $\mathbf{a} \times \mathbf{b}$ and $\mathbf{b} \times \mathbf{a}$ have *equal magnitudes and opposite directions*. Property 3 implies that $\mathbf{a} \times \mathbf{a} = \mathbf{0}$ for all $\mathbf{a}$. Property 5 states that $\mathbf{a} \times \mathbf{b}$ is orthogonal to both $\mathbf{a}$ and $\mathbf{b}$, and hence to every linear combination of $\mathbf{a}$ and $\mathbf{b}$.

Observe that

$$\mathbf{i} \times \mathbf{j} = \begin{vmatrix} \mathbf{i} & \mathbf{j} & \mathbf{k} \\ 1 & 0 & 0 \\ 0 & 1 & 0 \end{vmatrix} = 0\mathbf{i} + 0\mathbf{j} + 1\mathbf{k} = \mathbf{k}$$

Similarly,

$$\mathbf{j} \times \mathbf{k} = \mathbf{i}$$
$$\mathbf{k} \times \mathbf{i} = \mathbf{j}$$

Example 2. Find a unit vector $\mathbf{c}$ orthogonal to the vectors $\mathbf{a} = \mathbf{i} - \mathbf{j} + 2\mathbf{k}$ and $\mathbf{b} = 3\mathbf{j} - \mathbf{k}$.

SOLUTION. By Property 5 of the above theorem, $\mathbf{a} \times \mathbf{b}$ is orthogonal to $\mathbf{a}$ and $\mathbf{b}$.

$$\mathbf{a} \times \mathbf{b} = \begin{vmatrix} \mathbf{i} & \mathbf{j} & \mathbf{k} \\ 1 & -1 & 2 \\ 0 & 3 & -1 \end{vmatrix} = -5\mathbf{i} + \mathbf{j} + 3\mathbf{k}$$

Therefore,

$$\mathbf{c} = \frac{\mathbf{a} \times \mathbf{b}}{|\mathbf{a} \times \mathbf{b}|} = \frac{-5\mathbf{i} + \mathbf{j} + 3\mathbf{k}}{\sqrt{35}}$$

CHECK

$$\mathbf{a} \circ \mathbf{c} = \frac{1}{\sqrt{35}}(-5 - 1 + 6) = 0$$

$$\mathbf{b} \circ \mathbf{c} = \frac{1}{\sqrt{35}}(0 + 3 - 3) = 0$$

and

$$|\mathbf{c}|^2 \quad \frac{25 + 1 + 9}{35} = \frac{35}{35} = 1$$

Example 3. Given $\mathbf{a} = \mathbf{i} - 2\mathbf{j} + 3\mathbf{k}$. Find $\mathbf{b}$ and $\mathbf{c}$ so that $\mathbf{a}$, $\mathbf{b}$, and $\mathbf{c}$ are mutually orthogonal.

SOLUTION. Let $\mathbf{b} = b_1\mathbf{i} + b_2\mathbf{j} + b_3\mathbf{k}$. We require that
$$\mathbf{a} \circ \mathbf{b} = b_1 - 2b_2 + 3b_3 = 0$$
We may choose $b_2 = b_3 = 1$, so that $b_1 = -1$. Thus, $\mathbf{b} = -\mathbf{i} + \mathbf{j} + \mathbf{k}$. Let
$$\mathbf{c} = \mathbf{a} \times \mathbf{b} = \begin{vmatrix} \mathbf{i} & \mathbf{j} & \mathbf{k} \\ 1 & -2 & 3 \\ -1 & 1 & 1 \end{vmatrix} = -5\mathbf{i} - 4\mathbf{j} - \mathbf{k}$$

The student may verify that $\mathbf{a}$, $\mathbf{b}$, and $\mathbf{c}$ are mutually orthogonal.

EXERCISE 5

1. Compute $\mathbf{a} \times \mathbf{b}$ and $\mathbf{b} \times \mathbf{a}$ for each of the following pairs of vectors.
   (a) $\mathbf{a} = 2\mathbf{i} - \mathbf{j} + \mathbf{k}$, $\quad \mathbf{b} = \mathbf{i} + 2\mathbf{j} - \mathbf{k}$
   (b) $\mathbf{a} = \mathbf{i} - 4\mathbf{j}$, $\quad \mathbf{b} = \frac{1}{2}\mathbf{i} - \mathbf{k}$
   (c) $\mathbf{a} = \frac{2}{3}\mathbf{i} - \mathbf{j} + 4\mathbf{k}$, $\quad \mathbf{b} = \frac{1}{6}\mathbf{i} - \frac{1}{4}\mathbf{j} + \mathbf{k}$
   (d) $\mathbf{a} = \mathbf{i} + 2\mathbf{j} - \mathbf{k}$, $\quad \mathbf{b} = 3\mathbf{i} - \mathbf{j} + \mathbf{k}$
   (e) $\mathbf{a} = \mathbf{i} - \frac{1}{5}\mathbf{j} + \frac{2}{5}\mathbf{k}$, $\quad \mathbf{b} = -\frac{1}{2}\mathbf{i} + \frac{1}{10}\mathbf{j} - \frac{1}{5}\mathbf{k}$

2. For the given vector $\mathbf{a}$, find nonzero vectors $\mathbf{b}$ and $\mathbf{c}$ such that $\mathbf{a}$, $\mathbf{b}$, and $\mathbf{c}$ are mutually orthogonal. (See Example 3.)
   (a) $\mathbf{a} = \mathbf{i} + \mathbf{j} + \mathbf{k}$
   (b) $\mathbf{a} = 2\mathbf{i} - \mathbf{j}$
   (c) $\mathbf{a} = 4\mathbf{i} + \mathbf{j} - \mathbf{k}$
   (d) $\mathbf{a} = 2\mathbf{j} + 3\mathbf{k}$
   (e) $\mathbf{a} = \mathbf{i} - 2\mathbf{j} + 3\mathbf{k}$

3. Show by two methods that the vectors $\mathbf{a} = \frac{2}{3}\mathbf{i} - \frac{1}{6}\mathbf{j} + \frac{2}{3}\mathbf{k}$ and $\mathbf{b} = 2\mathbf{i} - \frac{1}{2}\mathbf{j} + 2\mathbf{k}$ are parallel.

4. Prove Parts 4, 5, 6, and 7 of the theorem.
   (*Hint:* To prove Part 7, square both sides, replace $\sin^2 \theta$ by $1 - \cos^2 \theta$, and use $|\mathbf{a}| \cdot |\mathbf{b}| \cos \theta = \mathbf{a} \circ \mathbf{b}$.)

5. Prove that for all vectors $\mathbf{a}$, $\mathbf{b}$, and $\mathbf{c}$
   (a) $\mathbf{a} \circ (\mathbf{b} \times \mathbf{c}) = \begin{vmatrix} a_1 & a_2 & a_3 \\ b_1 & b_2 & b_3 \\ c_1 & c_2 & c_3 \end{vmatrix}$

   (b) $\mathbf{a} \circ (\mathbf{b} \times \mathbf{c}) = (\mathbf{a} \times \mathbf{b}) \circ \mathbf{c}$

6. Compute $\mathbf{a} \circ \mathbf{b} \times \mathbf{c}$ for the following vectors. (See Problem 5.)
   (a) $\mathbf{a} = \mathbf{i} + \mathbf{j} + \mathbf{k}$, $\quad \mathbf{b} = -\mathbf{i} + 2\mathbf{j}$, $\quad \mathbf{c} = \mathbf{j} + \mathbf{k}$
   (b) $\mathbf{a} = 2\mathbf{i} - \mathbf{j} + \mathbf{k}$, $\quad \mathbf{b} = \mathbf{i} - 3\mathbf{j} - 2\mathbf{k}$, $\quad \mathbf{c} = \mathbf{i} - 2\mathbf{j}$
   (c) $\mathbf{a} = \mathbf{i} - 2\mathbf{j}$, $\quad \mathbf{b} = 2\mathbf{i} - 4\mathbf{j} + \mathbf{k}$, $\quad \mathbf{c} = -\mathbf{i} - \mathbf{j} - \mathbf{k}$

7. Use Problem 5 to prove that **a** and **b** are orthogonal to **a** × **b**.

*8. Prove that if $\mathbf{a} + \mathbf{b} + \mathbf{c} = \mathbf{0}$, then

$$\mathbf{a} \times \mathbf{b} = \mathbf{b} \times \mathbf{c} = \mathbf{c} \times \mathbf{a}$$

*9. Prove
   (a) $(\mathbf{a} \times \mathbf{b}) \times \mathbf{c} = (\mathbf{a} \circ \mathbf{c})\mathbf{b} - (\mathbf{b} \circ \mathbf{c})\mathbf{a}$
   (b) $\mathbf{a} \times (\mathbf{b} \times \mathbf{c}) = (\mathbf{a} \circ \mathbf{c})\mathbf{b} - (\mathbf{a} \circ \mathbf{b})\mathbf{c}$

*10. Prove that if **a** and **b** are nonparallel and **c** is orthogonal to both **a** and **b**, then $\mathbf{c} \parallel \mathbf{a} \times \mathbf{b}$. (Use Problem 9.)

chapter 9
Lines, Planes, Spheres, and Convex Sets

In this chapter, we shall use vectors to study *lines* and *planes* in three-dimensional space R^3. This will be followed by a brief treatment of *spheres*, *tangent planes* to spheres, and *convex sets*. Although the definitions and figures are motivated by solid Euclidean geometry, we make no use of the axioms and methods of geometry.

1. LINE SEGMENTS, RAYS, AND LINES

The definitions of *line segment*, *ray*, and *line* in R^3 have precisely the same form as in R^2.

Definition. The *line segment* with *endpoints* P_1 and P_2 is the set

$$\overline{P_1P_2} = \{P \mid \overrightarrow{P_1P} = t\overrightarrow{P_1P_2}, 0 \leq t \leq 1\} \quad \text{(Figure 9-1(a))}$$

The *ray* with initial point P_1 passing through $P_2 \neq P_1$ is the set

$$\text{ray } P_1P_2 = \{P \mid \overrightarrow{P_1P} = t\overrightarrow{P_1P_2}, t \geq 0\} \quad \text{(Figure 9-1(b))}$$

194

Line Segments, Rays, and Lines

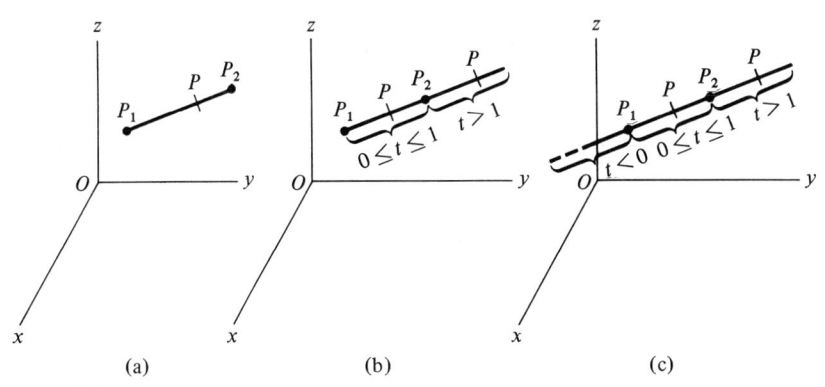

Figure 9-1

If $P_1 \neq P_2$, the *line* through P_1 and P_2 is the set

$$\ell = \{P \mid \overrightarrow{P_1P} = t\overrightarrow{P_1P_2}, t \in R\} \quad \text{(Figure 9-1(c))}$$

Observe that the definitions of *line segment, ray,* and *line* differ only in the range of values of the *parameter t*. All three are characterized by the same set of *parametric equations*

$$\begin{cases} x = x_1 + t(x_2 - x_1) \\ y = y_1 + t(y_2 - y_1) \\ z = z_1 + t(z_2 - z_1) \end{cases}$$

These are equivalent to the condition $\overrightarrow{P_1P} = t\overrightarrow{P_1P_2}$, where $P_1 = (x_1, y_1, z_1)$ and $P_2 = (x_2, y_2, z_2)$ are the given points, and $P(x, y, z)$ is any point on the segment, ray, or line. If $P(x, y, z)$ corresponds to a value of t with $0 < t < 1$, we say P lies between P_1 and P_2. It is easy to prove that $\overline{P_1P_2} = \overline{P_2P_1}$ (see Problem 5). The *midpoint* of $\overline{P_1P_2}$ (corresponding to $t = \frac{1}{2}$) is the point

$$M = \left(\frac{x_1 + x_2}{2}, \frac{y_1 + y_2}{2}, \frac{z_1 + z_2}{2} \right)$$

The *length* of the segment $\overline{P_1P_2}$ (also called the *distance between* P_1 and P_2) is defined by

$$|\overline{P_1P_2}| = |\overrightarrow{P_1P_2}| = [(x_2 - x_1)^2 + (y_2 - y_1)^2 + (z_2 - z_1)^2]^{1/2}$$

If $P_1 \neq P_2$ and P is on $\overline{P_1P_2}$, then $|\overrightarrow{P_1P}| = |t\overrightarrow{P_1P_2}|, 0 \leq t \leq 1$; so that

$$t = \frac{|\overrightarrow{P_1P}|}{|\overrightarrow{P_1P_2}|}$$

Thus, t is the ratio of the distance between P_1 and P to the length of the entire segment. $\overline{P_1P_2}$ is said to be *parallel* to $\overline{P_3P_4}$ ($\overline{P_1P_2} \parallel \overline{P_3P_4}$) iff $\overrightarrow{P_1P_2} \parallel \overrightarrow{P_3P_4}$.

Lines, Planes, Spheres, and Convex Sets

Example 1. Let $P_1 = (2, -3, 4)$ and $P_2 = (-1, 2, 3)$.
(a) Write the parametric equations of $\overrightarrow{P_1P_2}$.
(b) Find $|\overrightarrow{P_1P_2}|$.
(c) Find the midpoint of $\overrightarrow{P_1P_2}$.
(d) Find the point P which is $\frac{2}{3}$ the distance from P_2 along $\overrightarrow{P_1P_2}$.

SOLUTION

(a) $\quad x = 2 - 3t$
$\quad\quad y = -3 + 5t$
$\quad\quad z = 4 - t, \quad 0 \le t \le 1$
(b) $|\overrightarrow{P_1P_2}| = [(-3)^2 + 5^2 + (-1)^2]^{1/2} = \sqrt{35}$
(c) Letting $t = \frac{1}{2}$, we obtain the midpoint $M = (\frac{1}{2}, -\frac{1}{2}, \frac{7}{2})$.
(d) The desired point P is $\frac{1}{3}$ the distance from P_1 along $\overrightarrow{P_1P_2}$. Letting $t = \frac{1}{3}$, we have $P = (1, -\frac{4}{3}, \frac{11}{3})$.

Definition. The *distinct* points P_1, P_2, and P_3 are *collinear* iff they lie on the same line.

It is an immediate consequence of the definitions that P_1, P_2, and P_3 are collinear iff $\overrightarrow{P_1P_2} \parallel \overrightarrow{P_1P_3}$ (see Problem 7). The points P_1, P_2, and P_3 are said to be *noncollinear* iff they are *distinct* and none of the points is on the line joining the other two. Thus, the terms *collinear* and *noncollinear* apply only to sets of three (or more) points. The points $P_1, P_2, \ldots, P_n$ are said to be *collinear* iff they all lie on the same line.

Example 2. Determine whether the points $P_1(1, 2, -3)$, $P_2(2, 0, -4)$, and $P_3(4, -4, -6)$ are collinear.

SOLUTION. $\overrightarrow{P_1P_2} = \mathbf{i} - 2\mathbf{j} - \mathbf{k}$ and $\overrightarrow{P_1P_3} = 3\mathbf{i} - 6\mathbf{j} - 3\mathbf{k}$. Computing ratios of corresponding components, we obtain $\frac{1}{3} = \frac{-2}{-6} = \frac{-1}{-3}$. Therefore, $\overrightarrow{P_1P_2} \parallel \overrightarrow{P_1P_3}$ and P_1, P_2, and P_3 are collinear.

Since $P_1 \ne P_2$ in the definition of a line, the vector
$$\mathbf{v} = (x_2 - x_1)\mathbf{i} + (y_2 - y_1)\mathbf{j} + (z_2 - z_1)\mathbf{k}$$
is nonzero. If P_1 and $\mathbf{v}_1 \ne \mathbf{0}$ are given, the point P_2 such that $\overrightarrow{P_1P_2} = \mathbf{v}_1$ is determined, and so, therefore, is the line ℓ. Accordingly, a unique line is determined by a given point $P_1(x_1, y_1, z_1)$ and a given nonzero vector $\mathbf{v} = a\mathbf{i} + b\mathbf{j} + c\mathbf{k}$. Thus, $\ell = \{P | \overrightarrow{P_1P} = t\mathbf{v}, t \in R\}$ (Figure 9-2). We say ℓ is the line through P_1 parallel to $\mathbf{v}$, and write $\ell \parallel \mathbf{v}$. When $t = 0$, $P = P_1$; for $t \ne 0$, $P \ne P_1$ and $\overrightarrow{P_1P} \parallel \mathbf{v}$. It is easy to show that if $\mathbf{w} \parallel \mathbf{v}$, then the line through P_1 parallel to $\mathbf{w}$ is the same as ℓ, and we write $\ell \parallel \mathbf{w}$. (See Problem 8.)

Line Segments, Rays, and Lines

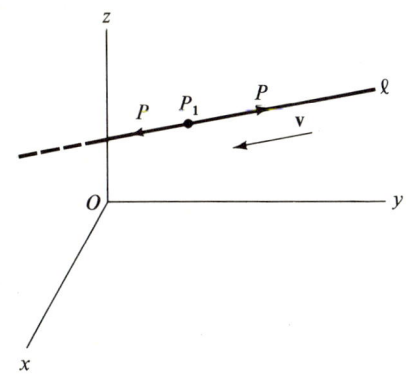

Figure 9-2

The parametric equations of ℓ may be written

$$\ell : \begin{cases} x = x_1 + at \\ y = y_1 + bt \\ z = z_1 + ct, \end{cases} \quad (t \in R)$$

where $\mathbf{v} = a\mathbf{i} + b\mathbf{j} + c\mathbf{k}$.

Example 3. Let $P_1 = (1, 2, 3)$ and $P_2 = (-3, 4, 5)$.

(a) Write the parametric equations of $\overrightarrow{P_1P_2}$, ray P_1P_2, and the line ℓ through P_1 and P_2.
(b) Find the points on ℓ corresponding to $t = 0, 1, \frac{1}{2}, \frac{3}{2}, -\frac{3}{2}$.

SOLUTION

(a) *Parametric equations*

$$\ell : \begin{cases} x = 1 - 4t \\ y = 2 + 2t \\ z = 3 + 2t \end{cases}$$

t	0	1	$\frac{1}{2}$	$\frac{3}{2}$	$-\frac{3}{2}$
P	$(1, 2, 3)$	$(-3, 4, 5)$	$(-1, 3, 4)$	$(-5, 5, 6)$	$(7, -1, 0)$

Example 4. Find the point of intersection (if it exists) for the following pair of lines.

$$\begin{cases} \ell_1: \text{through } P_1(4, 0, -2), \text{ parallel to } \mathbf{v}_1 = 2\mathbf{i} + 3\mathbf{k} \\ \ell_2: \text{through } P_2(-1, 2, -3), \text{ parallel to } \mathbf{v}_2 = 9\mathbf{i} + 4\mathbf{j} + \mathbf{k} \end{cases}$$

Lines, Planes, Spheres, and Convex Sets

SOLUTION

$$\ell_1 : \begin{cases} x = 4 + 2s \\ y = 0 + 3s \\ z = -2 + 0s \end{cases} \qquad \ell_2 : \begin{cases} x = -1 + 9t \\ y = 2 + 4t \\ z = -3 + t \end{cases}$$

The conditions for a point of intersection are

$$\begin{cases} 4 + 2s = -1 + 9t \\ 3s = 2 + 4t. \\ -2 = -3 + t \end{cases}$$

Solving the last two equations, we obtain $s = 2$ and $t = 1$. Since these also satisfy the first equation, ℓ_1 and ℓ_2 intersect. Substituting $s = 2$ in the equations for ℓ_1, we obtain the point $A(8, 6, -2)$. The same point is obtained when $t = 1$ is substituted in the equations for ℓ_2. (If the equations in s and t have no simultaneous solution, the lines do not intersect.)

EXERCISE 1

1. Write the parametric equations of the lines through the following pairs of points.
   (a) $P_1(2, -1, 3), P_2(0, 6, -2)$
   (b) $P_1(4, 2, 3), P_2(-2, 2, 3)$
   (c) $P_1(3, 4, 7), P_2(3, 4, -2)$
   (d) $P_1(0, 0, 0), P_2(2, -3, 4)$
   (e) $P_1(0, 1, 0), P_2(0, 5, 0)$

2. For each of the following pairs of points, find
   (a) the midpoint of $\overline{P_1 P_2}$
   (b) the point $\frac{1}{3}$ the distance from P_1 along $\overline{P_1 P_2}$
   (c) the point $\frac{3}{4}$ the distance from P_2 along $\overline{P_2 P_1}$
   (1) $P_1(2, -1, 0), P_2(1, 3, -4)$
   (2) $P_1(0, 0, 0), P_2(1, 2, 3)$
   (3) $P_1(1, 1, 1), P_2(1, -1, 2)$
   (4) $P_1(\frac{1}{2}, 2, -\frac{2}{3}), P_2(0, -1, 3)$

3. Determine whether the following sets of points are collinear.
   (a) $P_1(2, 1, 1), P_2(-1, 1, 2), P_3(-4, 1, 3)$
   (b) $P_1(-1, 2, -3), P_2(4, 3, 2), P_3(-5, 1, 3)$
   (c) $P_1(0, -2, 4), P_2(5, 2, -\frac{1}{2}), P_3(\frac{5}{4}, -1, \frac{23}{8})$
   (d) $P_1(1, -1, 1), P_2(3, 0, 2), P_3(0, -2, -3)$
   (e) $P_1(1, 0, 1), P_2(-3, 2, 0), P_3(-\frac{5}{3}, \frac{4}{3}, \frac{1}{3})$

4. Determine whether the following pairs of lines intersect. Find the point of intersection if it exists.
   (a) $\begin{cases} \ell_1 \text{ through } P_1(2, 1, 3), P_2(-1, 2, -4) \\ \ell_2 \text{ through } Q_1(5, 1, -2), Q_2(0, 4, 3) \end{cases}$

(b) $\ell_1: \begin{cases} x = 2 + 4t \\ y = -1 + 3t \\ z = -2t \end{cases}$ $\ell_2: \begin{cases} x = -1 + t \\ y = 3 + 7t \\ z = 5 + 3t \end{cases}$

(c) $\ell_1: \begin{cases} x = 6 + 4t \\ y = 2 - t \\ z = 3 + 2t \end{cases}$ $\ell_2: \begin{cases} x = 2t \\ y = 8 + t \\ z = -1 + 2t \end{cases}$

(d) $\begin{cases} \ell_1 \text{ through } P_1(1, -2, 4) \text{ parallel to } \mathbf{v} = -2\mathbf{i} + 3\mathbf{j} + 4\mathbf{k} \\ \ell_2 \text{ through } P_2(2, 0, 3) \text{ parallel to } \mathbf{w} = \mathbf{i} + 3\mathbf{j} + 7\mathbf{k} \end{cases}$

*5. Prove that $\overline{P_1 P_2} = \overline{P_2 P_1}$.
*6. Prove that if P_3 is on ray $P_1 P_2$, then ray $P_1 P_3 = $ ray $P_1 P_2$.
*7. Prove that P_1, P_2, and P_3 are collinear iff $\overrightarrow{P_1 P_2} \parallel \overrightarrow{P_1 P_3}$.
*8. (a) Let ℓ_1 be the line through P_1, P_2 and let Q_1, Q_2 be distinct points on ℓ_1. Prove that the line ℓ_2 through Q_1, Q_2 is equal to ℓ_1.
  (b) Let $\mathbf{v}$ and $\mathbf{w}$ be parallel vectors. Prove that the line through P_1 parallel to $\mathbf{v}$ is to the line through P_1 parallel to $\mathbf{w}$.

2. SYMMETRIC EQUATIONS OF A LINE; ANGLE BETWEEN TWO LINES

If the components of $\mathbf{v}$ are all nonzero, the parametric equations may each be solved for t. Since t is the same for all three equations, we have

$$\frac{x - x_1}{a} = \frac{y - y_1}{b} = \frac{z - z_1}{c}$$

These are called the *symmetric equations* for the line ℓ through the point $P_1(x_1, y_1, z_1)$, parallel to the vector $\mathbf{v} = a\mathbf{i} + b\mathbf{j} + c\mathbf{k}$. For example, the symmetric equations for the line of Example 1, Section 1 are

$$\frac{x - 2}{-3} = \frac{y + 3}{5} = \frac{z - 4}{-1}$$

If any of the components of $\mathbf{v}$ are zero, say $a = 0$ and $b, c \neq 0$, then $x = x_1$ for all points $P(x, y, z)$ on ℓ, and the symmetric equations of ℓ are defined to be

$$\frac{y - y_1}{b} = \frac{z - z_1}{c}, \quad x = x_1$$

There are three simultaneous linear equations involved in the symmetric equations, namely,

$$\begin{cases} \dfrac{x - x_1}{a} = \dfrac{y - y_1}{b} \\ \dfrac{x - x_1}{a} = \dfrac{z - z_1}{c} \\ \dfrac{y - y_1}{b} = \dfrac{z - z_1}{c} \end{cases} \quad (a, b, c \neq 0)$$

However, any two of these imply the third, so that only two are needed to characterize ℓ.

Example 1. Let ℓ be the line two of whose symmetric equations are equivalent to

$$\begin{cases} 2x - y + 4 = 0 \\ x + 3z - 1 = 0 \end{cases}$$

Find a set of symmetric equations for ℓ. Find a point on ℓ and a vector parallel to ℓ.

SOLUTION. We rearrange the first equation as follows:

$2x - y + 4 = 0$

$2(x + 2) = y$

$$\frac{x + 2}{1} = \frac{y}{2}$$

The second equation for ℓ must now be rearranged so that the term, $(x + 2)/1$, appears. Thus,

$x + 3z - 1 = 0$

$x + 2 = -3z + 3 = -3(z - 1)$

$$\frac{x + 2}{1} = \frac{z - 1}{-\frac{1}{3}}$$

One set of symmetric equations for ℓ is, therefore,

$$\ell : \frac{x + 2}{1} = \frac{y}{2} = \frac{z - 1}{-\frac{1}{3}}$$

A point on ℓ is $P_1(-2, 0, 1)$ and a vector parallel to ℓ is $\mathbf{v} = \mathbf{i} + 2\mathbf{j} - \frac{1}{3}\mathbf{k}$. Clearly, the symmetric equations for a line are not unique, since parametric equations are not.

Example 2. Write symmetric equations for the line ℓ through the points $P_1(2, 1, -3)$ and $P_2(-3, 1, 4)$.

SOLUTION

$$\mathbf{v} = \overrightarrow{P_1P_2} = -5\mathbf{i} + 0\mathbf{j} + 7\mathbf{k}$$

The symmetric equations for ℓ (using P_1) are

$$\frac{x - 2}{-5} = \frac{z + 3}{7}, \quad y = 1$$

Let $\mathbf{v} \parallel \ell$. The direction angles and cosines of $\mathbf{v}$ are also called direction angles and cosines of ℓ, whereas the components of $\mathbf{v}$ are called *direction numbers* of ℓ. Each line has two sets of direction angles and two sets of

Symmetric Equations of a Line; Angle Between Two Lines

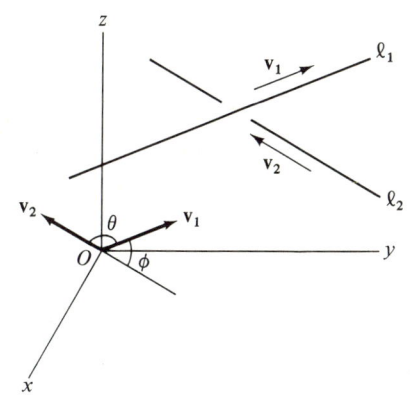

Figure 9-3

direction cosines. Any nonzero multiple of a set of direction numbers is also a set of direction numbers.

Now let $\mathbf{v}_1 \parallel \ell_1$ and $\mathbf{v} \parallel \ell_2$. We define the angle φ between ℓ_1 and ℓ_2 in terms of the angle θ between $\mathbf{v}_1$ and $\mathbf{v}_2$ as follows (see Figure 9-3):

$$\cos \varphi = |\cos \theta| = \frac{|\mathbf{v}_1 \circ \mathbf{v}_2|}{|\mathbf{v}_1| \cdot |\mathbf{v}_2|}, \quad 0 \leq \varphi \leq \frac{\pi}{2}$$

Note that the angle between two lines is defined whether or not the lines intersect. ℓ_1 and ℓ_2 are *parallel* or *perpendicular* according as $\mathbf{v}_1$ and $\mathbf{v}_2$ are parallel or perpendicular, respectively. If $\ell_1 \perp \ell_2$, we also write $\mathbf{v}_1 \perp \ell_2$. Two nonparallel lines that do not intersect are called *skew lines* (Figure 9-3).

Example 3. Show that the following lines are parallel and find the distance between them; that is, find the length of an arrow perpendicular to both and having an endpoint on each. (It can be shown that the lengths of any two such arrows are equal.)

$$\ell_1 : \begin{cases} x = 4 - 2s \\ y = -2 + s \\ z = 1 + 3s \end{cases} \qquad \ell_2 : \begin{cases} x = 2 - 3t \\ y = 0 + \frac{3}{2}t \\ z = 7 + \frac{9}{2}t \end{cases}$$

SOLUTION. Letting $\mathbf{v}_1 = -2\mathbf{i} + \mathbf{j} + 3\mathbf{k}$ and $\mathbf{v}_2 = -3\mathbf{i} + \frac{3}{2}\mathbf{j} + \frac{9}{2}\mathbf{k}$, we see that $\mathbf{v}_2 = \frac{3}{2}\mathbf{v}_1$, so that $\ell_2 \parallel \ell_1$. We choose $Q = (2, 0, 7)$ on ℓ_2. We seek a point $P(x, y, z)$ on ℓ_1 such that $\overrightarrow{QP} \perp \ell_1$. Now

$$\overrightarrow{QP} = (x - 2)\mathbf{i} + (y - 0)\mathbf{j} + (z - 7)\mathbf{k}$$
$$= (4 - 2s - 2)\mathbf{i} + (-2 + s - 0)\mathbf{j} + (1 + 3s - 7)\mathbf{k}$$

using the equations of ℓ_1. Since $\overrightarrow{QP} \circ \mathbf{v}_1 = 0$, we have

$$[(2 - 2s)\mathbf{i} + (-2 + s)\mathbf{j} + (-6 + 3s)\mathbf{k}] \circ [-2\mathbf{i} + \mathbf{j} + 3\mathbf{k}] = 0$$

Therefore,

$$-2(2 - 2s) + (-2 + s) + 3(-6 + 3s) = 0$$
$$14s - 24 = 0$$
$$s = \tfrac{12}{7}$$

The corresponding point P on ℓ_1 is $P(\tfrac{4}{7}, -\tfrac{2}{7}, \tfrac{43}{7})$. The distance between ℓ_1 and ℓ_2 is

$$d = |\overrightarrow{QP}| = [(\tfrac{4}{7} - 2)^2 + (-\tfrac{2}{7} - 0)^2 + (\tfrac{43}{7} - 7)^2]^{1/2}$$
$$= \tfrac{2}{7}\sqrt{35}$$

EXERCISE 2

1. Determine whether the point A lies on the line ℓ.

   (a) $A(-2, 3, 0);\ \ell: \begin{cases} x = 1 + 4t \\ y = -2 - t \\ z = 3 + 2t \end{cases}$

   (b) $A(1, -2, 3);\ \ell:$ through $P_1(-2, 4, -3)$ parallel to
   $\mathbf{v} = \mathbf{i} - 2\mathbf{j} + 2\mathbf{k}$

   (c) $A(-\tfrac{3}{2}, \tfrac{9}{2}, 2);\quad \ell: \begin{cases} x = -2 + t \\ y = 5 - t \\ z = 3 - 2t \end{cases}$

   (d) $A(-2, 1, 2);\ \ell:$ through $P_1(0, -4, 6)$ parallel to
   $\mathbf{v} = 2\mathbf{i} - 3\mathbf{j} + 4\mathbf{k}$

2. Write symmetric equations for the following lines.
   (a) through $P_1(0, -2, 4)$ and $P_2(\tfrac{2}{3}, -1, 3)$
   (b) through $P_1(-5, 6, 2)$, parallel to $\mathbf{v} = \mathbf{i} - 2\mathbf{j} + 4\mathbf{k}$
   (c) through $P_1(2, -3, 5)$ and $P_2(-6, -3, 2)$
   (d) $\ell: \begin{cases} x = -1 + 2t \\ y = 4 - 3t \\ z = 5t \end{cases}$
   (e) $\ell: \begin{cases} x + 2y - 2 = 0 \\ 3y - 2z + 1 = 0 \end{cases}$

   (Find a point on ℓ and a vector parallel to ℓ.)

   (f) $\ell: \begin{cases} y - 2x + 4 = 0 \\ z - 3x + 7 = 0 \end{cases}$

   (Find a point on ℓ and a vector parallel to ℓ.)
3. Find direction cosines for the following lines.
   (a) through $P_1(1, 2, -3)$ and $P_2(2, 0, -6)$

(b) through $P_1(6, -5, 1)$, parallel to $\mathbf{v} = 5\mathbf{i} - \mathbf{j} + 2\mathbf{k}$

(c) $\ell: \begin{cases} x = 5 - t \\ y = -6 + 3t \\ z = -t \end{cases}$

(d) $\ell: \dfrac{x + 2}{-1} = \dfrac{y - 3}{2} = \dfrac{z + 4}{6}$

(e) $\ell: \begin{cases} x = 4 - 2t \\ y = 6 + 3t \\ z = 5 \end{cases}$

4. Find the cosine of the angle between the following lines. State whether the lines of a given pair are parallel, perpendicular, or neither.

(a) $\begin{cases} \ell_1: \text{through } P_1(2, -1, 2) \text{ and } P_2(-1, 4, 7) \\ \ell_2: \text{through } Q_1(0, -2, 1) \text{ and } Q_2(-10, -12, 5) \end{cases}$

(b) $\begin{cases} \ell_1: \text{through } P_1(6, -2, 3), \text{ parallel to} \\ \qquad \mathbf{v} = -\mathbf{i} + 3\mathbf{j} - \mathbf{k} \\ \ell_2: \text{through } Q_1(1, -2, 6) \text{ and } Q_2(5, -14, 10) \end{cases}$

(c) $\ell_1: \begin{cases} x = 2t \\ y = -1 + 3t \\ z = 4 - t \end{cases} \quad \ell_2: \begin{cases} x = 6 - s \\ y = 2 + 2s \\ z = -1 + s \end{cases}$

5. In parts (a), (b), and (c) a point P_0 and line ℓ_1 are given.
   (1) Find the equations of the line through P_0
       (i) parallel to ℓ_1
       (ii) intersecting ℓ_1 orthogonally.
   (2) Find the distance from P_0 to ℓ_1.

(a) $P_0(-2, -2, 4); \ell_1: \begin{cases} x = -4 + 2t \\ y = 5 - t \\ z = 3 + 3t \end{cases}$

(b) $P_0(2, -3, 4); \ell_1:$ through $P_1(4, 3, -6)$, parallel to $\mathbf{v} = \mathbf{i} - \mathbf{j} + \mathbf{k}$

(c) $P_0(2, 5, -7); \ell_1: \dfrac{x + 3}{-2} = \dfrac{y - 4}{7} = \dfrac{z + 1}{1}$

6. Show that the lines of each pair are parallel and find the distance between them.

(a) $\ell_1: \begin{cases} x = 11 - 2t \\ y = 3 + t \\ z = -4 + 5t \end{cases} \quad \ell_2: \begin{cases} x = 7 - \frac{3}{5}s \\ y = 5 + \frac{3}{10}s \\ z = 1 + \frac{3}{2}s \end{cases}$

(b) $\ell_1: \dfrac{x + 1}{4} = \dfrac{y - 3}{1} = \dfrac{z + 7}{-3}$

$\ell_2: \dfrac{x - 7}{-12} = \dfrac{y + 1}{-3} = \dfrac{z - 2}{9}$

7. Given the lines

$$\ell_1: \frac{x-3}{2} = \frac{y+4}{6} = \frac{z-7}{-3}$$

$$\ell_2: \frac{x+4}{4} = \frac{y-2}{3} = \frac{z+2}{-1}$$

Find a line perpendicular to both ℓ_1 and ℓ_2.

8. Find the points in which the following line intersects the coordinate planes.

$$\ell: \begin{cases} x = 3 - 2t \\ y = 4 + t \\ z = 6 + 3t \end{cases}$$

9. Show that the following lines are *concurrent* (have a point in common), and find the point of intersection.

$$\ell_1: \begin{cases} x = 9 + 2t \\ y = 2 + t \\ z = -11 - 3t \end{cases} \quad \ell_2: \begin{cases} x = 2 + s \\ y = -1 + s \\ z = 3 + 2s \end{cases} \quad \ell_3: \begin{cases} x = 7 - 3u \\ y = -2 \\ z = -1 + u \end{cases}$$

3. PLANES

Although we call R^2 the "x–y plane," the notion of *plane* plays no role in R^2. However, as we shall see in Section 6, a plane in R^3 is essentially the same as R^2.

Definition. Let $\mathbf{v}_1$ and $\mathbf{v}_2$ be nonparallel vectors. The *plane* π determined by the point P_0 and $\mathbf{v}_1, \mathbf{v}_2$ is the set of points

$$\pi = \{P | \overrightarrow{P_0P} = s\mathbf{v}_1 + t\mathbf{v}_2, s, t \in R\}$$

A sum of the type $s\mathbf{v}_1 + t\mathbf{v}_2$ is called a *linear combination* of $\mathbf{v}_1$ and $\mathbf{v}_2$. Thus, the plane π consists of *all* points P such that the arrow $\overrightarrow{P_0P}$ represents a linear combination of $\mathbf{v}_1$ and $\mathbf{v}_2$ (Figure 9–4). When $s = t = 0$, $\overrightarrow{P_0P} = \mathbf{0}$ and $P = P_0$.

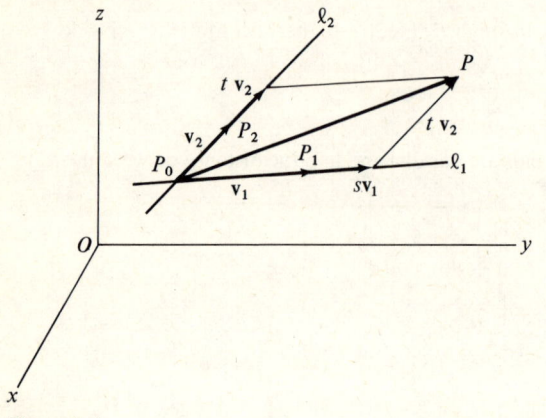

Figure 9-4

Thus, $P_0 \in \pi$. Setting $t = 0$, we obtain the line ℓ_1 through P_0 parallel to $\mathbf{v}_1$. Similarly, setting $s = 0$, we obtain the line ℓ_2 through P_0 parallel to $\mathbf{v}_2$. Thus, π contains the lines ℓ_1 and ℓ_2.

We now obtain a set of parametric equations for the plane. Let $P_0 = (x_0, y_0, z_0)$, $\mathbf{v}_1 = a_1\mathbf{i} + b_1\mathbf{j} + c_1\mathbf{k}$, and $\mathbf{v}_2 = a_2\mathbf{i} + b_2\mathbf{j} + c_2\mathbf{k}$. Then $P(x, y, z)$ lies in π iff for some $s, t \in R$, $\overrightarrow{P_0P} = s\mathbf{v}_1 + t\mathbf{v}_2$. This condition is equivalent to the set of equations

$$\pi: \begin{cases} x = x_0 + sa_1 + ta_2 \\ y = y_0 + sb_1 + tb_2 \\ z = z_0 + sc_1 + tc_2, \end{cases} \quad (s, t \in R)$$

These are called the *parametric equations* of π with parameters s and t. Note that a plane requires *two parameters* to characterize it, while a line requires only one.

We now show that the plane π may be represented by a single equation in x, y, and z. Since $\mathbf{v}_1 \nparallel \mathbf{v}_2$, the vector $\mathbf{n} = \mathbf{v}_1 \times \mathbf{v}_2$ is nonzero and is perpendicular to $\mathbf{v}_1$ and $\mathbf{v}_2$. Consequently, for every point $P(x, y, z)$ in π,

$$\overrightarrow{P_0P} \circ \mathbf{n} = (s\mathbf{v}_1 + t\mathbf{v}_2) \circ \mathbf{n} = s\mathbf{v}_1 \circ \mathbf{n} + t\mathbf{v}_2 \circ \mathbf{n} = 0$$

so $\overrightarrow{P_0P} \perp \mathbf{n}$. Conversely, if P is a point such that $\overrightarrow{P_0P} \perp \mathbf{n}$, then $\overrightarrow{P_0P} = s\mathbf{v}_1 + t\mathbf{v}_2$ for some $s, t \in R$, and so P lies in π. (See Problem 7.) Accordingly, the plane π is the set of all points $P(x, y, z)$ such that $\overrightarrow{P_0P} \circ \mathbf{n} = 0$. (This is often taken as the definition of a plane.) $\mathbf{n}$ is called a *normal vector* to π (Figure 9–5). Writing $\mathbf{n} = A\mathbf{i} + B\mathbf{j} + C\mathbf{k}$, we see that π has equation

$$\pi: A(x - x_0) + B(y - y_0) + C(z - z_0) = 0$$

This is called the *rectangular equation* of π. It may be written

$$\pi: Ax + By + Cz + D = 0$$

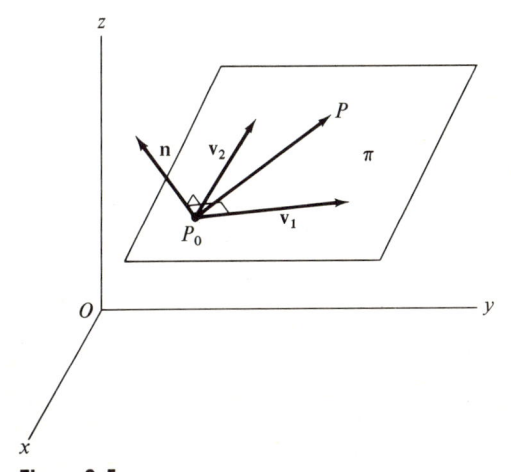

Figure 9–5

where

$$D = -(Ax_0 + By_0 + Cz_0)$$

The rectangular equation of a plane is *linear* in x, y, z, and *the coefficients of x, y, z are the components of a normal vector to the plane*. Thus, a plane is completely determined by a point P_0 and a normal vector $\mathbf{n}$.

Theorem. *Every linear equation, $Ax + By + Cz + D = 0$, with at least one of A, B, C nonzero represents a plane.*

PROOF. Let $P_0(x_0, y_0, z_0)$ and $P(x, y, z)$ be any points satisfying this equation. Then

$$Ax_0 + By_0 + Cz_0 + D = 0$$

and

$$Ax + By + Cz + D = 0$$

Subtracting, we obtain

$$A(x - x_0) + B(y - y_0) + C(z - z_0) = 0$$

which is the equation of the plane through P_0 having normal vector $\mathbf{n} = A\mathbf{i} + B\mathbf{j} + C\mathbf{k}$.

Example 1. Let $P_0 = (4, 1, -2)$, $\mathbf{v}_1 = 2\mathbf{i} - 3\mathbf{j} + \mathbf{k}$, and $\mathbf{v}_2 = 2\mathbf{j} - 4\mathbf{k}$. Give (a) parametric equations and (b) a rectangular equation of the plane π determined by P_0, $\mathbf{v}_1$, and $\mathbf{v}_2$.

SOLUTION

(a) $\pi: \begin{cases} x = 4 + 2s \\ y = 1 - 3s + 2t \\ z = -2 + s - 4t \end{cases}$

(b) $\mathbf{n} = \mathbf{v}_1 \times \mathbf{v}_2 = \begin{vmatrix} \mathbf{i} & \mathbf{j} & \mathbf{k} \\ 2 & -3 & 1 \\ 0 & 2 & -4 \end{vmatrix} = 10\mathbf{i} + 8\mathbf{j} + 4\mathbf{k}$

The rectangular equation of π is

$\pi: 10(x - 4) + 8(y - 1) + 4(z + 2) = 0,$

or, equivalently,

$\pi: 5x + 4y + 2z - 20 = 0.$

Observe that if $\mathbf{n}$ is normal to π, then so is $c\mathbf{n}$ ($c \neq 0$). Thus, any vector parallel to $\mathbf{n}$ together with any point in π may be used to write the equation of π. Any vector $\mathbf{v}$ or line ℓ parallel to $\mathbf{n}$ is said to be perpendicular to π and we write $\mathbf{v} \perp \pi$ or $\ell \perp \pi$. Any vector or line perpendicular to $\mathbf{n}$ is said to be parallel to π, and we write $\mathbf{v} \parallel \pi$ or $\ell \parallel \pi$.

The method of the following example may be used to show that any three noncollinear points determine a unique plane.

Example 2. Write the equation of the plane passing through $P_1(2, -1, 4)$, $P_2(-3, 0, 2)$, and $P_3(1, -4, 6)$.

SOLUTION. Let

$$\mathbf{v}_1 = \overrightarrow{P_1P_2} = -5\mathbf{i} + \mathbf{j} - 2\mathbf{k}$$

and

$$\mathbf{v}_2 = \overrightarrow{P_1P_3} = -\mathbf{i} - 3\mathbf{j} + 2\mathbf{k}$$

Then

$$\mathbf{n} = \mathbf{v}_1 \times \mathbf{v}_2 = -4\mathbf{i} + 12\mathbf{j} + 16\mathbf{k}$$

Using the point $P_1(2, -1, 4)$, we get the required equation

$$-4(x - 2) + 12(y + 1) + 16(z - 4) = 0$$

that is,

$$x - 3y - 4z + 11 = 0$$

Example 3. Find the parametric equations for the plane

$$\pi: 2x - 3y + z - 4 = 0$$

SOLUTION. The equation of π is $z = -2x + 3y + 4$. Let $P(x, y, z)$ be any point in π. Then parametric equations for π are

$$\pi : \begin{cases} x = x \\ y = y \\ z = -2x + 3y + 4 \end{cases}$$

with parameters x and y.

EXERCISE 3

1. Find parametric equations and a rectangular equation for the plane
   (a) determined by $P_0(2, -1, 3)$, $\mathbf{v}_1 = 2\mathbf{i} + \mathbf{j} + \mathbf{k}$, and $\mathbf{v}_2 = -\mathbf{i} + 2\mathbf{j} + 3\mathbf{k}$
   (b) through the points $P_1(2, 4, 1)$, $P_2(-2, -1, 3)$, and $P_3(0, 1, 4)$
   (c) containing $P_0(2, 1, -4)$ and the line

   $$\ell : \begin{cases} x = 6 + 4t \\ y = 2 - t \\ z = 3 + 2t \end{cases}$$

   (d) containing the lines

   $$\ell_1 : \begin{cases} x = 2 + 4s \\ y = -1 + 3s \\ z = -2s \end{cases} \quad \ell_2 : \begin{cases} x = -1 + t \\ y = 3 + 7t \\ z = 5 + 3t \end{cases}$$

(e) containing the lines
$$\ell_1: \begin{cases} x = -1 + 2s \\ y = -2 + s \\ z = 2 + 3s \end{cases} \quad \ell_2: \begin{cases} x = 3 + t \\ y = 1 + t \\ z = 9 + 2t \end{cases}$$

2. Write a set of parametric equations for the following planes.
   (a) $2x - 3y + 4z + 2 = 0$
   (b) $x + y - z = 0$
   (c) $3x + y = 0$
   (d) $4x = 8$

3. Using the methods of this section, write the equation of the plane (if it exists) determined by the following sets of points.
   (a) $P_1(2, -1, 3), P_2(1, 0, 4), P_3(-1, 4, 3)$
   (b) $P_1(0, \tfrac{1}{2}, 2), P_2(-4, 0, 6), P_3(2, -4, 3)$
   (c) $P_1(1, 2, -3), P_2(-1, 4, 6), P_3(0, 3, \tfrac{3}{2})$
   (d) $P_1(0, 2, \tfrac{1}{3}), P_2(1, 0, 4), P_3(-2, 6, -7)$

4. Given the plane $\pi: 2x - y + z = 4$ and the points $P_1(1, -1, 1)$ and $P_2(-2, 4, 12)$ in π.
   (a) Write the equations of the line ℓ_1 through P_1 and P_2 and show that every point on ℓ_1 lies in π.
   (b) Write the equations of the line through $A(5, -2, 3)$ parallel to ℓ_1.

*5. Let π be the plane $Ax + By + Cz + D = 0$. Let Q_1 and Q_2 be two distinct points in π. Show that π contains the line through Q_1 and Q_2.

6. Given the line
$$\ell_1: \begin{cases} x = 2 - 3t \\ y = -4 + t \\ z = 1 + 2t \end{cases}$$

and the plane $\pi: x - y + 2z - 3 = 0$. Show that there is a line in π which passes through $A(-4, -1, 3)$ and is parallel to ℓ_1.

*7. Prove that if $\mathbf{a} \not\parallel \mathbf{b}$ and $\mathbf{c} \perp \mathbf{a} \times \mathbf{b}$, then $\mathbf{c} = s\mathbf{a} + t\mathbf{b}$ for some scalars s and t. (*Hint:* Express the vector equation $\mathbf{c} = s\mathbf{a} + t\mathbf{b}$ as a system of three simultaneous equations in the unknowns s and t, and show that a solution exists.)

8. Let $\mathbf{n}$ be a normal vector to π. Show that if P_1 and P_2 are points in π, then $\overrightarrow{P_1P_2} \perp \mathbf{n}$ (and so $\overrightarrow{P_1P_2} \parallel \pi$).

4. THE ANGLE BETWEEN TWO PLANES; INTERSECTIONS OF PLANES

Let π_1 and π_2 be two planes with normals $\mathbf{n}_1 = A_1\mathbf{i} + B_1\mathbf{j} + C_1\mathbf{k}$ and $\mathbf{n}_2 = A_2\mathbf{i} + B_2\mathbf{j} + C_2\mathbf{k}$, respectively. The angle φ between π_1 and π_2 is defined in terms of the angle θ between $\mathbf{n}_1$ and $\mathbf{n}_2$ as follows:

The Angle Between Two Planes; Intersections of Planes

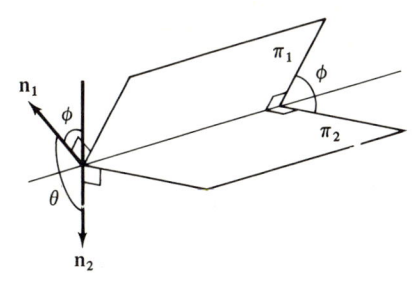

Figure 9-6

$$\cos \varphi = |\cos \theta| = \frac{|\mathbf{n}_1 \circ \mathbf{n}_2|}{|\mathbf{n}_1| \cdot |\mathbf{n}_2|} = \frac{|A_1 A_2 + B_1 B_2 + C_1 C_2|}{\sqrt{A_1^2 + B_1^2 + C_1^2} \sqrt{A_2^2 + B_2^2 + C_2^2}}$$

with $0 \le \varphi \le \frac{1}{2}\pi$ (Figure 9-6). We write $\pi_1 \perp \pi_2$ iff $\mathbf{n}_1 \perp \mathbf{n}_2$, and $\pi_1 \parallel \pi_2$ iff $\mathbf{n}_1 \parallel \mathbf{n}_2$. Thus $\pi_1 \parallel \pi_2$ iff $A_1, B_1,$ and C_1 are proportional to $A_2, B_2,$ and C_2.

Example 1. Determine whether the following pairs of planes are equal, parallel and unequal, or neither.

(a) $\begin{cases} \pi_1: 2x - 3y + 6 = 0 \\ \pi_2: 4x - 6y + 3z - 5 = 0 \end{cases}$

(b) $\begin{cases} \pi_1: 3x - 4y + 8z - 5 = 0 \\ \pi_2: \frac{3}{2}x - 2y + 4z - 6 = 0 \end{cases}$

(c) $\begin{cases} \pi_1: x - 2y + 4z - 6 = 0 \\ \pi_2: -3x + 6y - 12z + 18 = 0 \end{cases}$

SOLUTION
(a) Since $C_1 = 0$ and $C_2 = 3$, π_1 and π_2 are nonparallel, and hence unequal.

(b) $\dfrac{A_1}{A_2} = \dfrac{3}{\frac{3}{2}} = 2 = \dfrac{B_1}{B_2} = \dfrac{C_1}{C_2}$

However,

$$\frac{D_1}{D_2} = \frac{-5}{-6} = \frac{5}{6}$$

Hence π_1 and π_2 are parallel, but unequal.

(c) $\dfrac{A_1}{A_2} = \dfrac{1}{-3} = \dfrac{B_1}{B_2} = \dfrac{C_1}{C_2} = \dfrac{D_1}{D_2}$

Therefore, $\pi_1 = \pi_2$.

Lines, Planes, Spheres, and Convex Sets

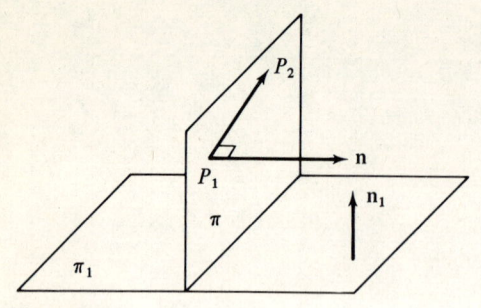

Figure 9-7

Example 2. Find the equation of the plane π passing through the points $P_1(2, -1, 3)$, $P_2(-1, -2, 1)$ and perpendicular to the plane

$$\pi_1 : x - 3y + 4z - 2 = 0$$

(Figure 9-7).

SOLUTION. Let $\mathbf{n}$ be a normal to π. $\mathbf{n}_1 = \mathbf{i} - 3\mathbf{j} + 4\mathbf{k}$ is a normal to π_1. Since P_1 and P_2 are in π, $\overrightarrow{P_1P_2} = -3\mathbf{i} - \mathbf{j} - 2\mathbf{k} \perp \mathbf{n}$. Now $\mathbf{n} \perp \mathbf{n}_1$ and $\mathbf{n} \perp \overrightarrow{P_1P_2}$. Therefore, $\mathbf{n} \parallel \mathbf{n}_1 \times \overrightarrow{P_1P_2}$. Hence $\mathbf{n}_1 \times \overrightarrow{P_1P_2}$ is a normal to π.

$$\mathbf{n}_1 \times \overrightarrow{P_1P_2} = \begin{vmatrix} \mathbf{i} & \mathbf{j} & \mathbf{k} \\ 1 & -3 & 4 \\ -3 & -1 & -2 \end{vmatrix} = 10\mathbf{i} - 10\mathbf{j} - 10\mathbf{k}$$

Thus, a suitable normal to π is $\mathbf{n} = \mathbf{i} - \mathbf{j} - \mathbf{k}$. Using P_1 to write the equation of π, we have

$$\pi : 1 \cdot (x - 2) - 1 \cdot (y + 1) - 1 \cdot (z - 3) = 0$$

that is,

$$\pi : x - y - z = 0$$

The student should verify that π has the required properties.

The fact that two nonparallel planes π_1 and π_2 with normals $\mathbf{n}_1$ and $\mathbf{n}_2$ intersect in a line parallel to $\mathbf{n}_1 \times \mathbf{n}_2$ may be shown as in the following example (since the line of intersection lies in both planes, it is perpendicular to both normals, hence parallel to $\mathbf{n}_1 \times \mathbf{n}_2$).

Example 3. Find the line of intersection ℓ of the planes

$$\pi_1 : x - 3y + z - 4 = 0 \quad \text{and} \quad \pi_2 : 2x + y - 3z + 1 = 0$$

SOLUTION. (Note that $\pi_1 \not\parallel \pi_2$.) A point $P(x, y, z)$ lies on ℓ iff it satisfies both equations. Since the determinant of the coefficients of x and y is nonzero, we may solve simultaneously for x and y in terms of z:

$$\begin{cases} x - 3y = -z + 4 \\ 2x + y = 3z - 1 \end{cases}$$

The solutions are

$$\begin{cases} x = \frac{1}{7} + \frac{8}{7}z \\ y = -\frac{9}{7} + \frac{5}{7}z \end{cases}$$

Thus, $P(x, y, z)$ lies on the intersection of π_1 and π_2 iff

$x = \frac{1}{7} + \frac{8}{7}z$
$y = -\frac{9}{7} + \frac{5}{7}z$
$z = 0 + 1 \cdot z$

These are the parametric equations (with parameter z) of the line ℓ through $P_0(\frac{1}{7}, -\frac{9}{7}, 0)$ parallel to $\mathbf{v} = \frac{8}{7}\mathbf{i} + \frac{5}{7}\mathbf{j} + \mathbf{k}$. Observe that

$$\mathbf{n}_1 \times \mathbf{n}_2 = \begin{matrix} \mathbf{i} & \mathbf{j} & \mathbf{k} \\ 1 & -3 & 1 \\ 2 & 1 & -3 \end{matrix} = 8\mathbf{i} + 5\mathbf{j} + 7\mathbf{k} = 7\mathbf{v}$$

so that $\mathbf{v} \parallel \mathbf{n}_1 \times \mathbf{n}_2$.

REMARK. A line in R^3 is often characterized as the intersection of two nonparallel planes, that is, as the set of simultaneous solutions of two linear equations in x, y, and z.

Traces in the Coordinate Planes—Intercept Form. Recall that the equations of the coordinate planes are $x = 0$, $y = 0$, and $z = 0$. Thus, the lines of intersection of the plane

$\pi: Ax + By + Cz + D = 0$

with the coordinate planes (if they exist) are

$Ax + By + Cz + D = 0 \quad (z = 0)$
$Ax + By + Cz + D = 0 \quad (y = 0)$
$Ax + By + Cz + D = 0 \quad (x = 0)$

They are called the *traces* of π in the coordinate planes. Since $z = 0$ in the x-y trace, we often write simply $Ax + By + D = 0$—x-y trace. The same holds for the other traces.

Example 4. Find the traces of the plane

$\pi: x + 2y + 3z - 12 = 0$

and sketch.

SOLUTION. Setting x, y, and z in turn equal to zero, we obtain (see Figure 9-8)

$2y + 3z - 12 = 0$—y-z trace
$x + 3z - 12 = 0$—x-z trace
$x + 2y - 12 = 0$—x-y trace

Lines, Planes, Spheres, and Convex Sets

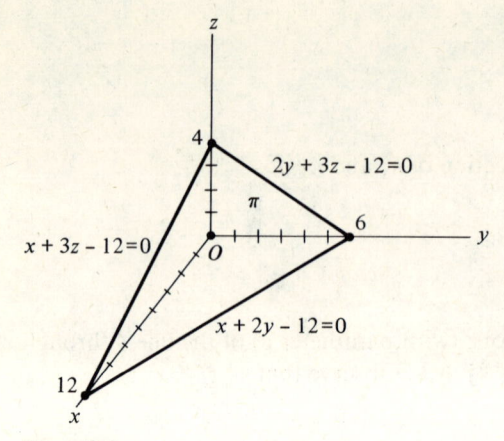

Figure 9-8

If the points $(a, 0, 0)$, $(0, b, 0)$, and $(0, 0, c)$ lie on the plane (and hence are the intersections of the plane with the coordinate axes), then a, b, and c are called, respectively, the x-, y-, and z-*intercepts* of the plane. In Example 4 they are, respectively, 12, 6, 4. It is not difficult to show that if a, b, and c are nonzero, the equation of π may be written in the following *intercept form* (see Problem 2):

$$\frac{x}{a} + \frac{y}{b} + \frac{z}{c} = 1$$

The intercepts (if they exist) are found by setting two of the variables equal to zero and solving for the third.

EXERCISE 4

1. Determine whether the following pairs of planes are parallel, equal, or neither. If neither, find the parametric equations of the line of intersection.

   (a) $\begin{cases} \pi_1 : x - 3y + 4z - 2 = 0 \\ \pi_2 : 2x - y + 1 = 0 \end{cases}$

   (b) $\begin{cases} \pi_1 : y - 4z + 3 = 0 \\ \pi_2 : 2y + z - 6 = 0 \end{cases}$

   (c) $\begin{cases} \pi_1 : x + y + z = 0 \\ \pi_2 : 2x - 3y + 4z = 5 \end{cases}$

   (d) $\begin{cases} \pi_1 : 2x - 4y + 3z = 6 \\ \pi_2 : 7x - 14y + \frac{21}{2}z = 3 \end{cases}$

2. (a) Derive the intercept form of the equation of a plane.
   (b) Write the equations of the following planes in intercept form and sketch.
      (1) $x + y + z + 4 = 0$

(2) $x - 2y + 3z = 6$
(3) $2x - 4y + 3z + 12 = 0$
(4) $2x - 4y - 6 = 0$
(5) $2y - 5z = 10$

3. Determine whether the following pairs of planes are parallel, perpendicular, or neither. If neither, find cos φ, where φ is the angle between the planes.

(a) $\begin{cases} \pi_1: 3x - 2y + z - 4 = 0 \\ \pi_2: x - 4y - 11z - 2 = 0 \end{cases}$

(b) $\begin{cases} \pi_1: 2x - 4y + 6z - 7 = 0 \\ \pi_2: x - 2y + 3z + 6 = 0 \end{cases}$

(c) $\begin{cases} \pi_1: x + y + z - 2 = 0 \\ \pi_2: 2x - y + 2z + 3 = 0 \end{cases}$

(d) $\begin{cases} \pi_1: 2x - y = 4 \\ \pi_2: 3x - y + 2z = 1 \end{cases}$

(e) $\begin{cases} \pi_1: 2y - z = 4 \\ \pi_2: x + 2z = 3 \end{cases}$

4. Find the equation of the plane satisfying the following condition.
   (a) through the point $A(4, 0, -2)$ and perpendicular to the line through $B(2, 1, -4)$ and $C(0, -2, 3)$
   (b) through the point $A(1, -2, 3)$ and parallel to the plane $4x - 2y + 3z = 2$
   (c) through the points $A(2, 1, 3)$, $B(4, 1, -1)$, and perpendicular to the plane $x - 2y + 3z = 1$.
   (d) perpendicular to the segment joining the points $A(2, -4, 3)$, $B(1, -3, 4)$, and through the midpoint of $\overline{AB}$
   (e) through the point $A(-1, 2, 3)$ and containing the line of intersection of the planes $2x - y + 3z - 4 = 0$ and $x + y - z - 2 = 0$
   (f) perpendicular to the planes $2x - y + 3z - 4 = 0$ and $x + y - z - 2 = 0$ and passing through the origin

5. For each of the following pairs of lines, find the line intersecting the two lines perpendicularly, and find the minimum distance between the lines. (If the lines intersect, find the common perpendicular through the point of intersection.)

(a) $\ell_1: \begin{cases} x = 1 - s \\ y = 5 + 3s \\ z = 1 - 2s \end{cases}$ ℓ_2: the x-axis

(b) $\ell_1: \begin{cases} x = -2 + 2s \\ y = -3 + s \\ z = 4 + s \end{cases}$ ℓ_2: the y-axis

(c) $\ell_1: \begin{cases} x = 6 + 2s \\ y = -2 - 3s \\ z = -2 + s \end{cases}$ $\ell_2: \begin{cases} x = 5 - t \\ y = -1 + 2t \\ z = -1 - 2t \end{cases}$

(d) $\ell_1: \begin{cases} x = 1 + s \\ y = 2 - 2s \\ z = -7 + 4s \end{cases}$ $\ell_2: \begin{cases} x = -3 + 2t \\ y = -7 + 3t \\ z = t \end{cases}$

5. THE NORMAL FORM OF THE EQUATION OF A PLANE; FAMILIES OF PLANES

For a given point $P_0(x_0, y_0, z_0)$ and plane $\pi: Ax + By + Cz + D = 0$, the line ℓ through P_0 and perpendicular to π has equations

$$\ell: \begin{cases} x = x_0 + At \\ y = y_0 + Bt \\ z = z_0 + Ct \end{cases}$$

ℓ intersects π in the point $P_1(x_1, y_1, z_1)$ given by (see Problem 8)

$$t_1 = -\frac{Ax_0 + By_0 + Cz_0 + D}{A^2 + B^2 + C^2}$$

The distance d from P_0 to π is defined to be $|\overline{P_0 P_1}|$. Thus,

$$d = |\overline{P_0 P_1}| = [(x_1 - x_0)^2 + (y_1 - y_0)^2 + (z_1 - z_0)^2]^{1/2}$$
$$= |t_1|(A^2 + B^2 + C^2)^{1/2}$$

$$d = \frac{|Ax_0 + By_0 + Cz_0 + D|}{\sqrt{A^2 + B^2 + C^2}}$$

Note that $d = 0$ iff P_0 lies in π.

The distance d from the *origin* to the plane is, thus, given by (see Figure 9-9)

$$d = \frac{|D|}{\sqrt{A^2 + B^2 + C^2}}$$

The line ℓ through the origin and perpendicular to π intersects π in the point

$$P_1\left(\frac{-AD}{A^2 + B^2 + C^2}, \frac{-BD}{A^2 + B^2 + C^2}, \frac{-CD}{A^2 + B^2 + C^2}\right)$$

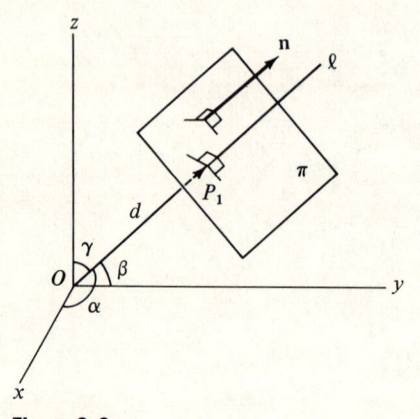

Figure 9-9

The Normal Form of the Equation of a Plane; Families of Planes

Assume now that π does not pass through the origin (that is, $D \neq 0$). Then the arrow $\overrightarrow{OP_1}$ has direction cosines (Figure 9–9)

$$\cos \alpha = \frac{A}{\sqrt{A^2 + B^2 + C^2}} \cdot \left(\frac{-D}{|D|}\right)$$

$$\cos \beta = \frac{B}{\sqrt{A^2 + B^2 + C^2}} \cdot \left(\frac{-D}{|D|}\right)$$

$$\cos \gamma = \frac{C}{\sqrt{A^2 + B^2 + C^2}} \cdot \left(\frac{-D}{|D|}\right)$$

The equation for π may be written

$$\pi: \frac{A}{\sqrt{A^2 + B^2 + C^2}} x + \frac{B}{\sqrt{A^2 + B^2 + C^2}} y + \frac{C}{\sqrt{A^2 + B^2 + C^2}} z + \frac{D}{\sqrt{A^2 + B^2 + C^2}} = 0$$

In terms of the direction cosines of $\overrightarrow{OP_1}$, we have

$$\pi: \left(\frac{|D|}{-D}\right)(\cos \alpha)x + \left(\frac{|D|}{-D}\right)(\cos \beta)y + \left(\frac{|D|}{-D}\right)(\cos \gamma)z + \frac{d}{|D|}D = 0$$

that is,

$$x \cos \alpha + y \cos \beta + z \cos \gamma + \left(\frac{-D}{|D|}\right)\left(\frac{D}{|D|}\right)d = 0$$

Since

$$\left(\frac{-D}{|D|}\right)\left(\frac{D}{|D|}\right) = \frac{-D^2}{D^2} = -1$$

we have

$$\pi: x \cos \alpha + y \cos \beta + z \cos \gamma - d = 0$$

This last equation is called the *normal form* of the equation of π. $\cos \alpha$, $\cos \beta$, and $\cos \gamma$ are the direction cosines of a *normal vector* with initial point O and terminal point on π, and d is the distance from O to π. Observe that the rectangular equation $Ax + By + Cz + D = 0$ may be reduced to the normal form by dividing through by $-\sqrt{A^2 + B^2 + C^2}$ if $D > 0$ and by $\sqrt{A^2 + B^2 + C^2}$ if $D < 0$. Note that if $D = 0$, there are two possible normal forms, obtained by dividing by $\sqrt{A^2 + B^2 + C^2}$ or $-\sqrt{A^2 + B^2 + C^2}$.

Example 1. Given the plane $\pi: 2x + y - z - 4 = 0$.
(a) Find the distance from $A(3, -1, 2)$ to π.
(b) Reduce to normal form and find the distance from the origin to π.

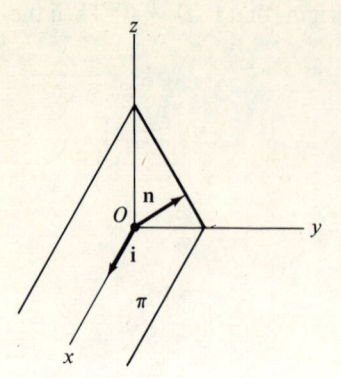

Figure 9-10 $By+CZ+D = 0\ (B, C \neq 0)$.

SOLUTION

(a) $d = \dfrac{|2\cdot 3 + (-1) - 2 - 4|}{\sqrt{2^2 + 1^2 + (-1)^2}} = \dfrac{1}{\sqrt{6}}$

(b) Since $D = -4 < 0$, we divide through by $\sqrt{2^2 + 1^2 + (-1)^2} = \sqrt{6}$. Thus, the normal form is $(2/\sqrt{6})x + y/\sqrt{6} - z/\sqrt{6} - 4/\sqrt{6} = 0$. The distance from the origin to π is $d = 4/\sqrt{6}$. $\cos\alpha = 2/\sqrt{6}$, $\cos\beta = 1/\sqrt{6}$, and $\cos\gamma = -1/\sqrt{6}$.

The following is a convenient device for obtaining the distance d from $P_0(x_0, y_0, z_0)$ to π: $Ax + By + Cz + D = 0$. Reduce the equation of π to normal form and substitute (x_0, y_0, z_0) for (x, y, z). Then d is the absolute value of the left-hand side of the equation.

Observe that if $A = 0$, $\cos\alpha = 0$ and so $\alpha = \frac{1}{2}\pi$. Thus, a normal $\mathbf{n}$ to π is perpendicular to the x-axis; that is, $\mathbf{n} \perp \mathbf{i}$. Accordingly, π is parallel to the x-axis (Figure 9-10). The case is similar for $B = 0$ or $C = 0$. Thus, *a plane is parallel to those coordinate axes for which the corresponding variable is absent from the equation of the plane.* For example, $2x - 3z + 4 = 0$ is the equation of a plane parallel to the y-axis.

Families of Planes. By a *family* of planes we mean a set of planes. Certain families can be characterized by writing a linear equation in x, y, and z with certain of the coefficients unspecified. A few examples will suffice to illustrate the idea.

Example 2. The family of all planes parallel to the plane

$\pi_1: 2x - 3y + z - 4 = 0$

has equation $2x - 3y + z + D = 0$, where D is arbitrary. (D is called a *parameter* for the family of planes.) That member of the family passing through the point $A(-2, 1, 3)$ is obtained by substituting the coordinates of A in the equation of the family and solving for D. Thus,

$$2(-2) - 3(1) + 3 + D = 0$$

Therefore,

$$D = 4$$

The desired member is $\pi: 2x - 3y + z + 4 = 0$.

Example 3. Write the equation of the family of all planes whose distance from the origin is 5. Find those members parallel to the plane

$$\pi_1: x + y + z + 4 = 0$$

SOLUTION. The equation of the family in normal form is

$$x \cos \alpha + y \cos \beta + z \cos \gamma - 5 = 0$$

The equation of π_1 in normal form is

$$\pi_1: \frac{x}{-\sqrt{3}} + \frac{y}{-\sqrt{3}} + \frac{z}{-\sqrt{3}} - \frac{4}{\sqrt{3}} = 0$$

A plane parallel to π_1 has normal vector with direction cosines

$$\left(-\frac{1}{\sqrt{3}}, -\frac{1}{\sqrt{3}}, -\frac{1}{\sqrt{3}}\right) \quad \text{or} \quad \left(\frac{1}{\sqrt{3}}, \frac{1}{\sqrt{3}}, \frac{1}{\sqrt{3}}\right)$$

Thus, there are two members of the family parallel to π_1. They are

$$-\frac{1}{\sqrt{3}}x - \frac{1}{\sqrt{3}}y - \frac{1}{\sqrt{3}}z - 5 = 0$$

and

$$\frac{1}{\sqrt{3}}x + \frac{1}{\sqrt{3}}y + \frac{1}{\sqrt{3}}z - 5 = 0$$

The family of planes containing the line of intersection of two given nonparallel planes is an especially interesting example. Let

$$\pi_1: A_1 x + B_1 y + C_1 z + D_1 = 0$$

and

$$\pi_2: A_2 x + B_2 y + C_2 z + D_2 = 0$$

be two nonparallel planes. The corresponding normals $\mathbf{n}_1$ and $\mathbf{n}_2$ are, thus, nonparallel vectors. Let h and k be two numbers not both zero. Consider the equation

$$h(A_1 x + B_1 y + C_1 z + D_1) + k(A_2 x + B_2 y + C_2 z + D_2) = 0 \qquad (1)$$

If $P_0(x_0, y_0, z_0)$ is any point on the line ℓ of intersection of π_1 and π_2, we have

$$h(A_1 x_0 + B_1 y_0 + C_1 z_0 + D_1) + k(A_2 x_0 + B_2 y_0 + C_2 z_0 + D_2)$$
$$= h \cdot 0 + k \cdot 0 = 0$$

Hence for all h and k, the graph of Equation (1) contains ℓ. Equation (1) may be written in the equivalent form

$$(hA_1 + kA_2)x + (hB_1 + kB_2)y + (hC_1 + kC_2)z + (hD_1 + kD_2) = 0 \quad (2)$$

Since Equation (2) is a linear equation, it represents a plane, provided the coefficients of x, y, and z are not all zero. However, it is easy to see that if all these coefficients are zero, then $\pi_1 \parallel \pi_2$, contrary to hypothesis. Thus, Equation (1) represents a family of planes containing the line of intersection of π_1 and π_2. (In fact, it contains all such planes.) Since $h \neq 0$ or $k \neq 0$, in practice we need only to determine the ratio h/k or k/h in order to select a particular member of the family.

Example 4. Given $\pi_1: x - 2y + 3z - 4 = 0$ and $\pi_2: 2x + y - 3z - 1 = 0$. Find the equation of the plane containing the line of intersection of π_1 and π_2 and having x-intercept 5.

SOLUTION. The equation of the family is

$$h(x - 2y + 3z - 4) + k(2x + y - 3z - 1) = 0$$

Substituting the coordinates of the point $(5, 0, 0)$, we obtain

$h + 9k = 0$
$h = -9k \quad (h, k \neq 0)$

Substituting for h in the equation of the family,

$$-9k(x - 2y + 3z - 4) + k(2x + y - 3z - 1) = 0$$

Dividing through by $k \neq 0$ and simplifying, we have

$$7x - 19y + 30z - 35 = 0$$

as the desired member of the family.

EXERCISE 5

1. Find the distance from the given point to the given plane.
   (a) $P_0(1, 1, 1); \pi: 3x - 2y + z - 4 = 0$
   (b) $P_0(1, 2, -1); \pi: x - y + 2z + 3 = 0$
   (c) $P_0(0, 1, -4); \pi: 2x - 3y = 4$
   (d) $P_0(2, -3, 0); \pi: 3x - y + 4z - 1 = 0$
   (e) $P_0(0, 0, 0); \pi: x + 3y - z + 2 = 0$

2. Reduce to normal form. Find the distance from the origin to the plane. Find the direction cosines of a normal vector with initial point at the origin and terminal point on the plane. Sketch the planes.
   (a) $6x - 3y + 5z - 30 = 0$
   (b) $x + 4y - 8 = 0$
   (c) $-3x + 2y - z = 0$

(d) $3y + 4z + 12 = 0$
(e) $2x + 4y - 3z = 12$
3. (a) Find the distance between the parallel planes $2x - y + 2z - 4 = 0$ and $6x - 3y + 6z + 2 = 0$.
  (b) Find the plane half way between the parallel planes $4x - 6y + 2z - 3 = 0$ and $2x - 3y + z - 1 = 0$.
4. Find the equation of a plane which passes through $P_1(0, 0, 1)$, $P_2(1, -1, 1)$ and makes an angle of $60°$ with the plane $x + z = 0$.
5. (a) Write the equation of the family of planes having x-intercept 5, y-intercept 2, and nonzero z-intercept. Find that member of the family perpendicular to the plane $3x - 2y + z = 4$.
  (b) Write the equation of the family of planes parallel to the plane $5x - 2y + 3z = 1$. Find that member of the family whose distance from the origin is $\sqrt{19}$.
6. Find the equation of the family of planes containing the line of intersection of the planes $2x + 2y - z - 4 = 0$ and $x - 3y + z - 2 = 0$. Find that member of the family
   (a) with z-intercept 3
   (b) passing through the origin
   (c) perpendicular to the plane $x + y + z = 7$
   (d) parallel to the line through $P_1(2, 4, 3)$ and $P_2(5, 2, 1)$.
7. The bisector of the angle between two nonparallel planes π_1 and π_2 is defined to be the plane π such that each point on π is equidistant from π_1 and π_2. There are two angle bisectors for each pair of nonparallel planes. Find the angle bisectors for each of the following pairs of planes.
   (a) $x - 2y + 3z - 5 = 0$; $3x - y + 2z - 6 = 0$
   (b) $3x + 5y - 4z - 2 = 0$; $x + z + 4 = 0$
8. Complete the details of the derivation of the formula for the distance from a point to a plane. That is, derive the formulas for t_1 and d.

* 6. THE RELATIONSHIP BETWEEN A PLANE AND R^2

It is common practice to regard a plane in R^3 as being essentially the same as R^2. We picture both as "flat surfaces," and tacitly assume that every statement that is true about a figure in R^2 is also true for the "same figure" in a plane in R^3. For instance, since the law of cosines holds for triangles in R^2, we assume it also holds for "triangles" in a plane in R^3. It is the purpose of this section to point out briefly why this practice is justified.

Consider first the x–y plane π_{xy}. Every point in this plane is of the form $(x, y, 0)$. There is a natural correspondence between points in π_{xy} and points in R^2, namely,

$P(x, y, 0) \leftrightarrow P'(x, y)$

This is a *one-to-one correspondence* having the following important property: If $P_1(x_1, y_1, 0)$ and $P_2(x_2, y_2, 0)$ are points in π_{xy} corresponding, respectively, to $P_1'(x_1, y_1)$ and $P_2'(x_2, y_2)$ in R^2, then

$$|\overline{P_1P_2}| = |\overline{P_1'P_2'}| = \sqrt{(x_2 - x_1)^2 + (y_2 - y_1)^2}$$

Thus, this correspondence preserves the distance between points and hence is an *isometry*. We say that π_{xy} and R^2 are *equivalent*. In a similar fashion, we may show that the coordinate planes π_{xz} and π_{yz} are equivalent to R^2. (See Problem 1.)

The problem of showing that an arbitrary plane π is equivalent to R^2 is more involved. We give only a brief discussion of it. Let π have equation

$$\pi: A(x - x_0) + B(y - y_0) + C(z - z_0) = 0$$

so that the point $P_0(x_0, y_0, z_0)$ lies in π. If we assume $C \neq 0$ (the other cases are similar), this equation may be written

$$\pi: z - z_0 = a(x - x_0) + b(y - y_0)$$

where $a = -A/C$ and $b = -B/C$. We seek a one-to-one distance and angle preserving correspondence between π and R^2. Because the derivation of such a correspondence is somewhat involved, we simply give the results. One such correspondence is the following. If $P(x, y, z)$ is any point in π, let its image $P'(x', y')$ in R^2 be given by the formulas

$$\begin{cases} x' = \sqrt{1 + a^2}\,(x - x_0) + \dfrac{ab}{\sqrt{1 + a^2}}(y - y_0) \\ y' = \dfrac{\sqrt{1 + a^2 + b^2}}{\sqrt{1 + a^2}}(y - y_0) \end{cases} \quad (1)$$

It is left as an exercise to show that the correspondence $P(x, y, z) \leftrightarrow P'(x', y')$ is one to one and preserves distances and dot products. (See Problem 2.) Note that P_0 corresponds to $O' = (0, 0)$ in R^2 (Figure 9-11).

Because of this isometry between π and R^2, we may transfer all the geometry and trigonometry of R^2 to any plane in R^3. For example, let A, B, and C be any three noncollinear points in R^3. These points determine a plane π (Figure

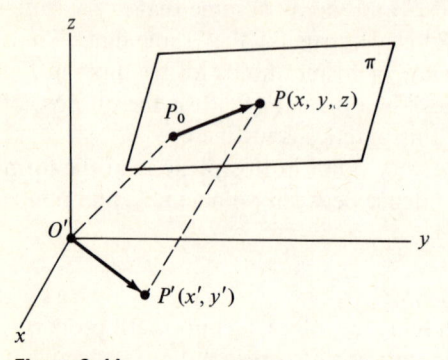

Figure 9-11

The Relationship Between a Plane and R^2

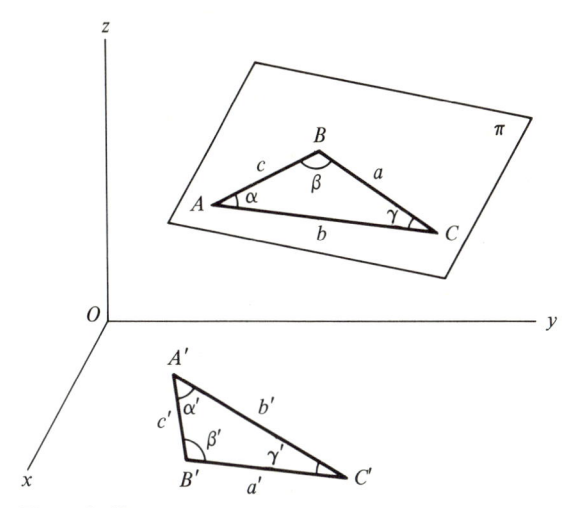

Figure 9-12

9-12). We define the triangle with vertices A, B, and C as in R^2, namely, $\triangle ABC = \overline{AB} \cup \overline{BC} \cup \overline{CA}$. $\triangle ABC$ is contained in π (why?). Let $A \leftrightarrow A'$, $B \leftrightarrow B'$, and $C \leftrightarrow C'$ under the isometry. Since the correspondence preserves distances, A', B', and C' are noncollinear, and so they are the vertices of a triangle in R^2 (Figure 9-12). We have $|\overline{AB}| = |\overline{A'B'}|, |\overline{BC}| = |\overline{B'C'}|$, and $|\overline{CA}| = |\overline{C'A'}|$. We say $\triangle ABC$ is *congruent* to $\triangle A'B'C'$, and we write $\triangle ABC \cong \triangle A'B'C'$. Furthermore, corresponding angles are equal (Figure 9-12). Denoting the side lengths by lower-case letters (Figure 9-12), we have, by the law of cosines,

$$a'^2 = b'^2 + c'^2 - 2b'c' \cos \alpha'$$

Hence for $\triangle ABC$ in π, we have

$$a^2 = b^2 + c^2 - 2bc \cos \alpha$$

Similar equations hold for angles β and γ. Thus, the *law of cosines* holds for triangles in R^3. Similarly, the *law of sines* and the *Pythagorean theorem* hold in R^3. We also have $0 < \alpha, \beta, \gamma < \pi$ and $\alpha + \beta + \gamma = \pi$.

EXERCISE 6

1. Show that there is a one-to-one correspondence between π_{xz} and R^2 which preserves distance. Do the same for π_{yz}.
*2. (a) Prove that Equations (1) establish a one-to-one correspondence between the plane π and R^2; that is, to each point $P(x, y, z)$ in π there corresponds exactly one point $P'(x', y')$ satisfying Equations (1), and vice versa.

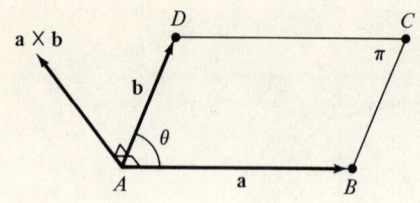

Figure 9-13

(b) Prove that the correspondence (Equation (1)) preserves distances; that is, if $P_1(x_1, y_1, z_1)$, $P_2(x_2, y_2, z_2)$ are in π and $P'_1(x'_1, y'_1)$, $P'_2(x'_2, y'_2)$ are the corresponding points in R^2, then $|\overline{P_1 P_2}| = |\overline{P'_1 P'_2}|$.

(c) Prove that the correspondence (Equations (1)) preserves dot products; that is, using the notation of Part (b), show that

$$\overrightarrow{P_0 P_1} \circ \overrightarrow{P_0 P_2} = \overrightarrow{O' P'_1} \circ \overrightarrow{O' P'_2}$$

where $O' = (0, 0)$ is the origin in R^2. (*Hint:* If $P(x, y, z)$ is in π, then $\overrightarrow{P_0 P} = (x - x_0)\mathbf{i} + (y - y_0)\mathbf{j} + [a(x - x_0) + b(y - y_0)]\mathbf{k}$.)

*3. Given the plane π: $3(x - 2) - 2(y + 3) + (z - 1) = 0$.
  (a) Obtain Equations (1) for π.
  (b) Find the points P'_1, P'_2, and P'_3 in R^2 corresponding to the following points in π by means of Equation (1): $P_1(0, 0, 13)$, $P_2(1, -1, 8)$, and $P_3(2, 1, 9)$. Verify that $\triangle P_1 P_2 P_3$ is congruent to $\triangle P'_1 P'_2 P'_3$.
  (c) Find the points P_1, P_2, and P_3 in π corresponding to the following points in R^2 by means of Equations (1): $P'_1(0, 0)$, $P'_2(1, -2)$, and $P'_3(-3, 4)$. Verify that $\triangle P_1 P_2 P_3$ is congruent to $\triangle P'_1 P'_2 P'_3$.

4. Let A, B, C, and D be the vertices of a parallelogram in the plane π. Let $\overrightarrow{AB} = \mathbf{a}$ and $\overrightarrow{AD} = \mathbf{b}$ (Figure 9-13). Prove that the area of the parallelogram is $|\mathbf{a} \times \mathbf{b}|$.

7. SPHERES AND TANGENT PLANES

Since the concepts presented here are direct extensions of those treated in Sections 5 to 7 of Chapter 4, we shall give only a brief discussion of them.

Definition. The *sphere* with *center* $P_0(x_0, y_0, z_0)$ and *radius* $r \geq 0$ is the set of points

$$S_r(P_0) = \{P(x, y, z) \, | \, |\overrightarrow{P_0 P}| = r\}$$

If $r = 0$, the sphere consists only of the point P_0, and is called a "point sphere." The point $P(x, y, z)$ is on the sphere iff

$$(x - x_0)^2 + (y - y_0)^2 + (z - z_0)^2 = r^2$$

This is called the *standard form* of the equation of the sphere. By squaring the binomials and collecting terms, we may put the equation in the form

$$Ax^2 + Ay^2 + Az^2 + Bx + Cy + Dz + E = 0 \qquad (A \neq 0)$$

which is called the *general form* of the equation of a sphere. The proof that any equation of this form (with $A \neq 0$) represents a sphere or the void set proceeds exactly as in the case of the circle. (See Chapter 4.)

Example 1. Reduce the following equation to standard form and find the center and radius (if the equation represents a sphere):

$$2x^2 + 2y^2 + 2z^2 - 4x + 6y + z + 15 = 0$$

SOLUTION. Dividing by 2 and completing the squares, we obtain

$$x^2 + y^2 + z^2 - 2x + 3y + \frac{z}{2} + \frac{15}{2} = 0$$

$$(x-1)^2 + \left(y + \frac{3}{2}\right)^2 + \left(z + \frac{1}{4}\right)^2 = -\frac{67}{16}$$

Since the right member is negative, the equation represents the empty set.

A *diameter* of a sphere is a line segment passing through the center and having endpoints on the sphere. Clearly, the length of a diameter is $d = 2r$ and its midpoint is the center of the sphere.

Example 2. Find the equation of the sphere having a diameter with endpoints $P_1(1, -3, 2)$ and $P_2(3, 1, -6)$.

SOLUTION. The center of the sphere is the midpoint M of $\overline{P_1 P_2}$. Thus, $M = (2, -1, -2)$. The radius is $r = |\overline{MP_1}| = \sqrt{21}$. The equation in standard form is

$$(x-2)^2 + (y+1)^2 + (z+2)^2 = 21$$

The general form is

$$x^2 + y^2 + z^2 - 4x + 2y + 4z - 12 = 0$$

Definition. Let $S_r(P_0)$ be the sphere with center P_0 and radius $r > 0$. The plane π is *tangent* to the sphere at P_1 iff P_1 lies on both the sphere and the plane, and the arrow $\overrightarrow{P_0 P_1}$ is perpendicular to π (Figure 9–14).

Note: It follows immediately from the definition that the plane π is tangent to the sphere $S_r(P_0)$ at some point iff the distance from P_0 to π equals r.

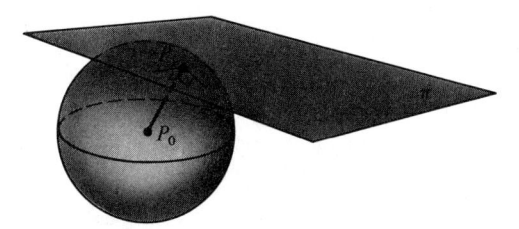

Figure 9-14

The following theorem gives an alternate characterization of a tangent plane.

Theorem. *The plane π is tangent to the sphere $S_r(P_0)$ $(r > 0)$ at P_1 iff P_1 is the only point common to the plane and the sphere.*

PROOF. (See Problem 9.)

Example 3. Find the equation of the sphere with center $P_0(2, 1, -1)$ and tangent to the plane
$$\pi: x - 2y + z + 7 = 0$$

SOLUTION. The radius of the sphere is the distance from P_0 to π. Thus,
$$r = \left|\frac{2 - 2\cdot 1 - 1 + 7}{\sqrt{6}}\right| = \sqrt{6}$$
and the equation of the sphere is
$$(x - 2)^2 + (y - 1)^2 + (z + 1)^2 = 6$$

EXERCISE 7

1. Write the equation of the sphere with the given center P_0 and radius r. Give both the standard form and the general form.
    (a) $P_0(5, -2, 1); r = 6$
    (b) $P_0(\frac{2}{3}, 4, -\frac{1}{2}); r = \sqrt{3}$
    (c) $P_0(-4, -2, -3); r = 0$
    (d) $P_0(\frac{2}{3}, -5, \sqrt{2}); r = \sqrt[3]{4}$

2. Reduce to standard form. Where appropriate give the center and radius of the sphere.
    (a) $x^2 + y^2 + z^2 - 2x + 6y - 2z + 9 = 0$
    (b) $9x^2 + 9y^2 + 9z^2 + 12x - 6y - 72z + 122 = 0$
    (c) $x^2 + y^2 + z^2 - 2x - 4y - 6z + 18 = 0$
    (d) $4x^2 + 4y^2 + 4z^2 + 4x - 12y + 20z + 34 = 0$

3. (a) Write the equation of the plane tangent to the sphere
    $$(x - 2)^2 + (y + 3)^2 + (z - 1)^2 = 5 \quad \text{at} \quad P_1(2, -4, 3)$$
    (b) Find the equation of the sphere with center at $P_0(4, 1, -6)$ and tangent to the plane $2x - 3y + 2z - 10 = 0$.
    *(c) Find the equation of the plane passing through the points $A(2, -1, 6), B(1, 0, 5)$ and tangent to the sphere $x^2 + y^2 + z^2 = 6$.

4. A *chord* of a sphere is a line segment whose endpoints are on the sphere. Prove that a line which passes through the center of a sphere and the midpoint of a chord is perpendicular to the chord.

5. Determine whether the given plane is tangent to the given sphere.
    (a) $x + 2y - 2z + 9 = 0; x^2 + y^2 + z^2 = 9$
    (b) $2x - y + 3z = 7; x^2 + y^2 + z^2 = 4$

6. Find the points of intersection (if any) of the given line and sphere.

(a) $\ell: \begin{cases} x = -4 + 2t \\ y = 1 + 2t; \\ z = 3 - t \end{cases} \quad x^2 + y^2 + z^2 = 17$

(b) $\ell: \begin{cases} x = t \\ y = -5 + 2t \\ z = 2t \end{cases} \quad x^2 + y^2 + z^2 - 5x - 8z + 29 = 0$

*7. A line ℓ is said to be tangent to the sphere $S_r(P_0)$ $(r > 0)$ at P_1 iff ℓ intersects $S_r(P_0)$ at P_1 and $\overrightarrow{P_0 P_1} \perp \ell$.
  (a) Prove that ℓ is tangent so $S_r(P_0)$ at P_1 iff P_1 is the only point at which ℓ intersects $S_r(P_0)$.
  (b) Find the equations of the line ℓ passing through the point $A(5, 0, 3)$, parallel to the plane $x + 2y + z = 4$, and tangent to the sphere $x^2 + y^2 + z^2 = 14$.

*8. Prove that the plane π tangent to the sphere $S_r(P_0)$ at P_1 is the union of all lines tangent to the sphere at P_1. (See Problem 7.)

*9. Prove the theorem of this section. (*Hint:* use Problems 8 and 7.)

★ 8. CONVEX SETS

The examples of convex sets presented here are direct analogues of those given in Section 7 of Chapter 4. We restate the definition for convenience.

Definition. A set $S \subseteq R^3$ is *convex* iff for all points $P_1, P_2 \in S$, $\overline{P_1 P_2} \subseteq S$ (Figure 9–15).

Thus, a set is convex iff it contains the line segment joining any two of its points. It is an immediate consequence of the definitions that the void set, sets consisting of a single point, line segments, rays, lines, and planes are convex sets.

The *solid sphere* with center P_0 and radius $r > 0$ is the set

$$\bar{S}_r(P_0) = \{P \mid |\overrightarrow{P_0 P}| \leq r\}$$

Thus, the solid sphere consists of all points "inside" and on the sphere $S_r(P_0)$.

Theorem 1. *A solid sphere is convex.*

PROOF. (See Problem 4.)

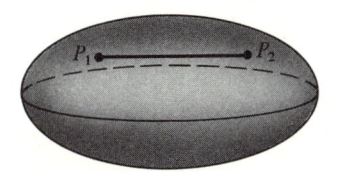

Figure 9–15

As another example of a convex set, consider the plane

$\pi: Ax + By + Cz + D = 0$

For every point $P(x, y, z) \in R^3$ one, and only one, of the following conditions holds:

$$Ax + By + Cz + D = 0 \qquad (1)$$
$$Ax + By + Cz + D > 0 \qquad (2)$$
$$Ax + By + Cz + D < 0 \qquad (3)$$

Let

$$S_1 = \{P(x, y, z) | Ax + By + Cz + D > 0\}$$
$$S_2 = \{P(x, y, z) | Ax + By + Cz + D < 0\}$$

S_1 and S_2 are called *sides* of π. Clearly π, S_1, and S_2 are nonvoid disjoint sets and $\pi \cup S_1 \cup S_2 = R^3$. Thus, π determines a *partition* of R^3 into three mutually disjoint sets. Moreover, this partition of R^3 is independent of the equation used to represent a given plane π.

Theorem 2. π, S_1, and S_2 are convex sets.

PROOF. (See Problem 5.)

The following theorem shows that a plane has no "thickness" and no "holes."

Theorem 3. Let $P_1(x_1, y_1, z_1) \in S_1$ and $P_2(x_2, y_2, z_2) \in S_2$. Then $\overline{P_1 P_2}$ intersects π in exactly one point (Figure 9-16).

PROOF. The proof is similar to that of Theorem 3 in Section 7 of Chapter 4. For arbitrary point $P(x, y, z) \in R^3$, let

$F(x, y, z) = F(P) = Ax + By + Cz + D$

Then

$F(P) = 0$ iff $P \in \pi$
$F(P) > 0$ iff $P \in S_1$
$F(P) < 0$ iff $P \in S_2$

Accordingly, $F(P_1) > 0$ and $F(P_2) < 0$. We seek a point $P(x, y, z) \in \overline{P_1 P_2}$ such that $F(P) = 0$. These conditions are equivalent to

$$\begin{cases} x = (1-t)x_1 + tx_2 \\ y = (1-t)y_1 + ty_2 \\ z = (1-t)z_1 + tz_2, \qquad 0 < t < 1 \end{cases}$$

and $Ax + By + Cz + D = 0$.

The desired point $P(x, y, z)$ is the point on $\overline{P_1 P_2}$ corresponding to

$$t = \frac{F(P_1)}{F(P_1) - F(P_2)}$$

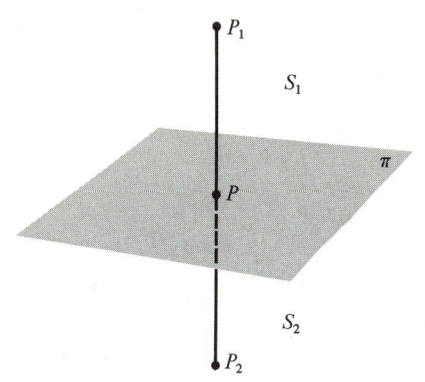

Figure 9-16

The student should verify that $0 < t < 1$ and that $F(P) = 0$.

REMARK. We say that two points lie on the same side of π iff they are both in S_1 or both in S_2. If $P_1 \in S_1$ and $P_2 \in S_2$, we say P_1 and P_2 lie on opposite sides of π (Figure 9-16).

Example. Given $\pi: 2x - 4y + 2z - 3 = 0$ and the points $A(-3, 1, 4)$ and $B(2, 0, 3)$. (a) Show that $F(A) < 0$ and $F(B) > 0$. (b) Find the point at which $\overline{AB}$ intersects π.

SOLUTION

$$F(P) = F(x, y, z) = 2x - 4y + 2z - 3$$

(a) $F(A) = F(-3, 1, 4) = -5$
$F(B) = F(2, 0, 3) = 7$

(b) The required point $P(x, y, z)$ is given by

$$t = \frac{F(B)}{F(B) - F(A)} = \frac{7}{12}$$

Thus,

$x = \frac{5}{12}(2) + \frac{7}{12}(-3) = -\frac{11}{12}$
$y = \frac{5}{12}(0) + \frac{7}{12}(1) = \frac{7}{12}$
$z = \frac{5}{12}(3) + \frac{7}{12}(4) = \frac{43}{12}$

CHECK

$F(-\frac{11}{12}, \frac{7}{12}, \frac{43}{12}) = 2(-\frac{11}{12}) - 4(\frac{7}{12}) + 2(\frac{43}{12}) - 3$
$= 0$

so

$P(-\frac{11}{12}, \frac{7}{12}, \frac{43}{12}) \in \pi$

EXERCISE 8

1. Given $\pi: 3x - 2y - z + 4 = 0$, $P_1(1, -2, 12)$ and $P_2(-2, -3, 1)$.
   (a) Show that P_1 and P_2 are on opposite sides of π.
   (b) Find the point $P(x, y, z)$ at which $\overline{P_1 P_2}$ intersects π.
2. Given $\pi: -6x + y - 2z + 4 = 0$, $P_1(0, 2, 1)$, and $P_2(4, 0, -3)$.
   (a) Show that P_1 and P_2 are on opposite sides of π.
   (b) Find the point $P(x, y, z)$ at which $\overline{P_1 P_2}$ intersects π.
3. Prove that if the plane π is tangent to the sphere $S_r(P_0)$ at the point P_1, then all points of $\overline{S_r(P_0)}$ (except P_1) lie on the same side of π. (*Hint:* Use Theorem 3.)
4. Prove Theorem 1. (*Hint:* See Theorem 4, Section 7 of Chapter 4.)
5. Prove Theorem 2. (*Hint:* See Theorem 2 Section 7 of Chapter 4.)

chapter 10
Surfaces and Curves

1. SURFACES

In Chapter 9, we encountered two examples of a *surface:* the *plane* and the *sphere*. Recall that a plane is a set of points satisfying an equation of the form $Ax + By + Cz + D = 0$, whereas a sphere has equation $x^2 + y^2 + z^2 + ax + by + cz + d = 0$. Both equations have the form $F(x, y, z) = 0$, where F is a function of three variables.

On the other hand, we saw that the plane was also characterized by the *parametric equations*
$$\begin{cases} x = x_0 + sa_1 + ta_2 \\ y = y_0 + sb_1 + tb_2 \\ z = z_0 + sc_1 + tc_2, \quad s, t \in R \end{cases}$$
In Section 5 we shall see how parametric equations for the sphere may be obtained. We shall use *parametric equations* to *define* a general surface, and then show how a surface is related to a *rectangular* equation $F(x, y, z) = 0$.

Definition. A surface S is a set of points $P(x, y, z)$ determined by the equations
$$\begin{cases} x = f(r, s) \\ y = g(r, s) \\ z = h(r, s), \quad r \in I_1, s \in I_2 \end{cases} \tag{1}$$

where I_1 and I_2 are intervals on the real line, and f, g, h are functions of two variables. The variables r and s are called *parameters*, and Equations (1) are called *parametric equations* of the surface S.

If there exists a function F of three variables (someimes written $F(x, y, z)$) such that

$$F(f(r, s), g(r, s), h(r, s)) = 0$$

for all (r, s) in the domain of f, g, and h, the equation

$$F(x, y, z) = 0 \tag{2}$$

is called an *eliminant* of Equations (1). The process of determining Equation (2) from Equation (1) is called *eliminating the parameters*. The set of points $P(x, y, z)$ satisfying Equation (2) (that is, the graph of Equation (2)) is *also* called a surface and it contains S. If this set equals S, Equation (2) is called the *rectangular equation of S*. In this section, we shall study certain surfaces by means of the rectangular equation. The discussion of parametric equations will be resumed in Section 5.

We often refer to the sketch of a surface as the "graph" of the surface. Our sketches will consist of traces in various planes. An important aspect of surface sketching is *symmetry* of the surface.

Symmetry. If $F(x, y, z) = 0$ implies $F(-x, y, z) = 0$, the surface is said to be *symmetric with respect to (or about) the y–z plane*. Symmetry with respect to the other coordinate planes is defined similarly. If $F(x, y, z) = 0$ implies $F(x, -y, -z) = 0$, the surface is said to be *symmetric with respect to (or about) the x-axis*. Symmetry with respect to the other axes is defined similarly. The surface is said to be *symmetric about the origin* if $F(x, y, z) = 0$ implies $F(-x, -y, -z) = 0$

Traces. The *trace* of a surface in a plane is the intersection of the surface with the plane. Thus, the trace of the surface S with equation $F(x, y, z) = 0$ in the vertical plane $\pi: x = c$ (c constant) is the set

$$T = \{P(x, y, z) | F(x, y, z) = 0 \text{ and } x = c\}$$
$$= \{P(c, y, z) | F(c, y, z) = 0\}$$

Now $F(c, y, z) = 0$ is an equation in the *two* variables y, z and so it represents a *curve* in R^2. We shall show how the trace T may be associated with this curve.

Every point P in the plane π is of the form $P(c, y, z)$, where y and z are arbitrary. Consider the mapping φ from π to R^2 defined by

$$\varphi(c, y, z) = (y, z)$$

Then φ is an isometry from π onto R^2 (see Chapter 6, Section 3); that is, φ is an onto mapping which preserves distances (see Problem 8). We say π is equivalent to R^2. Accordingly, we *identify* the trace T with its image under φ, namely, the curve

$$C = \{(y, z) | F(c, y, z) = 0\}$$

Surfaces

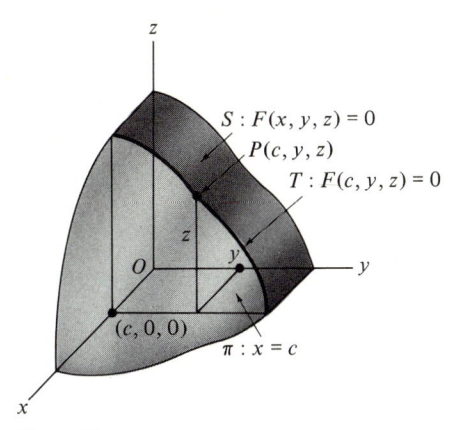

![Figure 10-1: Surface S: F(x,y,z)=0 with point P(c,y,z), trace T: F(c,y,z)=0 in plane π: x=c, passing through (c,0,0).]

Figure 10-1

in R^2. We sketch the curve C in the plane π, regarding π as R^2 (Figure 10–1). Thus, *the equation of the curve C is obtained from the equation of the surface S by substituting c for x.* In a similar manner, we may obtain the trace of S in the vertical plane $y = c$ and the horizontal plane $z = c$.

It is often helpful in sketching a surface S to obtain the trace in the vertical plane $\pi: y = mx$ ($m > 0$) as shown in Figure 10–2. This trace may be identified with a curve in R^2 as follows: Since every point P in π is of the form (x, mx, z), we define a mapping G from π to R^2 by

$$G(x, mx, z) = (x', y')$$

where

$$\begin{cases} x' = \sqrt{1 + m^2} \cdot x \\ y' = z \end{cases}$$

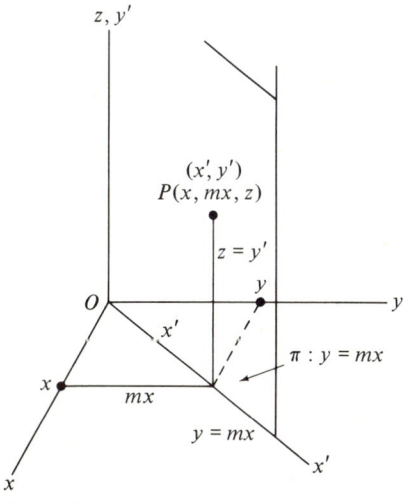

![Figure 10-2: Plane π: y = mx in 3-space, with point P(x, mx, z) mapped to (x', y') where z = y' and x' axis along y = mx.]

Figure 10-2

The mapping G is an isometry which maps $(0, 0, 0)$ into $(0, 0)$. (See Problem 9.) In Figure 10-2, the x'-axis is the line $y = mx$ in the x–y plane, and the y'-axis is the z-axis. The trace of the surface $S: F(x, y, z) = 0$ in the plane $\pi: y = mx$ is the set

$$T = \{P(x, y, z) | F(x, y, z) = 0 \text{ and } y = mx\}$$
$$= \{P(x, mx, z) | F(x, mx, z) = 0\}$$

The image of T under the mapping $G: \pi \to R^2$ is the set

$$C = \{(x', y') | x' = \sqrt{1 + m^2} \cdot x, \ y' = z \text{ and } F(x, mx, z) = 0\}$$

Since $x = x'/\sqrt{1 + m^2}$ and $z = y'$, we have

$$C = \{(x', y') | F\left(\frac{x'}{\sqrt{1 + m^2}}, \frac{mx'}{\sqrt{1 + m^2}}, y'\right) = 0\}$$

It is this curve in R^2 which we shall sketch in π. Thus, *the equation of the curve C is obtained from the equation of the surface S by substituting $x'/\sqrt{1 + m^2}$ for x, $mx'/\sqrt{1 + m^2}$ for y, and y' for z.*

Example 1. Discuss and sketch the surface

$$S: F(x, y, z) = \frac{x^2}{4} + \frac{y^2}{4} + \frac{z^2}{16} - 1 = 0$$

DISCUSSION

Symmetry. $F(x, y, z) = 0$ implies

$$F(-x, y, z) = \frac{(-x)^2}{4} + \frac{y^2}{4} + \frac{z^2}{16} - 1 = 0$$

So S is symmetric with respect to the y–z plane. Similarly, S is symmetric about the other coordinate planes. Since $F(x, y, z) = 0$ implies

$$F(x, -y, -z) = \frac{x^2}{4} + \frac{(-y)^2}{4} + \frac{(-z)^2}{16} - 1 = 0$$

S is symmetric about the x-axis; similarly S is symmetric about the other axes and the origin.

Traces. The traces in the coordinate planes have equations

$\frac{y^2}{4} + \frac{z^2}{16} = 1$ (ellipse in the y–z plane)

$\frac{x^2}{4} + \frac{z^2}{16} = 1$ (ellipse in the x–z plane)

$x^2 + y^2 = 4$ (circle in the x–y plane)

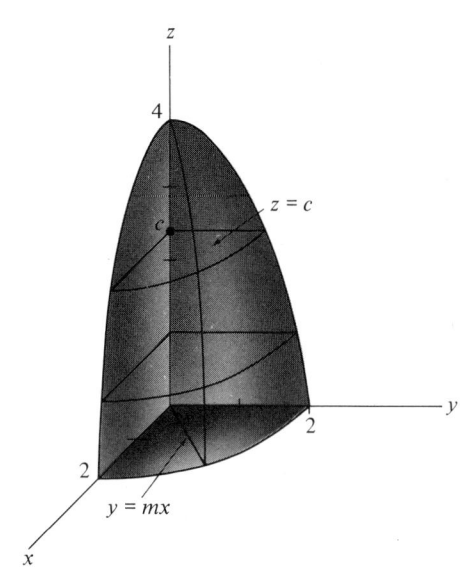

Figure 10-3

Since

$$\frac{x^2}{4} + \frac{y^2}{4} = \frac{16 - z^2}{16}$$

the trace in the plane $z = c \,(-4 < c < 4)$ is a circle. The equation of the trace in the plane $y = mx$ is

$$F\left(\frac{x'}{\sqrt{1 + m^2}}, \frac{mx'}{\sqrt{1 + m^2}}, y'\right) = 0$$

that is,

$$\frac{x'^2}{4(1 + m^2)} + \frac{m^2 x'^2}{4(1 + m^2)} + \frac{y'^2}{16} - 1 = 0$$

which reduces to

$$\frac{x'^2}{4} + \frac{y'^2}{16} = 1$$

Thus, the trace (which in this case is independent of m) is an ellipse with semimajor axis $a = 4$ and semiminor axis $b = 2$. This surface is a special case of an *ellipsoid* (Figure 10-3).

Example 2. Discuss and sketch the surface

$$S: F(x, y, z) = x^2 + y^2 - z^2 = 0$$

DISCUSSION

Symmetry. The surface is symmetric about each coordinate plane, each coordinate axis, and the origin.

Traces. The traces in the coordinate planes have equations

$z = \pm y$ (lines in the y–z plane)
$z = \pm x$ (lines in the x–z plane)
$x^2 + y^2 = 0$ (the origin in the x–y plane)

The trace in the plane $z = c$ (constant) has equation

$x^2 + y^2 = c^2$ (circle of radius $|c|$)

The equation of the trace in the plane $y = mx$ is

$$F\left(\frac{x'}{\sqrt{1+m^2}}, \frac{mx'}{\sqrt{1+m^2}}, y'\right) = 0$$

that is,

$$\frac{x'^2}{1+m^2} + \frac{m^2 x'^2}{1+m^2} - y'^2 = 0$$

which reduces to

$$y'^2 = x'^2$$

or

$$y' = \pm x'$$

Thus, the traces are lines and are independent of m. This surface is called a *circular cone* (Figure 10–4).

A surface is often specified by a condition which the points on the surface

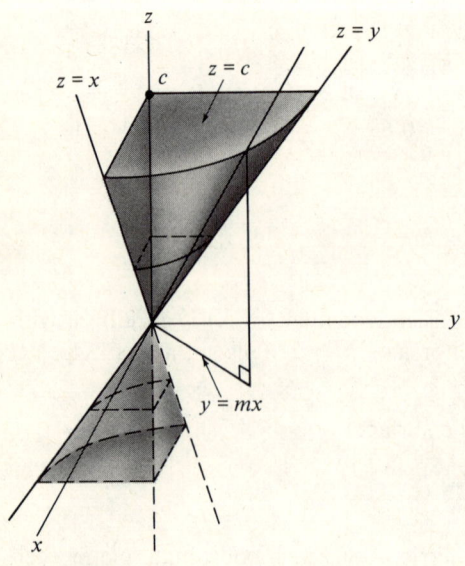

Figure 10-4

must satisfy. The problem of finding the equation of the surface is referred to as a *locus problem*.

Example 3. Find the equation of the set of points the difference of whose distances from O and $A(0, 6, 0)$ is equal to 2.

SOLUTION. Let $P(x, y, z)$ be an arbitrary point of the surface. Then

$$\sqrt{x^2 + y^2 + z^2} - \sqrt{x^2 + (y-6)^2 + z^2} = \pm 2$$
$$\sqrt{x^2 + y^2 + z^2} = \sqrt{x^2 + (y-6)^2 + z^2} \pm 2$$

After squaring and simplifying, the equation reduces to

$$\frac{x^2}{8} - \frac{(y-3)^2}{1} + \frac{z^2}{8} = -1$$

DISCUSSION

Symmetry. The graph is *symmetric* with respect to the x–y plane, the y–z plane, and the y-axis.

Traces.

y–z plane: $\dfrac{(y-3)^2}{1} - \dfrac{z^2}{8} = 1$ (hyperbola)

x–y plane: $\dfrac{(y-3)^2}{1} - \dfrac{x^2}{8} = 1$ (hyperbola)

Since

$$(y-3)^2 = \frac{x^2}{8} + \frac{z^2}{8} + 1 \geq 1$$

we see that $|y - 3| \geq 1$; that is, $y \geq 4$ or $y \leq 2$. The traces in the planes $y = c$, $c \geq 4$ or $c \leq 2$ are circles

$$x^2 + z^2 = 8[(c-3)^2 - 1]$$

The surface is shown in Figure 10–5. It is called a *hyperboloid of two sheets*.

EXERCISE 1

1. Discuss and sketch the following surfaces.
    (a) $2x - 3y + 6z - 12 = 0$
    (b) $4x^2 + 4y^2 + 4z^2 - 8x + 16y - 12z - 7 = 0$
    (c) $x^2 + z^2 = 9$
    (d) $x^2 - y^2 + z^2 = 0$
    (e) $x^2 + y^2 - z = 0$
2. Discuss and sketch the following surfaces.
    (a) $y^2 - x = 0$

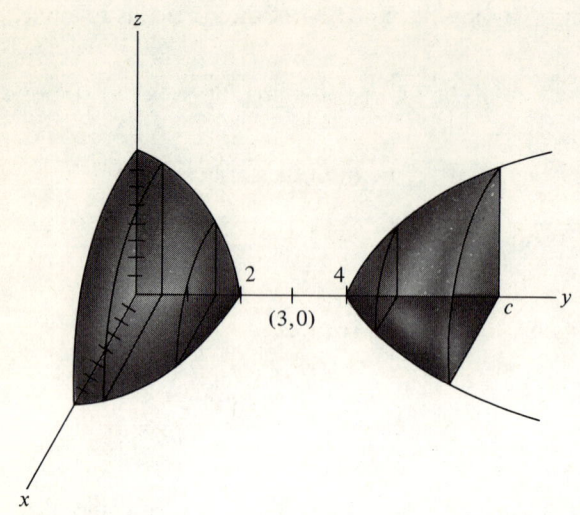

Figure 10-5

(b) $\dfrac{x^2}{4} + \dfrac{y^2}{1} + \dfrac{z^2}{9} = 1$

(c) $\dfrac{x^2}{4} + \dfrac{y^2}{9} = z$

(d) $\dfrac{x^2}{4} + \dfrac{y^2}{9} - \dfrac{z^2}{16} = -1$

(e) $\dfrac{x^2}{4} + \dfrac{y^2}{9} - \dfrac{z^2}{16} = 1$

3. Find the equation for each of the following sets of points.
    (a) the set of all points $P(x, y, z)$ such that the distance from P to $(1, -1, 1)$ is twice the distance from P to $(8, 1, 2)$
    (b) the set of all points $P(x, y, z)$ equidistant from the origin and the plane $x + y + z = 4$
    (c) the set of all points $P(x, y, z)$ equidistant from the origin and the plane $y = 5$
    (d) the set of all points $P(x, y, z)$ equidistant from $A(2, 0, -4)$ and $B(3, 2, 1)$
    (e) the set of all points equidistant from the planes $2x - y + 3z = 4$ and $x - y + z = 1$
4. Consider the set of all line segments with one endpoint $(-2, 0, 1)$ and the other endpoint on the sphere with center $(2, -1, 3)$ and radius 3. Find the equation of the set of midpoints of these segments.
5. The distance from a point to a sphere is the distance to the nearest point of the sphere. This is the distance along a line joining the point to the center of the sphere.
    (a) Find the equation of the set of points $P(x, y, z)$ equidistant from the sphere $x^2 + y^2 + z^2 = 1$ and the plane $x + y + 9 = 0$.
    (b) Find the equation of the set of points $P(x, y, z)$ equidistant from the sphere $x^2 + y^2 + z^2 = 4$ and the point $A(0, 6, 0)$. (Compare this problem with Example 3.)

6. Find the equation of the set of points $P(x, y, z)$ the difference of whose distances from $A(0, 0, 4)$ and $B(0, 0, -4)$ is numerically equal to 6.
7. Find the equation of the set of points $P(x, y, z)$ the sum of whose distances from $A(0, 0, 4)$ and $B(0, 0, -4)$ is equal to 10.
*8. Prove that the mapping $\varphi: \pi \to R^2$ defined by $\varphi(c, y, z) = (y, z)$ is an isometry (that is, φ is onto and preserves distances).
*9. Prove that the mapping $G: \pi \to R^2$ defined in this section is an isometry (G is onto and preserves distances).

2. CYLINDERS

Consider a curve C in the x–y plane having equation $f(x, y) = 0$. Let $\mathbf{v} = a\mathbf{i} + b\mathbf{j} + c\mathbf{k}$ be a vector not prallel to the x–y plane, so that $c \neq 0$. A line ℓ through a point $(X, Y, 0)$ on C parallel to $\mathbf{v}$ has parametric equations

$$\ell : \begin{cases} x = X + at \\ y = Y + bt \\ z = ct \end{cases}$$

If $P(x, y, z)$ is any point on ℓ, then $t = z/c$ and so

$$X = x - \frac{a}{c}z, \qquad Y = y - \frac{b}{c}z$$

Since $f(X, Y) = 0$, the point $P(x, y, z)$ must satisfy the equation

$$f\left(x - \frac{a}{c}z, y - \frac{b}{c}z\right) = 0$$

Conversely, if $P(x, y, z)$ satisfies this equation, the point $(X, Y, 0)$, where $X = x - (a/c)z$, $Y = y - (b/c)z$, lies on C, since $f(X, Y) = 0$. Letting $t = z/c$, we have

$$X = x - at, \quad Y = y - bt, \quad \text{and } z = ct$$

Accordingly, $x = X + at$, $y = Y + bt$, $z = ct$, and so the point $P(x, y, z)$ lies on the line ℓ through $(X, Y, 0)$ parallel to $\mathbf{v}$. The union of all such lines is called a *cylindrical surface*, or simply a *cylinder*, and each line ℓ is called an *element*, or *ruling*, of the cylinder. The curve C is called a *directrix* of the cylinder (Figure 10–6).

Definition. Let C be a curve in the x–y plane with equation $f(x, y) = 0$, and let $\mathbf{v} = a\mathbf{i} + b\mathbf{j} + c\mathbf{k}$ be a vector with $c \neq 0$. The *cylinder* with *directrix* C and *elements* parallel to $\mathbf{v}$ is the union of all lines, each of which passes through some point of C and is parallel to $\mathbf{v}$.

From the above discussion, we see that the equation of the cylinder is

$$F(x, y, z) = f\left(x - \frac{a}{c}z, y - \frac{b}{c}z\right) = 0$$

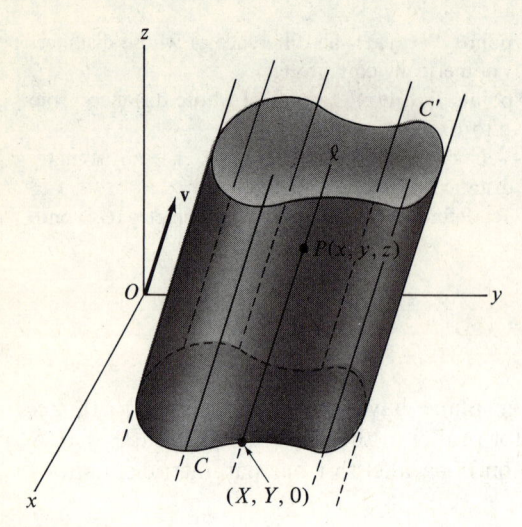

Figure 10-6

Observe that the trace of the cylinder in the plane $z = k$ (that is, a plane parallel to the x–y plane) is the curve (see Figure 10–6)

$$C': f\left(x - \frac{a}{c}k, y - \frac{b}{c}k\right) = 0$$

Since C' is merely a *translate* of C, it is clear that all cross sections of the cylinder in planes parallel to the x–y plane are "carbon copies" of C.

In an analogous fashion, we may formulate definitions of cylinders with directrices in the x–z plane, the y–z plane, or any plane (see Problem 3). However, every cylinder may be regarded as a cylinder with directrix in a *coordinate* plane. (One takes the trace in that plane as directrix.)

Example 1. Find the equation of the cylinder with directrix $C: z - x^2 = 0$ and having elements parallel to $\mathbf{v} = \mathbf{i} + 2\mathbf{j} + 3\mathbf{k}$.

SOLUTION. The directrix C is in the x–z plane and its equation is

$$g(x, z) = z - x^2 = 0$$

Since $b = 2 \neq 0$, $\mathbf{v}$ is not parallel to the x–z plane; so the required cylinder exists. We shall derive the equation of the cylinder from the definition. Let ℓ be the line through $(X, 0, Z)$ on C parallel to $\mathbf{v}$. (See Figure 10–7.) Then

$$\ell: \begin{cases} x = X + t \\ y = 2t \\ z = Z + 3t \end{cases}$$

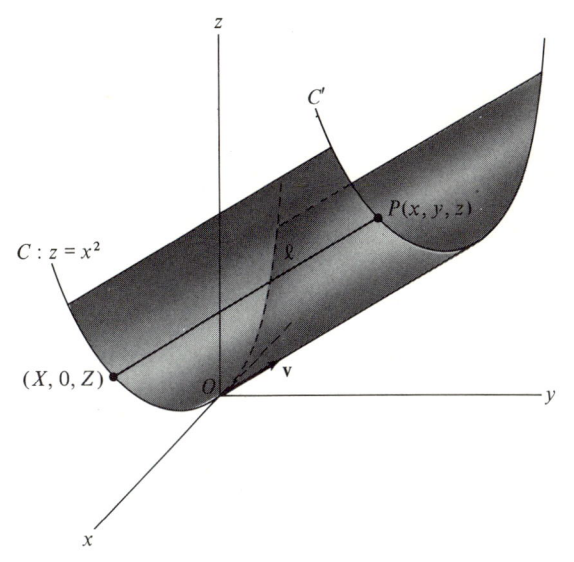

Figure 10-7

Thus, $t = y/2$, $X = x - y/2$, and $Z = z - \frac{3}{2}y$. Since $g(X, Z) = Z - X^2 = 0$, the equation of the cylinder is $g(x - y/2, z - \frac{3}{2}y) = 0$; that is,

$$z - \tfrac{3}{2}y = \left(x - \frac{y}{2}\right)^2$$

Note that the trace of the parabolic cylinder in a plane parallel to the x-z plane ($y = k$) has equations

$$z - \tfrac{3}{2}k = \left(x - \frac{k}{2}\right)^2, \quad y = k$$

This is a translate of the directrix $z = x^2$. Thus, all traces C' in planes parallel to the plane of the directrix are congruent parabolas (Figure 10–7).

Right Cylinders. A cylinder is said to be a *right cylinder with respect to the plane π* if π is perpendicular to its elements. The name of the cylinder is often taken from the name of the directrix in π. The equation of a right cylinder with respect to a coordinate plane is found as follows. If the directrix is $C: f(x, y) = 0$ in the x-y plane and $\mathbf{v} = 0\mathbf{i} + 0\mathbf{j} + c\mathbf{k}$ ($c \neq 0$), then the equation of the cylinder is

$$f\left(x - \frac{0 \cdot z}{c}, y - \frac{0 \cdot z}{c}\right) = 0$$

that is, $F(x, y, z) = f(x, y) = 0$. Accordingly, *a right cylinder with respect to a coordinate plane has precisely the same equation as the directrix in that plane.*

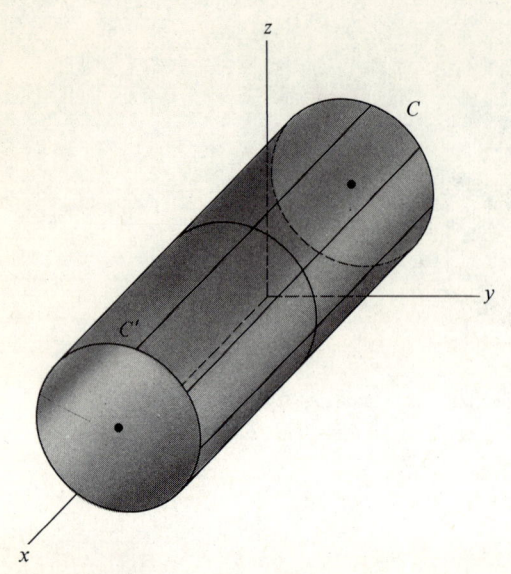

Figure 10-8

Example 2. Find the equation of the right cylinder whose directrix is the circle in the y–z plane with center $(0, 2, 3)$ and radius 2.

SOLUTION. The directrix has equation

$$(y - 2)^2 + (z - 3)^2 = 4$$

This is also the equation of the cylinder (Figure 10-8).

EXERCISE 2

1. Find the equation of the cylinder with the given directrix and elements parallel to the given vector. Derive each equation from the definition of a cylinder. Sketch the cylinder.
   (a) $x^2 + y^2 = 9$; $\mathbf{v} = \mathbf{i} - 2\mathbf{j} + \mathbf{k}$
   (b) $z^2 = y$; $\mathbf{v} = 2\mathbf{i}$
   (c) $x + z = 4$; $\mathbf{v} = 2\mathbf{j} - \mathbf{k}$
   (d) $\dfrac{x^2}{4} + \dfrac{y^2}{9} = 1$; $\mathbf{v} = \mathbf{i} + \mathbf{j} + \mathbf{k}$
   (e) $\dfrac{x^2}{4} - \dfrac{y^2}{9} = 1$; $\mathbf{v} = \mathbf{k}$

2. Sketch the following cylinders.
   (a) $4y^2 - 8y - z + 6 = 0$
   (b) $2y - 3z + 6 = 0$
   (c) $x^2 + y^2 - 4x - 6y + 11 = 0$
   (d) $z = \sqrt{x - 1}$

(e) $|y| + |z| = 1$

*3. Formulate the definition of a cylinder with directrix $g(x, z) = 0$ in the x–z plane and elements parallel to $\mathbf{v} = a\mathbf{i} + b\mathbf{j} + c\mathbf{k}$ ($b \neq 0$). Obtain the equation of the cylinder in terms of g.

3. CURVES

In Chapter 9, we learned that a line is characterized by *three* parametric equations

$$x = x_0 + at$$
$$y = y_0 + bt$$
$$z = z_0 + ct \qquad (t \in R)$$

A line is an example of a *curve* in space, according to the following definition.

Definition. Let f, g, and h be functions of one variable defined on an interval I. The curve C determined by f, g, h (in that order) is the set of points

$$C = \{P(x, y, z) \mid x = f(t),\ y = g(t),\ z = h(t),\ t \in I\}$$

In the special case of the line, $f(t) = x_0 + at$, $g(t) = y_0 + bt$, $h(t) = z_0 + ct$, and $I = R$. The equations defining a curve are called the *parametric equations* of the curve, and the variable t is the *parameter*.

Example 1. Let C be the curve whose parametric equations are

$$C: \begin{cases} x = 5 \cos t \\ y = 5 \sin t \\ z = 3t, \quad t \in R \end{cases}$$

Sketch the curve.

DISCUSSION. For every point $P(x, y, z)$ on C, $x^2 + y^2 = 25$. Hence C lies on the circular cylinder $x^2 + y^2 = 25$. We compute several points on C.

$t = 0;\ P = (5, 0, 0)$
$t = \frac{1}{2}\pi;\ P = (0, 5, \frac{3}{2}\pi)$
$t = \pi;\ P = (-5, 0, 3\pi)$

As t increases, the point P "moves" on the cylinder in the counterclockwise sense and spirals upward. As t decreases, the curve spirals downward in the clockwise sense. The curve is called the *cylindrical helix* (Figure 10–9).

In the above example, the cylinder $x^2 + y^2 = 25$ is called the *projecting cylinder* of the curve C with respect to the x–y plane. It is a right cylinder with respect to the x–y plane, each of whose elements contains a point of the curve. Observe that it was obtained by eliminating the parameter t from the parametric equations for x and y. Similarly, by eliminating the parameter

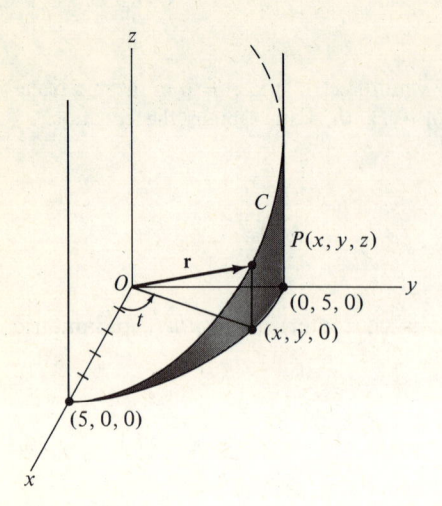

Figure 10-9

from the equations for x and z, we obtain the projecting cylinder with respect to the x–z plane. Since $t = z/3$, we have $x = 5\cos(z/3)$. This is the equation of the projecting cylinder whose directrix is the cosine curve in the x–z plane. The projecting cylinder with respect to the y–z plane is $y = 5\sin(z/3)$. In general, *eliminating the parameter from a pair of parametric equations yields the equation of a cylinder which contains the projecting cylinder with respect to a coordinate plane.* In the following examples and exercises, however, the student may assume that the cylinders obtained in this manner are *precisely* the projecting cylinders.

Example 2. Find the projecting cylinders of the curve

$$C: \begin{cases} x = 2 + 3t \\ y = 1 + t \\ z = 3 + 2t, \quad t \in R \end{cases}$$

SOLUTION. C is the line through $(2, 1, 3)$ parallel to $\mathbf{v} = 3\mathbf{i} + \mathbf{j} + 2\mathbf{k}$. Eliminating t, we have

$$\frac{x-2}{3} = \frac{y-1}{1} = \frac{z-3}{2}.$$

The projecting cylinders are

$$\frac{x-2}{3} = \frac{y-1}{1}$$

$$\frac{y-1}{1} = \frac{z-3}{2}$$

$$\frac{x-2}{3} = \frac{z-3}{2}$$

Thus, *the symmetric equations of a line are the equations of its projecting cylinders.*

A curve C is often characterized as the intersection of two surfaces:

$$C: \begin{cases} f(x, y, z) = 0 \\ g(x, y, z) = 0 \end{cases}$$

In this case, we may obtain the projecting cylinders for C as follows. Let $\varphi(x, y)$ be a function with the property that whenever $f(x, y, z) = 0$ and $g(x, y, z) = 0$, it follows that $\varphi(x, y) = 0$. That is, for each point $P(x, y, z)$ lying on the curve, the x-, y-coordinates satisfy the equation

$$\varphi(x, y) = 0$$

This equation is called an *eliminant* of the equations for C. It is the equation of a right cylinder perpendicular to the x–y plane, and it contains the curve C. Hence it contains a projecting cylinder for the curve. The equation $\varphi(x, y) = 0$ is usually obtained by *eliminating* z between the two equations for C, hence the term *eliminant*. In a similar manner, we may obtain the other projecting cylinders.

Example 3. Show that the curve of intersection of the sphere $x^2 + y^2 + z^2 = 9$ and the plane $x + y + z = 1$ is a circle. Find the projecting cylinders.

SOLUTION. We shall find the center and radius of the circle by assuming that the curve of intersection is a circle and that the line from the center of the sphere perpendicular to the plane passes through the center of the circle. We will then verify that the curve of intersection is indeed the circle obtained.

The line ℓ through the center of the sphere and perpendicular to the plane has equations

$$\ell : \begin{cases} x = t \\ y = t \\ z = t \end{cases}$$

The center of the circle is the intersection of ℓ with the plane. Solving simultaneously, we obtain $t = \frac{1}{3}$; so the center is $A(\frac{1}{3}, \frac{1}{3}, \frac{1}{3})$. For any point P on the circle, $|\overline{OP}|^2 = |\overline{OA}|^2 + |\overline{AP}|^2$ (Figure 10–10). Now $|\overline{OP}| = 3$, since

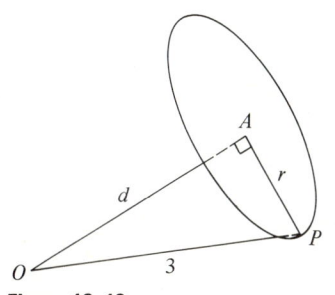

Figure 10-10

P is on the sphere. The distance from O to the plane is $d = |\overline{OA}| = 1/\sqrt{3}$, whereas the radius of the circle is $r = |\overline{AP}|$. Accordingly, $r = \sqrt{26/3}$.

We now verify that any point $P(x, y, z)$ lying on both the plane and the sphere is at a distance $\sqrt{26/3}$ from A. Thus,

$$\begin{aligned}|\overrightarrow{AP}|^2 &= (x - \tfrac{1}{3})^2 + (y - \tfrac{1}{3})^2 + (z - \tfrac{1}{3})^2 \\ &= x^2 + y^2 + z^2 - \tfrac{2}{3}(x + y + z) + \tfrac{1}{3} \\ &= 9 - \tfrac{2}{3} \cdot 1 + \tfrac{1}{3} \\ &= \tfrac{26}{3}\end{aligned}$$

since $x^2 + y^2 + z^2 = 9$ and $x + y + z = 1$.

Conversely, any point $P(x, y, z)$ in the plane, which is at a distance $\sqrt{26/3}$ from A, satisfies the equations

$$\begin{cases}(x - \tfrac{1}{3})^2 + (y - \tfrac{1}{3})^2 + (z - \tfrac{1}{3})^2 = \tfrac{26}{3} \\ x + y + z = 1\end{cases}$$

From the first equation, we obtain

$$x^2 + y^2 + z^2 - \tfrac{2}{3}(x + y + z) + \tfrac{1}{3} = \tfrac{26}{3}$$

which, by virtue of the second equation, reduces to

$$x^2 + y^2 + z^2 = 9$$

that is, $P(x, y, z)$ lies on the sphere. Thus, the curve of intersection is precisely the circle with the stated center and radius.

We obtain the equation of the projecting cylinder perpendicular to the x–y plane by eliminating z between the two given equations. We have $z = 1 - (x + y)$ from the equation of the plane. Substituting this in the equation of the sphere, we obtain

$$x^2 + y^2 + (1 - (x + y))^2 = 9$$

which reduces to

$$2x^2 + 2xy + 2y^2 - 2x - 2y - 8 = 0$$

This is the equation of an elliptic cylinder.

EXERCISE 3

1. Find the projecting cylinders and sketch the following curves.

   (a) $\begin{cases} x = 4 \sin t \\ y = 4 \cos t \\ z = 2t \end{cases}$ (circular helix)

   (b) $\begin{cases} x = 2 \sin t \\ y = 4 \cos t \\ z = t \end{cases}$ (elliptical helix)

(c) $\begin{cases} x = t \cos t \\ y = t \sin t \\ z = t \quad \text{(conical helix)} \end{cases}$

(d) $\begin{cases} x = 2 - t \\ y = 3 + 2t \\ z = 4t \end{cases}$

2. Find the projecting cylinders and sketch the following curves. If the directrices of the cylinders are conic sections, identify them.

(a) $\begin{cases} x^2 + y^2 + z^2 = 9 \\ x^2 + y^2 = z^2 \end{cases}$

(b) $\begin{cases} x^2 + y^2 + z^2 = 9 \\ x^2 + y^2 + (z - 3)^2 = 9 \end{cases}$

(c) $\begin{cases} x^2 + y^2 + z^2 = 16 \\ x + y + z = 0 \end{cases}$

(d) $\begin{cases} (x - 1)^2 + (y - 2)^2 + (z - 3)^2 = 4 \\ x + y + z - 7 = 0 \end{cases}$

(e) $\begin{cases} \dfrac{x^2}{16} + \dfrac{y^2}{9} + \dfrac{z^2}{4} = 1 \\ x + y - z = 0 \end{cases}$

(f) $\begin{cases} \dfrac{x^2}{16} + \dfrac{y^2}{9} + \dfrac{z^2}{4} = 1 \\ x^2 + y^2 = z^2 \end{cases}$

(g) $\begin{cases} x^2 + y^2 = 16 \\ z - x = 0 \end{cases}$

3. Assume that the curve of intersection of the given sphere and plane is a circle. Find its center and radius. Verify that the intersection is a circle, as in Example 3. Draw a figure.

(a) $\begin{cases} x^2 + y^2 + z^2 = 16 \\ 2y + 3z = 6 \end{cases}$

(b) $\begin{cases} (x - 1)^2 + (y - 2)^2 + (z - 3)^2 = 9 \\ x + y + z - 3 = 0 \end{cases}$

(c) $\begin{cases} (x + 1)^2 + (y - 2)^2 + (z - 4)^2 = 16 \\ x + 2y + 3z - 15 = 0 \end{cases}$

*4. Show that the curve of intersection of the sphere $x^2 + y^2 + z^2 = r^2$ and the plane $Ax + By + Cz + D = 0$ is a circle, provided

$$\left| \frac{D}{\sqrt{A^2 + B^2 + C^2}} \right| < r$$

4. SURFACES OF REVOLUTION

Consider the curve $C: Z = \sqrt{y}$, $y \geq 0$, $x = 0$, in the y–z plane. The *surface of revolution* obtained by revolving the curve about the y-axis is the set of points determined as follows.

Imagine that each point $(0, Y, Z)$ on C is revolved about the y-axis along a circle of radius Z in a plane perpendicular to the y-axis at $(0, Y, 0)$ (Figure 10–11). Now a point $P(x, y, z)$ lies on such a circle iff $y = Y$ and $x^2 + z^2 = Z^2$. Since $Z = \sqrt{Y} = \sqrt{y}$, the equation of the surface is

$$x^2 + z^2 = (\sqrt{y})^2 = y$$

Observe that the surface is symmetric about the y–z plane, the x–y plane, and the y-axis. The trace in a plane perpendicular to the positive y-axis is a circle with center on the y-axis. The curve C is called the *generator* of the surface and the y-axis is called the *axis of revolution*. This suggests the following definition.

Definition. Let $C: Z = f(y)$, $x = 0$, be a curve in the y–z plane. The *surface of revolution* with C as *generator* and the y-axis as *axis of revolution* is the set of points $P(x, y, z)$ satisfying the equation

$$x^2 + z^2 = [f(y)]^2$$

Similarly, if the curve $C: y = g(z)$, $x = 0$, in the y–z plane is revolved about the z-axis, the surface of revolution has equation

$$x^2 + y^2 = [g(z)]^2$$

(Although we may define surfaces of revolution with generators in planes other than coordinate planes, and axes other than the coordinate axes, we shall not do so.) Observe that the equation of a surface of revolution, as given by the above definition, always has the following form: *The sum of*

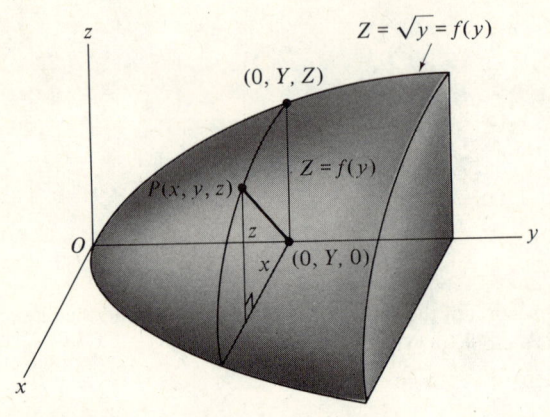

Figure 10–11

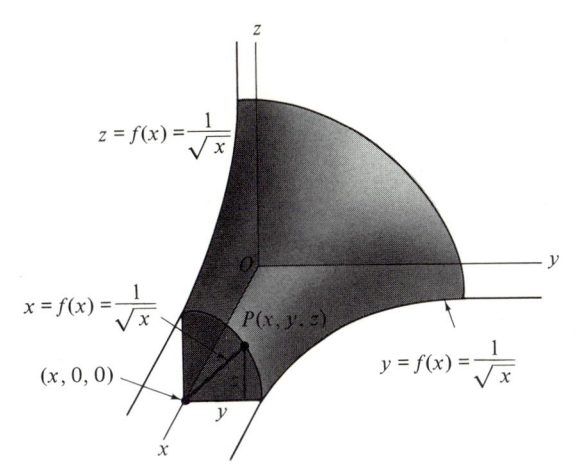

Figure 10-12

the squares of two of the coordinate variables equals the square of the function of the third coordinate variable. This function is the one associated with the generator C and must always be written in the form $u = f(v)$, where u and v are x, y, or z, and the v-axis is the axis of revolution. The generator is then a curve in the u–v plane.

Example. Show that the surface

$$S: x(y^2 + z^2) = 1$$

is a surface of revolution. Find a generator and the axis of revolution.

SOLUTION. The equation may be written

$$y^2 + z^2 = \frac{1}{x} \quad \text{(since } x \neq 0\text{)}$$

This is in the form $y^2 + z^2 = [f(x)]^2$, where $f(x) = 1/\sqrt{x}$ ($x > 0$). Accordingly, this is a surface of revolution, and the x-axis is the axis of revolution. As generator we may take the curve $y = f(x) = 1/\sqrt{x}$ in the x–y plane, or $z = f(x) = 1/\sqrt{x}$ in the x–z plane (Figure 10–12).

EXERCISE 4

1. Find the equation of the surface obtained by revolving the given curve about the indicated axis. Draw a figure.
   (a) $x^2 + y^2 = 4$; (i) x-axis, (ii) y-axis
   (b) $z = y^2$; (i) y-axis, (ii) z-axis
   (c) $z = \sin x$ $\quad (0 \leq x \leq \pi)$; x-axis
   (d) $y = 2x$ $\quad (x \geq 0)$; (i) x-axis, (ii) y-axis

(e) $\dfrac{z^2}{4} + \dfrac{x^2}{9} = 1$; (i) x-axis, (ii) z-axis

2. (Same as Problem 1.)
   (a) $(y - 2)^2 + z^2 = 1$; (i) y-axis, (ii) z-axis
   (b) $2x - y = 2 \quad (2 \leq y \leq 4)$; y-axis
   (c) $\dfrac{(y - 4)^2}{4} + \dfrac{z^2}{1} = 1$; (i) y-axis, (ii) z-axis
   (d) $y = x^2 + 1$; (i) x-axis, (ii) y-axis
   (e) $\dfrac{y^2}{4} - \dfrac{z^2}{1} = 1$; (i) y-axis, (ii) z-axis

3. The following are surfaces of revolution. Determine an axis of revolution and a generating curve. Draw a figure.
   (a) $x^2 + z^2 - y = 0$
   (b) $y^2 - x + z^2 = 0$
   (c) $x^2 + y^2 + z^2 = 25$
   (d) $4x^2 + 25y^2 + 25z^2 = 100$
   (e) $y^2 - 4x^2 - 4z^2 = 4$
   (f) $x^2 + y^2 - 4z^2 = 4$
   (g) $x^2y^2 + y^2z^2 = 1$
   (h) $x^2z^2 + y^2z^2 = 1$

5. CYLINDRICAL AND SPHERICAL COORDINATES

If the ordered triple of real numbers (r, θ, z) satisfies the equations

$$\begin{cases} x = r \cos \theta \\ y = r \sin \theta \\ z = z \end{cases} \tag{1}$$

then (r, θ, z) is called a set of *cylindrical coordinates* for the point $P(x, y, z)$ (Figure 10–13). x, y, and z are called the *rectangular coordinates* of the point $P'(r, \theta, z)$. We may express r, θ, and z in terms of x, y, and z as follows:

$$\begin{cases} r^2 = x^2 + y^2 \\ \cos \theta = \dfrac{x}{\sqrt{x^2 + y^2}}; \sin \theta = \dfrac{y}{\sqrt{x^2 + y^2}} \quad (x^2 + y^2 \neq 0) \\ \quad \text{or } \tan \theta = \dfrac{y}{x} \quad (x \neq 0) \\ z = z \end{cases} \tag{2}$$

If (r, θ, z) are given, x, y, and z are uniquely determined by Equations (1).

Cylindrical and Spherical Coordinates

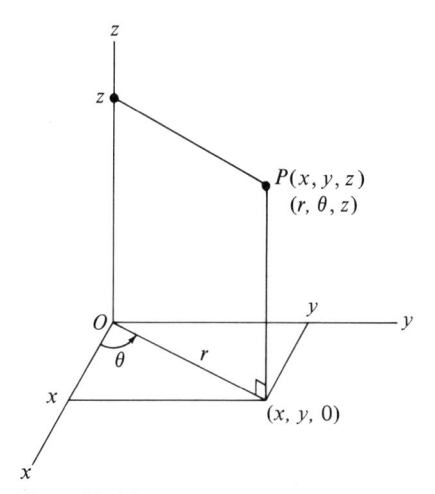

Figure 10-13

However, if x, y, and z are given, infinitely many triples (r, θ, z) are determined by Equations (2). For points $(0, 0, z)$ on the z-axis, $r = 0$ and θ is not defined. The reason for the name "cylindrical coordinates" is that the surface $r = c > 0$ (c constant) is the circular cylinder $x^2 + y^2 = c^2$. Cylindrical coordinates are frequently used to simplify the equations appearing in physical problems. Note that r and θ are the polar coordinates of the point (x, y) in R^2.

Example 1. Find the rectangular coordinates of the point whose cylindrical coordinates are $(5, \frac{2}{3}\pi, -3)$.

SOLUTION

$r = 5, \quad \theta = \frac{2}{3}\pi, \quad z = -3.$

Therefore, $\begin{cases} x = 5 \cos \frac{2}{3}\pi = -\frac{5}{2} \\ y = 5 \sin \frac{2}{3}\pi = \frac{5}{2}\sqrt{3} \\ z = -3 \end{cases}$

Example 2. Convert the following equation in rectangular coordinates to an equivalent equation in cylindrical coordinates: $(x^2 + y^2)z - 2xy = 0$.

SOLUTION. Substituting from Equations (1), we have

$r^2 z - 2r^2 \sin \theta \cos \theta = 0$

that is, $r^2(z - \sin 2\theta) = 0$

Example 3. Express the following equation in rectangular coordinates: $z = r \cos^2 2\theta$.

SOLUTION. For $r \neq 0$, we have

$z = r(\cos^2 \theta - \sin^2 \theta)^2$

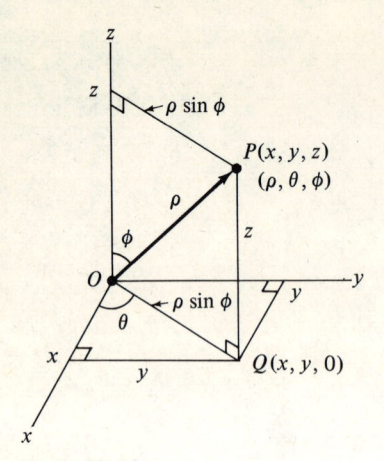

$$= r\left(\frac{x^2 - y^2}{r^2}\right)^2 = \frac{(x^2 - y^2)^2}{\pm(x^2 + y^2)^{3/2}}$$

Finally,

$$\pm(x^2 + y^2)^{3/2} z = (x^2 - y^2)^2$$
$$(x^2 + y^2)^3 z^2 = (x^2 - y^2)^4$$

Spherical Coordinates. If the ordered triple of real numbers (ρ, θ, φ) satisfies the equations

$$\begin{cases} x = \rho \sin \varphi \cos \theta \\ y = \rho \sin \varphi \sin \theta \\ z = \rho \cos \varphi \end{cases} \quad (3)$$

where $\rho \geq 0$, $0 \leq \varphi \leq \pi$, and θ is arbitrary, then (ρ, θ, φ) is called a set of *spherical coordinates* for the point $P(x, y, z)$. x, y, and z are called *rectangular coordinates* of the point $P'(\rho, \theta, \varphi)$ (Figure 10–14). Observe that $\rho = |\overrightarrow{OP}|$, while φ is the angle between the positive z-axis and $\overrightarrow{OP}$. Moreover, letting $Q = (x, y, 0)$, we see that θ is the angle between the positive x-axis and $\overrightarrow{OQ}$.

From Equations (3), we obtain

$$\begin{cases} \rho = \sqrt{x^2 + y^2 + z^2} \\ \cos \theta = \dfrac{x}{\sqrt{x^2 + y^2}}; \sin \theta = \dfrac{y}{\sqrt{x^2 + y^2}} \quad (x^2 + y^2 \neq 0) \\ \qquad \text{or } \tan \theta = \dfrac{y}{x} \quad (x \neq 0) \\ \cos \varphi = \dfrac{z}{\sqrt{x^2 + y^2 + z^2}} \quad (0 \leq \varphi \leq \pi; x^2 + y^2 + z^2 \neq 0) \end{cases} \quad (4)$$

As in the case of polar coordinates, care must be exercised to assure that θ is chosen in the proper quadrant. The name *spherical coordinates* derives from the fact that the surface $\rho = c > 0$ (c constant) is a sphere with center at the origin and radius c.

Example 4. Find the rectangular coordinates of the point with spherical coordinates $(6, \frac{3}{4}\pi, \frac{2}{3}\pi)$.

SOLUTION. $\rho = 6, \theta = \frac{3}{4}\pi$, and $\varphi = \frac{2}{3}\pi$.

Therefore,
$$\begin{cases} x = 6 \sin \frac{2}{3}\pi \cos \frac{3}{4}\pi = -\frac{3}{2}\sqrt{6} \\ y = 6 \sin \frac{2}{3}\pi \sin \frac{3}{4}\pi = \frac{3}{2}\sqrt{6} \\ z = 6 \cos \frac{2}{3}\pi = -3 \end{cases}$$

Example 5. Find a set of spherical coordinates for the point with rectangular coordinates $\left(-\frac{5}{2}\sqrt{6}, -\frac{5}{2}\sqrt{2}, -5\sqrt{2}\right)$.

SOLUTION. $x = -\frac{5}{2}\sqrt{6}, y = -\frac{5}{2}\sqrt{2}$, and $z = -5\sqrt{2}$.

Therefore,
$$\rho = \sqrt{\tfrac{75}{2} + \tfrac{25}{2} + 50} = 10$$

$$\tan \theta = \frac{-\frac{5}{2}\sqrt{2}}{-\frac{5}{2}\sqrt{6}} = \frac{1}{\sqrt{3}}$$

Therefore,

$\theta = \frac{7}{6}\pi$, since θ is in the third quadrant.

$$\cos \varphi = \frac{-5\sqrt{2}}{10} = -\frac{\sqrt{2}}{2}$$

Therefore,

$\varphi = \frac{3}{4}\pi$

So one set of spherical coordinates is $(10, \frac{7}{6}\pi, \frac{3}{4}\pi)$.

Example 6. Convert the following to an equivalent equation in spherical coordinates:

$$(x^2 + y^2 + z^2)^{3/2}(x^2 + y^2) = 4xyz$$

SOLUTION. Exclusive of points on the z-axis, the given equation is equivalent to

$$x^2 + y^2 + z^2 = \frac{4xyz}{(x^2 + y^2)\sqrt{x^2 + y^2 + z^2}}$$

By Equations (4), this is equivalent to

$$\rho^2 = 4\cos\theta \sin\theta \cos\varphi$$

Therefore,

$$\rho^2 = 2\sin 2\theta \cos\varphi$$

Example 7. Find the rectangular equation corresponding to the spherical equation $\rho \tan\theta = \sec\varphi$.

SOLUTION

$$\rho \tan\theta = \sec\varphi$$

Therefore,

$$\rho \frac{\sin\theta}{\cos\theta} = \frac{1}{\cos\varphi}$$

$$\rho \cos\varphi \sin\theta = \cos\theta$$

Substituting from Equation (4), we have

$$z \cdot \frac{y}{\sqrt{x^2 + y^2}} = \frac{x}{\sqrt{x^2 + y^2}}$$

Therefore,

$$zy = x$$

Parametric Equations for a Surface. The equation in spherical coordinates of the sphere with center at O and radius $a > 0$ is simply $\rho = a$. From Equations (3), we have

$$\begin{cases} x = a\sin\varphi \cos\theta = f(\varphi, \theta) \\ y = a\sin\varphi \sin\theta = g(\varphi, \theta) \\ z = a\cos\varphi = h(\varphi, \theta) \end{cases} \quad (0 \le \theta < 2\pi, 0 \le \varphi \le \pi)$$

These are *parametric equations* for the sphere with parameters θ and φ (see Section 1). When φ is 0 or π, z is a or $-a$, in which case, we *assign* the value 0 to x and y (since θ is undefined). As with the sphere, we may use Equations (3) to obtain a set of parametric equations of a surface with given rectangular equation.

Example 8. Find a set of parametric equations for the surface $xz = y$ with parameters θ and φ.

SOLUTION. Substituting from Equations (3), we have

$$\rho \sin\varphi \cos\theta \, \rho \cos\varphi = \rho \sin\varphi \sin\theta$$

Solving for ρ, we obtain

$$\rho = \frac{\tan\theta}{\cos\varphi}$$

provided ρ, $\sin \varphi$, $\cos \theta$, $\cos \varphi \neq 0$.

Substituting this expression for ρ in Equation (3), we obtain the parametric equations

$$\begin{cases} x = \sin \theta \tan \varphi \\ y = \sin \theta \tan \theta \tan \varphi \\ z = \tan \theta \end{cases}$$

with $0 \leq \varphi \leq \pi$, $\theta \neq (2n + 1)\pi/2$, $\varphi \neq \pi/2$.

Parametric equations for the above surface with parameters x and y are

$$\begin{cases} x = x \\ y = y \\ z = \dfrac{y}{x} \end{cases} \quad (x \neq 0)$$

EXERCISE 5

1. Find cylindrical coordinates for the following points given in rectangular coordinates.
   (a) $(-1, \sqrt{3}, 7)$
   (b) $(-5\sqrt{2}, -5\sqrt{2}, 6)$
   (c) $(2, \sqrt{5}, -2)$

2. Find spherical coordinates for the following points given in rectangular coordinates.
   (a) $(-\sqrt{3}, 3, 2)$
   (b) $(\frac{5}{2}\sqrt{6}, -\frac{5}{2}\sqrt{6}, -5)$
   (c) $(-\frac{24}{5}\sqrt{2}, \frac{18}{5}\sqrt{2}, -3)$

3. Find the rectangular coordinates of the following points given in cylindrical coordinates.
   (a) $(5, \frac{5}{6}\pi, -6)$
   (b) $(3, \frac{3}{2}\pi, 3)$
   (c) $(6, \cos^{-1}(-\frac{2}{3}), 0)$

4. Find the rectangular coordinates of the following points given in spherical coordinates.
   (a) $\left(7, \dfrac{\pi}{6}, \dfrac{\pi}{4}\right)$
   (b) $(8, \frac{4}{3}\pi, \frac{2}{3}\pi)$
   (c) $(9, \cos^{-1}\frac{1}{3}, \sin^{-1}(-\frac{1}{3}))$

5. Convert the following equations to equivalent equations in cylindrical coordinates and spherical coordinates.
   (a) $x^2 + y^2 = 9$
   (b) $x^2 + y^2 + z^2 = 4$
   (c) $x^2 + y^2 = z^2$

(d) $2x + 3y - z = 4$
(e) $x^2 + y^2 = z$

6. Same as Problem 5.
   (a) $\dfrac{x^2}{4} + \dfrac{y^2}{9} = z$
   (b) $\dfrac{x^2}{4} + \dfrac{y^2}{9} = z^2$
   (c) $(x^2 + y^2 + z^2)^{3/2} = z$
   (d) $x^2 y = z^2$
   (e) $z(x^2 + y^2) = xy$

7. Convert the following equations given in cylindrical coordinates to equivalent equations in rectangular coordinates.
   (a) $r = \sin 2\theta$
   (b) $z = r \tan \theta$
   (c) $z^2 + \cos^2 \theta = r^2$
   (d) $z = r \tan 2\theta$
   (e) $z = r \sin 3\theta$

8. Convert the following equations given in spherical coordinates to equivalent equations in rectangular coordinates.
   (a) $\rho = \cos \varphi$
   (b) $\rho = 3 \sin \varphi \sin \theta$
   (c) $\rho \cos(\theta + \varphi) = 3$
   (d) $\rho = \cot \varphi \csc \varphi$
   (e) $\rho^2 = \dfrac{4}{\sin^2 \varphi \sin 2\theta}$

9. Give a set of equations for transforming cylindrical coordinates to spherical coordinates and vice versa.

10. Write a set of parametric equations for each of the following surfaces with parameters θ and φ as in Example 8.
    (a) $z = x^2 + y^2$
    (b) $x^2 - z^2 = y$
    (c) $x^2 + y^2 + z^2 = 9$
    (d) $z = \dfrac{x^2}{4} + \dfrac{y^2}{9}$
    (e) $z = \dfrac{1}{x^2 + y^2}$

*chapter 11
Isometries in Space and Quadric Surfaces

The purpose of this chapter is to investigate the general second-degree equation in three variables:

$$Ax^2 + By^2 + Cz^2 + 2Dxy + 2Exz + 2Fyz + Gx + Hy + Iz + J = 0$$

where at least one of the coefficients of the second-degree terms is nonzero. (The factor 2 is introduced to simplify a matrix associated with this equation.) The graph of this equation in R^3 is called a *quadric surface*. As in the case of the second-degree equation in two variables, we shall employ isometries to simplify the equation. Our methods will be analogous to those used in Chapters 6 and 7. Properties of matrices and determinants used in Sections 2, 3, and 4 are given in the Appendix.

1. QUADRIC SURFACES

A few quadric surfaces have already been studied—planes and spheres in Chapter 9, cylinders and cones in Chapter 10. In this section, we shall sketch the graphs of certain quadric surfaces which are analogous to the conic sections. They are the *paraboloid*, *ellipsoid*, and *hyperboloid*. For the purpose of sketching, the most important aspects of these surfaces are *intercepts*, *symmetry*, and *traces* in various planes.

Example 1

$$\frac{x^2}{4} + \frac{y^2}{9} - z = 0$$

DISCUSSION. The equation may be written

$$z = \frac{x^2}{4} + \frac{y^2}{9}$$

Since $z \geq 0$ for all (x, y), the surface lies above the x–y plane.

Intercepts. $(0, 0, 0)$

Symmetry. About the y–z plane and the x–z plane

Traces. x–z plane: $x^2 = 4z$ (parabola)
y–z plane: $y^2 = 9z$ (parabola)
x–y plane: The origin, $(0, 0, 0)$
the planes, $z = c > 0$: $x^2/4 + y^2/9 = c$ (ellipse)

Since two of the above types of traces are parabolas and one is an ellipse, the surface is called an *elliptic paraboloid* (Figure 11–1).

Example 2

$$\frac{x^2}{9} + \frac{y^2}{25} + \frac{z^2}{16} = 1$$

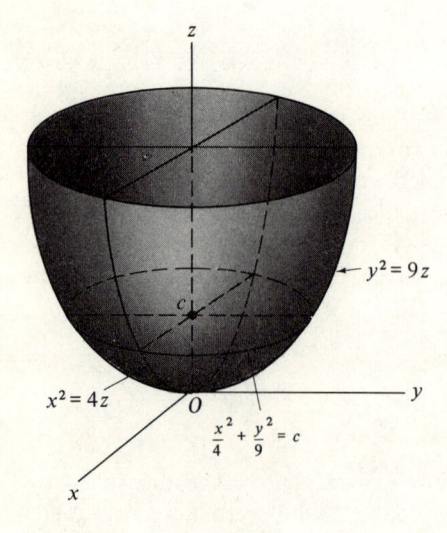

Figure 11–1 Elliptic paraboloid: $x^2/4 + y^2/9 - z = 0$.

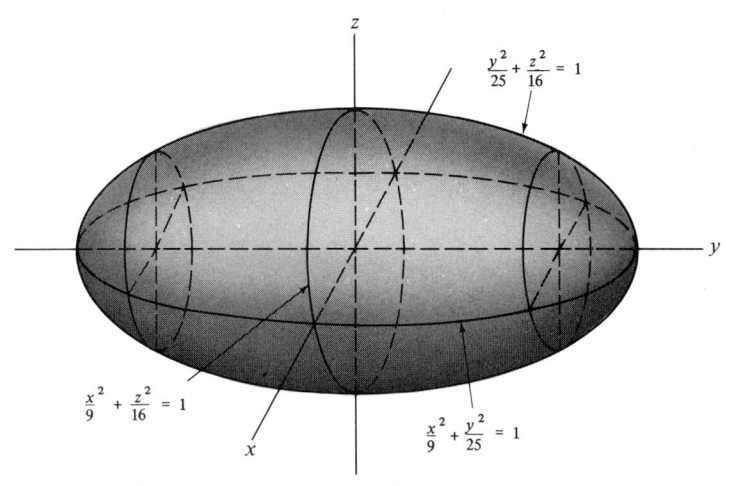

Figure 11-2 Ellipsoid: $x^2/9 + y^2/25 + z^2/16 = 1$.

DISCUSSION

Intercepts. $(\pm 3, 0, 0)$, $(0, \pm 5, 0)$, $(0, 0, \pm 4)$

Symmetry. About all coordinate planes and the origin

Traces. x–y plane: $x^2/9 + y^2/25 = 1$ (ellipse)
x–z plane: $x^2/9 + z^2/16 = 1$ (ellipse)
y–z plane: $y^2/25 + z^2/16 = 1$ (ellipse)

Since all the above traces are ellipses, the surface is called an *ellipsoid* (Figure 11-2).

Example 3

$$\frac{x^2}{9} + \frac{y^2}{16} - \frac{z^2}{25} = 1$$

DISCUSSION

Intercepts. $(\pm 3, 0, 0)$, $(0, \pm 4, 0)$

Symmetry. About all coordinate planes and the origin

Traces. x–y plane: $x^2/9 + y^2/16 = 1$ (ellipse)
y–z plane: $y^2/16 - z^2/25 = 1$ (hyperbola)
x–z plane: $x^2/9 - z^2/25 = 1$ (hyperbola)

Since two of the above traces are hyperbolas and one is an ellipse, the surface is called an *elliptic hyperboloid*. Because the surface consists of "one piece," it is called an elliptic hyperboloid *of one sheet* (Figure 11-3).

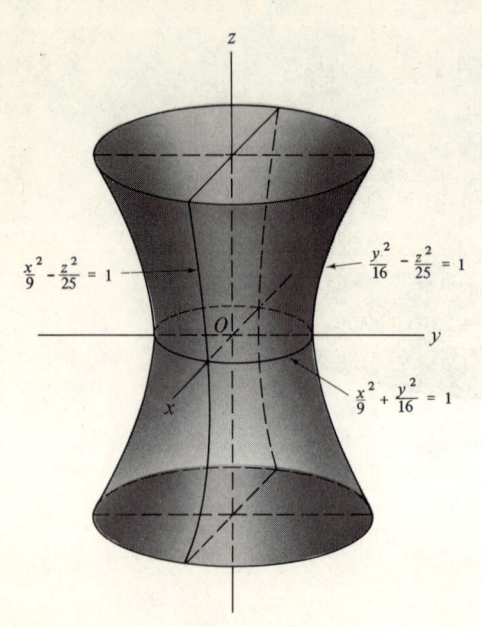

Figure 11-3 Elliptic hyperboloid of one sheet: $x^2/9 + y^2/16 - z^2/25 = 1$.

Example 4

$$\frac{x^2}{9} - \frac{y^2}{16} + \frac{z^2}{4} = -1$$

DISCUSSION

Intercepts. $(0, \pm 4, 0)$

Symmetry. About all coordinate planes and the origin

Traces. x–y plane: $y^2/16 - x^2/9 = 1$ (hyperbola)
y–z plane: $y^2/16 - z^2/4 = 1$ (hyperbola)
x–z plane: no trace

Solving the equation for y^2, we obtain

$$y^2 = 16\left(\frac{x^2}{9} + \frac{z^2}{4} + 1\right) \geq 16$$

Therefore, $|y| \geq 4$. Traces in the planes $y = c$ with $|c| \geq 4$ are ellipses

$$\frac{x^2}{9} + \frac{z^2}{4} = \frac{c^2}{16} - 1$$

The surface is called an *elliptic hyperboloid* of *two sheets* (Figure 11–4).

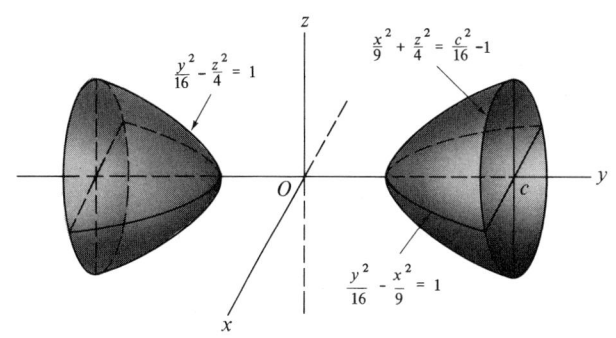

Figure 11-4 Elliptic hyperboloid of two sheets: $x^2/9 - y^2/16 + z^2/4 = -1$.

Example 5

$$\frac{y^2}{9} - \frac{x^2}{4} - z = 0$$

DISCUSSION. The equation may be written

$$z = \frac{y^2}{9} - \frac{x^2}{4}$$

Intercepts. $(0, 0, 0)$

Symmetry. About y–z plane and x–z plane

Traces. x–y plane: $y^2/9 - x^2/4 = 0$; that is $y = \pm\frac{3}{4}x$ (two lines)
y–z plane: $y^2 = 9z$ (parabola)
x–z plane: $x^2 = 4z$ (parabola)

Traces in the planes $x = c$ are parabolas opening upward:

$$y^2 = 9\left(z + \frac{c^2}{4}\right)$$

Traces in the planes $y = c$ are parabolas opening downward:

$$x^2 = -4\left(z - \frac{c^2}{9}\right)$$

Traces in the planes $z = c \neq 0$ are hyperbolas:

$$\frac{y^2}{9} - \frac{x^2}{4} = c$$

The transverse axis is parallel to the y-axis if $c > 0$, and parallel to the x-axis if $c < 0$. The surface is called a *hyperbolic paraboloid*. (It sometimes is called a *saddle surface*. See Figure 11–5.)

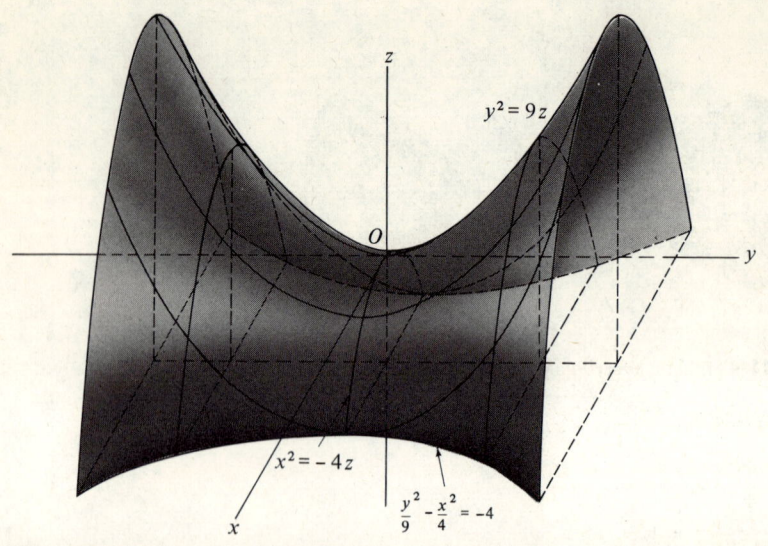

Figure 11-5 Hyperbolic paraboloid: $y^2/9 - x^2/4 - z = 0$.

EXERCISE 1

Discuss and sketch the following quadric surfaces.

1. $\dfrac{x^2}{9} + \dfrac{y^2}{4} + z = 0$

2. $\dfrac{x^2}{4} + \dfrac{y^2}{4} + \dfrac{z^2}{25} = 1$

3. $\dfrac{x^2}{9} - \dfrac{y^2}{16} + \dfrac{z^2}{9} = 1$

4. $\dfrac{x^2}{4} + \dfrac{y^2}{9} - \dfrac{z^2}{16} = -1$

5. $\dfrac{x^2}{9} - \dfrac{y^2}{16} + z = 0$

6. $\dfrac{x^2}{4} - \dfrac{y^2}{9} - z = 0$

7. $x^2 + y^2 - z^2 = 0$

8. $x^2 + y - z = 0$

9. $x - \dfrac{y^2}{9} + \dfrac{z^2}{16} = 1$

10. $y^2 - x^2 - z^2 = 9$

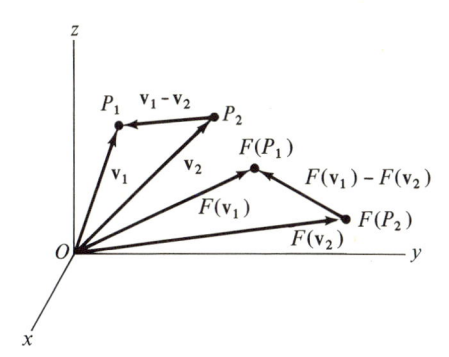

Figure 11-6

2. ISOMETRIES OF R^3

The following definition is similar to that given in Chapter 6.

Definition. An isometry F of R^3 is a mapping of R^3 onto itself with the property that for all $P_1, P_2 \in R^3$

$$|\overline{P_1 P_2}| = |\overline{F(P_1)F(P_2)}|$$

Since an isometry preserves the distance between points, it is also called a *distance-preserving mapping* (or *transformation*) of R^3. By the argument used for an isometry of R^2, we may conclude that an isometry of R^3 is one to one, and that the inverse of an isometry is also an isometry. (See Chapter 6.) If $P = (x, y, z)$ and $\mathbf{v} = x\mathbf{i} + y\mathbf{j} + z\mathbf{k}$, so that $\mathbf{v}$ is the position vector of P, the value of F at P will be denoted by $F(P)$, $F(x, y, z)$, or $F(\mathbf{v})$. However, while $F(P)$ and $F(x, y, z)$ denote a point in R^3, $F(\mathbf{v})$ denotes a *vector*, namely, the position vector of $F(P)$. In vector notation, F is an isometry iff for all vectors $\mathbf{v}_1$ and $\mathbf{v}_2$ (see Figure 11-6)

$$|\mathbf{v}_1 - \mathbf{v}_2| = |F(\mathbf{v}_1) - F(\mathbf{v}_2)|$$

In the following propositions, we establish important properties of a special kind of isometry which will allow us to identify the isometry with a *rotation matrix*, that is, an *orthogonal matrix whose determinant has value* $+1$. This, in turn, will be used to define a *rotation of axes* in R^3. Since a rotation must map the origin into itself, we shall investigate those isometries G with the property that $G(\mathbf{0}) = \mathbf{0}$. Throughout this section, G will always denote such an isometry. In several of the proofs, we make important use of the fact that for any vector $\mathbf{w}$, $|\mathbf{w}|^2 = \mathbf{w} \circ \mathbf{w}$.

Lemma 1. *For any vector* $\mathbf{v}$, $|G(\mathbf{v})| = |\mathbf{v}|$. *(G preserves the length of a vector.)*

PROOF

$$\begin{aligned}
|G(\mathbf{v})| &= |G(\mathbf{v}) - \mathbf{0}| \\
&= |G(\mathbf{v}) - G(\mathbf{0})| \quad \text{(by hypothesis)} \\
&= |\mathbf{v} - \mathbf{0}| \quad \text{(by definition of isometry)} \\
&= |\mathbf{v}|
\end{aligned}$$

Therefore,

$|G(\mathbf{v})| = |\mathbf{v}|$

Corollary. *For all vectors* $\mathbf{v}$, $G(\mathbf{v}) \circ G(\mathbf{v}) = \mathbf{v} \circ \mathbf{v}$.

Lemma 2. *For all vectors* $\mathbf{v}$ *and* $\mathbf{w}$, $G(\mathbf{v}) \circ G(\mathbf{w}) = \mathbf{v} \circ \mathbf{w}$. *(G preserves dot products.)*

PROOF. Since G is an isometry,

$|G(\mathbf{v}) - G(\mathbf{w})|^2 = |\mathbf{v} - \mathbf{w}|^2$

Therefore,

$[G(\mathbf{v}) - G(\mathbf{w})] \circ [G(\mathbf{v}) - G(\mathbf{w})] = (\mathbf{v} - \mathbf{w}) \circ (\mathbf{v} - \mathbf{w})$

Therefore,

$G(\mathbf{v}) \circ G(\mathbf{v}) - 2G(\mathbf{v}) \circ G(\mathbf{w}) + G(\mathbf{w}) \circ G(\mathbf{w}) = \mathbf{v} \circ \mathbf{v} - 2\mathbf{v} \circ \mathbf{w} + \mathbf{w} \circ \mathbf{w}$

By the Corollary to Lemma 1, $G(\mathbf{v}) \circ G(\mathbf{v}) = \mathbf{v} \circ \mathbf{v}$ and $G(\mathbf{w}) \circ G(\mathbf{w}) = \mathbf{w} \circ \mathbf{w}$. Canceling terms and simplifying, we obtain

$G(\mathbf{v}) \circ G(\mathbf{w}) = \mathbf{v} \circ \mathbf{w}$

Corollary 1. *For all vectors* $\mathbf{v}$ *and* $\mathbf{w}$.

$G(\mathbf{v}) \perp G(\mathbf{w})$ *iff* $\mathbf{v} \perp \mathbf{w}$

Corollary 2. *The vectors* $G(\mathbf{i})$, $G(\mathbf{j})$, *and* $G(\mathbf{k})$ *form an orthonormal set; that is,* $|G(\mathbf{i})| = |G(\mathbf{j})| = |G(\mathbf{k})| = 1$, *and* $G(\mathbf{i}) \circ G(\mathbf{j}) = G(\mathbf{i}) \circ G(\mathbf{k}) = G(\mathbf{j}) \circ G(\mathbf{k}) = 0$.

PROOF. The vectors $\mathbf{i}$, $\mathbf{j}$, and $\mathbf{k}$ form an orthonormal set. Therefore, by Lemma 1 and Corollary 1 above, $G(\mathbf{i})$, $G(\mathbf{j})$, and $G(\mathbf{k})$ form an orthonormal set.

Lemma 3. *For all numbers r and vectors* $\mathbf{v}$,

$G(r\mathbf{v}) = rG(\mathbf{v})$

(A mapping with this property is said to be homogeneous.)

PROOF. We shall prove that $G(r\mathbf{v}) - rG(\mathbf{v}) = \mathbf{0}$. Now

$$\begin{aligned}|G(r\mathbf{v}) - rG(\mathbf{v})|^2 &= [G(r\mathbf{v}) - rG(\mathbf{v})] \circ [G(r\mathbf{v}) - rG(\mathbf{v})] \\ &= G(r\mathbf{v}) \circ G(r\mathbf{v}) - 2rG(r\mathbf{v}) \circ G(\mathbf{v}) + r^2 G(\mathbf{v}) \circ G(\mathbf{v}) \\ &= (r\mathbf{v}) \circ (r\mathbf{v}) - 2r(r\mathbf{v}) \circ \mathbf{v} + r^2 \mathbf{v} \circ \mathbf{v} \quad \text{(by Lemma 2)} \\ &= r^2 \mathbf{v} \circ \mathbf{v} - 2r^2 \mathbf{v} \circ \mathbf{v} + r^2 \mathbf{v} \circ \mathbf{v} \\ &= 0\end{aligned}$$

Therefore,

$G(r\mathbf{v}) - rG(\mathbf{v}) = \mathbf{0}$

Lemma 4. *For all vectors* **v** *and* **w**,

$G(\mathbf{v} + \mathbf{w}) = G(\mathbf{v}) + G(\mathbf{w})$

(A mapping with this property is said to be additive.)

PROOF. The proof is similar to that of Lemma 3 and is left as an exercise. (See Problem 8.)

Definition. A mapping Φ is called *linear* iff for all numbers r, s and vectors **v**, **w**,

$\Phi(r\mathbf{v} + s\mathbf{w}) = r\Phi(\mathbf{v}) + s\Phi(\mathbf{w})$

Lemmas 3 and 4 imply the following important theorem:

Theorem. *If G is an isometry with $G(\mathbf{0}) = \mathbf{0}$, then G is linear.*

PROOF

$$G(r\mathbf{v} + s\mathbf{w}) = G(r\mathbf{v}) + G(s\mathbf{w}) \quad \text{(by Lemma 4)}$$
$$= rG(\mathbf{v}) + sG(\mathbf{w}) \quad \text{(by Lemma 3)}$$

Therefore G is linear.

Corollary 1. *If $r_1, r_2, \ldots, r_n$ are real numbers, and $\mathbf{v}_1, \mathbf{v}_2, \ldots, \mathbf{v}_n$ are vectors, then*

$G(r_1\mathbf{v}_1 + r_2\mathbf{v}_2 + \ldots + r_n\mathbf{v}_n) = r_1 G(\mathbf{v}_1) + r_2 G(\mathbf{v}_2) + \ldots + r_n G(\mathbf{v}_n)$

PROOF. (See Problem 9.)

Corollary 2. *If $\mathbf{v} = x\mathbf{i} + y\mathbf{j} + z\mathbf{k}$, then*

$G(\mathbf{v}) = xG(\mathbf{i}) + yG(\mathbf{j}) + zG(\mathbf{k})$

PROOF. Apply Corollary 1 of the Theorem.

Now let

$$\begin{cases} G(\mathbf{i}) = p_{11}\mathbf{i} + p_{12}\mathbf{j} + p_{13}\mathbf{k} \\ G(\mathbf{j}) = p_{21}\mathbf{i} + p_{22}\mathbf{j} + p_{23}\mathbf{k} \\ G(\mathbf{k}) = p_{31}\mathbf{i} + p_{32}\mathbf{j} + p_{33}\mathbf{k} \end{cases} \quad (1)$$

where the p_{st} ($s, t = 1, 2, 3$) are real numbers. Let

$$P = \begin{pmatrix} p_{11} & p_{12} & p_{13} \\ p_{21} & p_{22} & p_{23} \\ p_{31} & p_{32} & p_{33} \end{pmatrix}$$

P is called the *coefficient matrix* for the system (1) and is uniquely determined by G. Since $G(\mathbf{i})$, $G(\mathbf{j})$, and $G(\mathbf{k})$ form an orthonormal set, the rows of P (regarded as vectors) do also. Thus, for $s, t = 1, 2, 3$,

$$p_{s1}p_{t1} + p_{s2}p_{t2} + p_{s3}p_{t3} = \begin{cases} 0 & \text{if } s \neq t \\ 1 & \text{if } s = t \end{cases} \quad (2)$$

Letting P^t denote the transpose of P, we have from Equation (2)

$$PP^t = I_3 \tag{3}$$

where

$$I_3 = \begin{pmatrix} 1 & 0 & 0 \\ 0 & 1 & 0 \\ 0 & 0 & 1 \end{pmatrix}$$

the 3×3 identity matrix. But Equation (3) implies that P and P^t are inverses of each other; that is,

$$PP^t = P^t P = I_3$$

so that P is an *orthogonal matrix*. Substituting $P = (P^t)^t$, we see that P^t is an orthogonal matrix, so that the *columns of P also form an orthonormal set*. From Equation (3) and the product theorem for matrices, we have

$$|P| \cdot |P^t| = |I_3| = 1$$

But $|P^t| = |P|$. Therefore, $|P|^2 = 1$, and $|P| = \pm 1$. Also $|P^t| = \pm 1$. An orthogonal matrix P with $|P| = 1$ is called a *rotation matrix*.

We shall now show how the values of G may be computed by matrix operations involving P, the result being that G may be identified with P.

Let $\mathbf{v} = x\mathbf{i} + y\mathbf{j} + z\mathbf{k}$ be the position vector of the point $P(x, y, z)$, and let

$$G(\mathbf{v}) = x'\mathbf{i} + y'\mathbf{j} + z'\mathbf{k} \tag{4}$$

From Corollary (2) of the Theorem,

$$G(\mathbf{v}) = xG(\mathbf{i}) + yG(\mathbf{j}) + zG(\mathbf{k})$$

Substituting from Equations (1), we have

$$G(\mathbf{v}) = x(p_{11}\mathbf{i} + p_{12}\mathbf{j} + p_{13}\mathbf{k}) + y(p_{21}\mathbf{i} + p_{22}\mathbf{j} + p_{23}\mathbf{k}) \\ + z(p_{31}\mathbf{i} + p_{32}\mathbf{j} + p_{33}\mathbf{k})$$

Performing the multiplications, regrouping terms, and comparing with Equation (4), we see that

$$\begin{cases} x' = p_{11}x + p_{21}y + p_{31}z \\ y' = p_{12}x + p_{22}y + p_{32}z \\ z' = p_{13}x + p_{23}y + p_{33}z \end{cases} \tag{5}$$

In matrix form, Equations (5) may be written

$$\begin{pmatrix} x' \\ y' \\ z' \end{pmatrix} = \begin{pmatrix} p_{11} & p_{21} & p_{31} \\ p_{12} & p_{22} & p_{32} \\ p_{13} & p_{23} & p_{33} \end{pmatrix} \begin{pmatrix} x \\ y \\ z \end{pmatrix} \tag{6}$$

With

$$X' = \begin{pmatrix} x' \\ y' \\ z' \end{pmatrix}, \text{ and } X = \begin{pmatrix} x \\ y \\ z \end{pmatrix}$$

Equation (6) takes the form

$$X' = P^t X \tag{7}$$

Thus, *to find the image of a point (x, y, z) under the isometry G, we multiply the column matrix* $\begin{pmatrix} x \\ y \\ z \end{pmatrix}$ *by the matrix* P^t. Multiplying Equation (7) on the left by P and using the fact that $PP^t = I_3$, we obtain

$$X = PX' \tag{8}$$

If P is a rotation matrix, then G is called a *rotation of axes* (or simply *rotation*). Moreover, Equation (7) can be used to define an isometry G with $G(0) = 0$. (We omit the proof.) Accordingly, G will be identified with P, and Equation (7) will be called *the equation of the rotation of axes determined by P*. Although Equation (8) is equivalent to Equation (7) (and hence is also called *the equation of the rotation*), it is important to note that *to obtain the image X' of the point X under this rotation, one must multiply X by P^t as in Equation (7)*.

As in R^2, we regard "one point" as having "two sets of coordinates," one called its x-, y-, z- and the other its x'-, y'-, z'-coordinates, related by Equations (7) or (8). We use the same dot to represent both (Figure 11-7). We define the positive x'-, y'-, z'-axes to be the rays with initial point O passing through the points $(x' = 1, y' = 0, z' = 0)$, $(x' = 0, y' = 1, z' = 0)$,

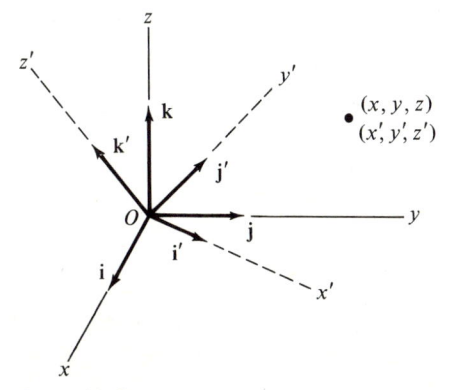

Figure 11-7 Rotation of axes.

and $(x' = 0, y' = 0, z' = 1)$, respectively. The x-, y-, z-coordinates of these points are obtained from Equation (8). We have

$$P \cdot \begin{pmatrix} 1 \\ 0 \\ 0 \end{pmatrix} = \begin{pmatrix} p_{11} \\ p_{21} \\ p_{31} \end{pmatrix}$$

$$P \cdot \begin{pmatrix} 0 \\ 1 \\ 0 \end{pmatrix} = \begin{pmatrix} p_{12} \\ p_{22} \\ p_{32} \end{pmatrix}$$

$$P \cdot \begin{pmatrix} 0 \\ 0 \\ 1 \end{pmatrix} = \begin{pmatrix} p_{13} \\ p_{23} \\ p_{33} \end{pmatrix}$$

Thus, the columns of P, regarded as vectors, are unit orthogonal vectors along the x'-, y'-, z'-axes, and will be denoted by $\mathbf{i'}$, $\mathbf{j'}$, and $\mathbf{k'}$, respectively. Hence the x'-, y'-, z'-axes are mutually perpendicular.

A *translation of axes* in R^3 is defined in the same manner as a translation of axes in R^2. Let (l, m, n) be a fixed point and let $\mathbf{v}_0 = l\mathbf{i} + m\mathbf{j} + n\mathbf{k}$. The translation of axes to the point (l, m, n) is the transformation $T_{\mathbf{v}_0}$ of R^3 defined as follows. For each $\mathbf{v} = x\mathbf{i} + y\mathbf{j} + z\mathbf{k}$,

$$T_{\mathbf{v}_0}(\mathbf{v}) = \mathbf{v} - \mathbf{v}_0 = (x - l)\mathbf{i} + (y - m)\mathbf{j} + (z - n)\mathbf{k} \tag{9}$$

Letting $T_{\mathbf{v}_0}(\mathbf{v}) = x'\mathbf{i} + y'\mathbf{j} + z'\mathbf{k}$, we see that

$$\begin{cases} x' = x - l \\ y' = y - m \\ z' = z - n \end{cases} \tag{10}$$

Equations (10) are called *the equations for a translation of axes to the point (l, m, n)*. Note that if $(x, y, z) = (l, m, n)$, then $(x', y', z') = (0, 0, 0)$. For this reason, the point (l, m, n) is called the "origin" in the x'-, y'-, z'-coordinate system. The x'-, y'-, z'-axes are drawn through the point $O' = (l, m, n)$ and parallel to the x-, y-, z-axes, respectively (Figure 11–8). As before, a point is considered as having two sets of coordinates: (x, y, z) and (x', y', z'). The same dot in Figure 11–8 represents both. In matrix form, Equation (10) becomes

$$\begin{pmatrix} x' \\ y' \\ z' \end{pmatrix} = \begin{pmatrix} x \\ y \\ z \end{pmatrix} - \begin{pmatrix} l \\ m \\ n \end{pmatrix} \tag{11}$$

A translation is an isometry of R^3. (See Problem 6.) It is a simple matter to show that *every isometry F is the composite of a translation and an isometry G with $G(\mathbf{0}) = \mathbf{0}$.*

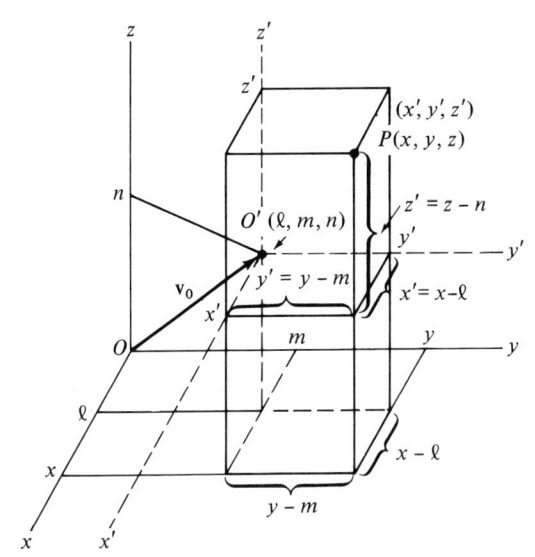

Figure 11-8 Translation of axes.

Example. Let

$$P = \begin{pmatrix} \frac{2}{3} & -\frac{2}{3} & -\frac{1}{3} \\ \frac{1}{3} & \frac{2}{3} & -\frac{2}{3} \\ \frac{2}{3} & \frac{1}{3} & \frac{2}{3} \end{pmatrix}$$

(a) Show that P is a rotation matrix.

(b) Sketch the x'-, y'-, z'-axes for the rotation of axes given by

$$\begin{pmatrix} x' \\ y' \\ z' \end{pmatrix} = P^t \cdot \begin{pmatrix} x \\ y \\ z \end{pmatrix}$$

(c) Compute the x'-, y'-, z'-coordinates of the point $(3, -\frac{3}{2}, 2)$.

SOLUTION

(a) By direct calculation, we have $P \cdot P^t = I_3$, so P is orthogonal. Moreover, $|P| = 1$; so P is a rotation matrix.

(b) The columns of P are unit vectors along the positive x'-, y'-, z'-axes (Figure 11-9).

(c) $\begin{pmatrix} x' \\ y' \\ z' \end{pmatrix} = \begin{pmatrix} \frac{2}{3} & \frac{1}{3} & \frac{2}{3} \\ -\frac{2}{3} & \frac{2}{3} & \frac{1}{3} \\ -\frac{1}{3} & -\frac{2}{3} & \frac{2}{3} \end{pmatrix} \begin{pmatrix} 3 \\ -\frac{3}{2} \\ 2 \end{pmatrix} = \begin{pmatrix} \frac{17}{6} \\ -\frac{7}{3} \\ \frac{4}{3} \end{pmatrix}$

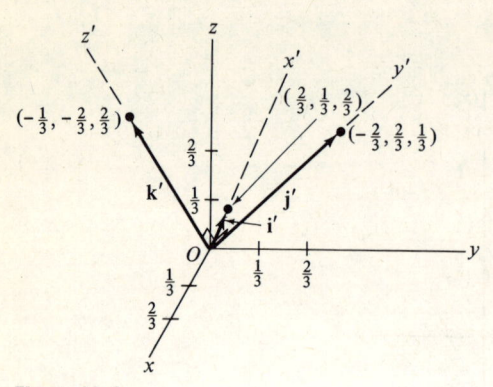

Figure 11-9

EXERCISE 2

1. Let

$$P = \begin{pmatrix} -\frac{1}{\sqrt{2}} & \frac{1}{\sqrt{2}} & 0 \\ \frac{1}{\sqrt{3}} & \frac{1}{\sqrt{3}} & \frac{1}{\sqrt{3}} \\ \frac{1}{\sqrt{6}} & \frac{1}{\sqrt{6}} & -\frac{2}{\sqrt{6}} \end{pmatrix}$$

   (a) Show that P is orthogonal.
   (b) Show that $|P| = 1$.
   (c) Sketch the x'-, y'-, z'-axes under a rotation of axes by the matrix P; that is,

$$\begin{pmatrix} x' \\ y' \\ z' \end{pmatrix} = P^t \cdot \begin{pmatrix} x \\ y \\ z \end{pmatrix}$$

   (d) Find the x'-, y'-, z'-coordinates of the point $(2, \sqrt{2}, -1)$. Plot the point.

2. Let (x', y', z') denote the coordinates of the point (x, y, z) after a translation of axes. Sketch the x'-, y'-, z'-axes for the following translations of axes. Compute the x'-, y'-, z'-coordinates of the point $(2, -1, 3)$ for each translation. Plot the point.

   (a) $\begin{pmatrix} x' \\ y' \\ z' \end{pmatrix} = \begin{pmatrix} x \\ y \\ z \end{pmatrix} - \begin{pmatrix} 2 \\ -3 \\ 5 \end{pmatrix}$

   (b) $\begin{pmatrix} x \\ y \\ z \end{pmatrix} = \begin{pmatrix} x' \\ y' \\ z' \end{pmatrix} + \begin{pmatrix} 3 \\ -5 \\ 7 \end{pmatrix}$

3. Let (x', y', z') denote the coordinates of the point (x, y, z) after a rotation of axes. Sketch the x'-, y'-, z'-axes for the following rotations of axes. Prove that they are indeed rotations. Compute the x'-, y'-, z'-coordinates of this point $(\sqrt{2}, -\sqrt{3}, -\sqrt{2})$

for each rotation, and plot the point.

(a) $\begin{pmatrix} x' \\ y' \\ z' \end{pmatrix} = \begin{pmatrix} \frac{1}{\sqrt{2}} & 0 & \frac{1}{\sqrt{2}} \\ 0 & -1 & 0 \\ \frac{1}{\sqrt{2}} & 0 & -\frac{1}{\sqrt{2}} \end{pmatrix} \begin{pmatrix} x \\ y \\ z \end{pmatrix}$

(b) $\begin{pmatrix} x \\ y \\ z \end{pmatrix} = \begin{pmatrix} -\frac{1}{\sqrt{2}} & \frac{1}{\sqrt{2}} & 0 \\ \frac{1}{\sqrt{3}} & \frac{1}{\sqrt{3}} & \frac{1}{\sqrt{3}} \\ \frac{1}{\sqrt{6}} & \frac{1}{\sqrt{6}} & -\frac{2}{\sqrt{6}} \end{pmatrix} \begin{pmatrix} x' \\ y' \\ z' \end{pmatrix}$

4. Let G be an isometry with $G(0) = 0$ such that
$$\begin{cases} G(\mathbf{i}) = -\frac{2}{\sqrt{5}}\mathbf{i} + \frac{1}{\sqrt{5}}\mathbf{j} \\ G(\mathbf{j}) = \frac{1}{\sqrt{5}}\mathbf{i} + \frac{2}{\sqrt{5}}\mathbf{j} \\ G(\mathbf{k}) = -\mathbf{k} \end{cases}$$

(a) Find the matrix P which represents G.
(b) Verify that P is a rotation matrix.
(c) Write the matrix equation corresponding to G.
(d) Compute $G(\sqrt{5}, 0, -1)$.

*5. (a) Prove that if A and B are orthogonal 3×3 matrices, then AB is orthogonal.
(b) Prove that if A and B are 3×3 rotation matrices, then AB is a rotation matrix.
(c) Prove directly from the definition of isometry that if F and G are isometries, then the composite $F(G)$ is an isometry. Moreover, if $F(0) = G(0) = 0$, then $F(G)(0) = 0$.

6. Prove that a translation of axes is an isometry.

7. (a) Show that a translation of axes to the point (l, m, n) may be written in the form
$$\begin{pmatrix} x' \\ y' \\ z' \\ 1 \end{pmatrix} = \begin{pmatrix} 1 & 0 & 0 & -l \\ 0 & 1 & 0 & -m \\ 0 & 0 & 1 & -n \\ 0 & 0 & 0 & 1 \end{pmatrix} \begin{pmatrix} x \\ y \\ z \\ 1 \end{pmatrix}$$

(b) Show that a rotation of axes may be written in the form
$$\begin{pmatrix} x' \\ y' \\ z' \\ 1 \end{pmatrix} = \left(\begin{array}{c|c} P^t & \begin{matrix} 0 \\ 0 \\ 0 \end{matrix} \\ \hline 0 \ 0 \ 0 & 1 \end{array} \right) \begin{pmatrix} x \\ y \\ z \\ 1 \end{pmatrix}$$

where P is a 3×3 rotation matrix.

*8. Prove Lemma 4.

9. Prove Corollary 1 of the Theorem.

†*10. As in the text, let

$$\mathbf{i}' = P \cdot \begin{pmatrix} 1 \\ 0 \\ 0 \end{pmatrix}, \mathbf{j}' = P \cdot \begin{pmatrix} 0 \\ 1 \\ 0 \end{pmatrix}, \mathbf{k}' = P \cdot \begin{pmatrix} 0 \\ 0 \\ 1 \end{pmatrix}$$

(a) Prove that if $|P| = 1$, then

$$\mathbf{i}' \times \mathbf{j}' = \mathbf{k}', \mathbf{j}' \times \mathbf{k}' = \mathbf{i}', \mathbf{k}' \times \mathbf{i}' = \mathbf{j}'$$

(*Hint:* Use the fact that $\mathbf{i}' \times \mathbf{j}' \circ \mathbf{k}' = |P|$. See Problem 5, Exercise 5 of Chapter 8.) (We say that $\mathbf{i}', \mathbf{j}', \mathbf{k}'$ form a *right-handed system*.)

(b) Prove that if $|P| = -1$, then

$$\mathbf{i}' \times \mathbf{j}' = -\mathbf{k}', \mathbf{j}' \times \mathbf{k}' = -\mathbf{i}', \mathbf{k}' \times \mathbf{i}' = -\mathbf{j}'$$

(In this case, we say that $\mathbf{i}', \mathbf{j}', \mathbf{k}'$ form a *left-handed coordinate system*.)

*11. Prove that if G is a rotation, then G preserves cross products; that is,

$$G(\mathbf{a} \times \mathbf{b}) = G(\mathbf{a}) \times G(\mathbf{b}) \quad (\text{use Problem 10(a)})$$

3. THE GENERAL QUADRATIC EQUATION

We now examine the general quadratic equation

$$Ax^2 + By^2 + Cz^2 + 2Dxy + 2Exz + 2Fyz + Gx + Hy + Iz + J = 0 \tag{1}$$

This may be written in matrix form, as was done in the case of two variables:

$$(x \ y \ z) \begin{pmatrix} A & D & E \\ D & B & F \\ E & F & C \end{pmatrix} \begin{pmatrix} x \\ y \\ z \end{pmatrix} + (G \ H \ I) \begin{pmatrix} x \\ y \\ z \end{pmatrix} + J = 0 \tag{2}$$

(See Problem 1.)
Letting

$$X = \begin{pmatrix} x \\ y \\ z \end{pmatrix}, \quad Q = \begin{pmatrix} A & D & E \\ D & B & F \\ E & F & C \end{pmatrix}, \quad S = (G \ H \ I)$$

we may write Equation (2) in the form

$$X'QX + SX + J = 0 \tag{3}$$

The first term of Equation (3) contains all the second-degree terms of Equation (1), the second term contains all the linear terms, while the third term J is the constant term. The symmetric matrix $Q \neq (0)$ is called the *symmetric matrix associated with the quadratic equation*.

Now let P be a rotation matrix and consider the rotation of axes

$$\begin{pmatrix} x' \\ y' \\ z' \end{pmatrix} = P^t \cdot \begin{pmatrix} x \\ y \\ z \end{pmatrix} \tag{4}$$

or equivalently,

$$\begin{pmatrix} x \\ y \\ z \end{pmatrix} = P \cdot \begin{pmatrix} x' \\ y' \\ z' \end{pmatrix} \tag{5}$$

(Recall that $P^{-1} = P^t$.) Equation (5) may be written

$$X = PX' \tag{6}$$

where

$$X' = \begin{pmatrix} x' \\ y' \\ z' \end{pmatrix}$$

Substitution of (6) in (3) yields

$$X'^t(P^tQP)X' + (SP)X' + J = 0 \tag{7}$$

which is called the *transform* of Equation (3), or of Equation (1), under a rotation of axes. Letting $Q' = P^tQP$, we may write the transform of Equation (1) as

$$X'^tQ'X' + (SP)X' + J = 0 \tag{8}$$

Equation (8) is a quadratic equation in x', y', z', with $X'^tQ'X'$ comprising the quadratic part, $(SP)X'$ the linear part, and J the constant term. Observe that $Q'^t = P^tQ^t(P^t)^t = P^tQP$, since Q is symmetric and $(P^t)^t = P$. Thus, $Q'^t = Q'$, so that Q' is the symmetric matrix associated with Equation (8). Moreover,

$$|Q'| = |P^tQP| = |P^t| \cdot |Q| \cdot |P| = |Q|$$

Thus, *the determinant of the symmetric matrix is invariant under a rotation of axes.*

As before, we examine the quadratic part $X'^tQ'X'$ in an attempt to eliminate the product terms $x'y'$, $x'z'$, and $y'z'$. (Note that $Q' \neq (0)$, since $Q \neq (0)$.) Our problem, then, is to find a rotation matrix P such that

$$X'^t(P^tQP)X' = \lambda_1 x'^2 + \lambda_2 y'^2 + \lambda_3 z'^2$$

$$= (x' \ y' \ z') \begin{pmatrix} \lambda_1 & 0 & 0 \\ 0 & \lambda_2 & 0 \\ 0 & 0 & \lambda_3 \end{pmatrix} \begin{pmatrix} x' \\ y' \\ z' \end{pmatrix}$$

Thus, P must satisfy

$$P^t Q P = \begin{pmatrix} \lambda_1 & 0 & 0 \\ 0 & \lambda_2 & 0 \\ 0 & 0 & \lambda_3 \end{pmatrix} \qquad (9)$$

for some scalars λ_1, λ_2, and λ_3. Let the unknown matrix P be written

$$P = \begin{pmatrix} p_{11} & p_{12} & p_{13} \\ p_{21} & p_{22} & p_{23} \\ p_{31} & p_{32} & p_{33} \end{pmatrix}$$

and let X_1, X_2, and X_3 be the columns of P, called *column matrices* or *column vectors*. So we may write

$$P = (X_1 \ X_2 \ X_3)$$

Since $P^t = P^{-1}$, Equation (9) takes the form

$$Q \cdot (X_1 \ X_2 \ X_3) = (X_1 \ X_2 \ X_3) \cdot \begin{pmatrix} \lambda_1 & 0 & 0 \\ 0 & \lambda_2 & 0 \\ 0 & 0 & \lambda_3 \end{pmatrix}$$

Thus, the columns of P must satisfy

$$QX_1 = \lambda_1 X_1; \qquad QX_2 = \lambda_2 X_2; \qquad QX_3 = \lambda_3 X_3$$

We see, then, that λ_1, λ_2, and λ_3 are *eigenvalues* of Q, and X_1, X_2, and X_3 are corresponding *eigenvectors*, according to the following definition.

Definition. Let Q be a 3×3 matrix. A scalar λ for which there exists a vector $X \neq 0$ such that $QX = \lambda X$ is called an *eigenvalue*, or *characteristic value*, of Q, and the vector X is called an *eigenvector*, or *characteristic vector*, of Q corresponding to (or *belonging to*) λ.

Now let

$$X = \begin{pmatrix} p_1 \\ p_2 \\ p_3 \end{pmatrix}$$

denote an arbitrary column vector and λ an arbitrary scalar. Our problem is to determine λ such that for some $X \neq 0$, we have $QX - \lambda X = (0)$; that is,

$$(Q - \lambda I_3)X = (0) \qquad (10)$$

$Q - \lambda I_3$ is called the *characteristic matrix* of Q. Equation (10) is equivalent to the following system of homogeneous linear equations:

$$\begin{cases} (A - \lambda)p_1 + Dp_2 + Ep_3 = 0 \\ Dp_1 + (B - \lambda)p_2 + Fp_3 = 0 \\ Ep_1 + Fp_2 + (C - \lambda)p_3 = 0 \end{cases} \qquad (11)$$

Equation (11) has a solution $X \neq 0$ iff the determinant of the matrix of coefficients is zero; that is,

$$|Q - \lambda I_3| = 0 \tag{12}$$

(The proof of this fact proceeds by the elementary process of elimination and will be omitted.) Equation (12) is called the *characteristic equation* of Q. It is a cubic equation in λ and always has three real roots, not necessarily distinct (proof omitted), which are the eigenvalues of Q. If the three roots are distinct, the corresponding eigenvectors are mutually orthogonal (see the Theorem of Chapter 7, Section 6). If Equation (12) has a repeated root, we must find orthogonal eigenvectors belonging to that eigenvalue. After three mutually orthogonal eigenvectors have been found, we normalize them (that is, convert them to unit vectors). These normalized vectors are then the columns of the required rotation matrix P. The columns of P must be arranged so that $|P| = 1$ (rather than -1). That such a matrix P always exists is guaranteed by the following theorem, which we state without proof.

Theorem. *If Q is a 3×3 symmetric matrix, there exists a rotation matrix P such that*

$$P'QP = \begin{pmatrix} \lambda_1 & 0 & 0 \\ 0 & \lambda_2 & 0 \\ 0 & 0 & \lambda_3 \end{pmatrix}$$

for some scalars λ_1, λ_2, and λ_3. Moreover, λ_1, λ_2, and λ_3 are eigenvalues of Q, and the columns of P are mutually orthogonal unit eigenvectors of Q belonging to λ_1, λ_2, and λ_3, respectively. If $Q \neq (0)$, at least one of the eigenvalues is nonzero.

Example 1. Rotate axes so as to eliminate the product terms from the following equation. Find its transform.

$$2x^2 + y^2 + 2z^2 + 2xy - 2yz - 4 = 0$$

SOLUTION. In matrix form, the equation is

$$(x \quad y \quad z) \begin{pmatrix} 2 & 1 & 0 \\ 1 & 1 & -1 \\ 0 & -1 & 2 \end{pmatrix} \begin{pmatrix} x \\ y \\ z \end{pmatrix} - 4 = 0$$

Thus,

$$Q = \begin{pmatrix} 2 & 1 & 0 \\ 1 & 1 & -1 \\ 0 & -1 & 2 \end{pmatrix}$$

We first solve the characteristic equation

$$|Q - \lambda I_3| = \begin{vmatrix} 2-\lambda & 1 & 0 \\ 1 & 1-\lambda & -1 \\ 0 & -1 & 2-\lambda \end{vmatrix} = -\lambda^3 + 5\lambda^2 - 6\lambda = 0$$

Equivalently,

$$\lambda^3 - 5\lambda + 6\lambda = 0$$
$$\lambda(\lambda - 2)(\lambda - 3) = 0$$

The roots of this equation (that is, the eigenvalues of Q) are $\lambda = 0, 2, 3$. To find corresponding eigenvectors, we consider the eigenvalues separately.

$\lambda = 0$. An eigenvector $\begin{pmatrix} p_1 \\ p_2 \\ p_3 \end{pmatrix}$ must satisfy

$$\begin{pmatrix} 2 & 1 & 0 \\ 1 & 1 & -1 \\ 0 & -1 & 2 \end{pmatrix} \begin{pmatrix} p_1 \\ p_2 \\ p_3 \end{pmatrix} = 0 \cdot \begin{pmatrix} p_1 \\ p_2 \\ p_3 \end{pmatrix}$$

This is equivalent to the system of equations

$$\begin{cases} 2p_1 + p_2 & = 0 \\ p_1 + p_2 - p_3 & = 0 \\ -p_2 + 2p_3 & = 0 \end{cases}$$

One solution (there are infinitely many) is $p_1 = 1$, $p_2 = -2$, and $p_3 = -1$. A *unit* eigenvector is $(1/\sqrt{6}, -2/\sqrt{6}, -1/\sqrt{6})$.

$\lambda = 2$. We must solve

$$\begin{pmatrix} 2 & 1 & 0 \\ 1 & 1 & -1 \\ 0 & -1 & 2 \end{pmatrix} \begin{pmatrix} p_1 \\ p_2 \\ p_3 \end{pmatrix} = 2 \cdot \begin{pmatrix} p_1 \\ p_2 \\ p_3 \end{pmatrix}$$

The system of equations to be solved is

$$\begin{cases} 2p_1 + p_2 & = 2p_1 \\ p_1 + p_2 - p_3 & = 2p_2 \\ -p_2 + 2p_3 & = 2p_3 \end{cases}$$

A solution is $p_1 = 1$, $p_2 = 0$, and $p_3 = 1$, so that a unit eigenvector is $(1/\sqrt{2}, 0, 1/\sqrt{2})$.

$\lambda = 3$. The equation to be solved is

$$\begin{pmatrix} 2 & 1 & 0 \\ 1 & 1 & -1 \\ 0 & -1 & 2 \end{pmatrix} \begin{pmatrix} p_1 \\ p_2 \\ p_3 \end{pmatrix} = 3 \cdot \begin{pmatrix} p_1 \\ p_2 \\ p_3 \end{pmatrix}$$

that is,
$$\begin{cases} 2p_1 + p_2 = 3p_1 \\ p_1 + p_2 - p_3 = 3p_2 \\ -p_2 + 2p_3 = 3p_3 \end{cases}$$

A solution is $p_1 = 1$, $p_2 = 1$, and $p_3 = -1$; a unit eigenvector is $(1/\sqrt{3}, 1/\sqrt{3}, -1/\sqrt{3})$.

There are several ways in which the matrix P may be constructed, using the above unit eigenvectors as columns. Any arrangement of columns is suitable, provided that $|P| = 1$. One such arrangement is

$$P = (X_1 \ X_2 \ X_3) = \begin{pmatrix} \dfrac{1}{\sqrt{3}} & \dfrac{1}{\sqrt{2}} & \dfrac{1}{\sqrt{6}} \\ \dfrac{1}{\sqrt{3}} & 0 & \dfrac{-2}{\sqrt{6}} \\ \dfrac{-1}{\sqrt{3}} & \dfrac{1}{\sqrt{2}} & \dfrac{-1}{\sqrt{6}} \end{pmatrix}$$

so that $\lambda_1 = 3$, $\lambda_2 = 2$, and $\lambda_3 = 0$. Accordingly, Q' will be
$$\begin{pmatrix} 3 & 0 & 0 \\ 0 & 2 & 0 \\ 0 & 0 & 0 \end{pmatrix}$$

Any other arrangement of columns simply permutes λ_1, λ_2, and λ_3 in the main diagonal of Q'. As a check on our calculations, we have

$$Q' = P^t Q P = \begin{pmatrix} \dfrac{1}{\sqrt{3}} & \dfrac{1}{\sqrt{3}} & \dfrac{-1}{\sqrt{3}} \\ \dfrac{1}{\sqrt{2}} & 0 & \dfrac{1}{\sqrt{2}} \\ \dfrac{1}{\sqrt{6}} & \dfrac{-2}{\sqrt{6}} & \dfrac{-1}{\sqrt{6}} \end{pmatrix} \begin{pmatrix} 2 & 1 & 0 \\ 1 & 1 & -1 \\ 0 & -1 & 2 \end{pmatrix} \begin{pmatrix} \dfrac{1}{\sqrt{3}} & \dfrac{1}{\sqrt{2}} & \dfrac{1}{\sqrt{6}} \\ \dfrac{1}{\sqrt{3}} & 0 & \dfrac{-2}{\sqrt{6}} \\ \dfrac{-1}{\sqrt{3}} & \dfrac{1}{\sqrt{2}} & \dfrac{-1}{\sqrt{6}} \end{pmatrix}$$

$$= \begin{pmatrix} \sqrt{3} & \sqrt{3} & -\sqrt{3} \\ \sqrt{2} & 0 & \sqrt{2} \\ 0 & 0 & 0 \end{pmatrix} \begin{pmatrix} \dfrac{1}{\sqrt{3}} & \dfrac{1}{\sqrt{2}} & \dfrac{1}{\sqrt{6}} \\ \dfrac{1}{\sqrt{3}} & 0 & \dfrac{-2}{\sqrt{6}} \\ \dfrac{-1}{\sqrt{3}} & \dfrac{1}{\sqrt{2}} & \dfrac{-1}{\sqrt{6}} \end{pmatrix}$$

$$= \begin{pmatrix} 3 & 0 & 0 \\ 0 & 2 & 0 \\ 0 & 0 & 0 \end{pmatrix}$$

The transform of the given equation is

$X''Q'X' - 4 = 0$

$(x' \ y' \ z') \begin{pmatrix} 3 & 0 & 0 \\ 0 & 2 & 0 \\ 0 & 0 & 0 \end{pmatrix} \begin{pmatrix} x' \\ y' \\ z' \end{pmatrix} - 4 = 0$

$3x'^2 + 2y'^2 - 4 = 0$ (elliptic cylinder)

(Note that the coefficients of the squared terms in the transform are the eigenvalues of Q in the order in which they appear in the main diagonal of Q'.)

If the given quadratic equation contains linear terms, we may rotate axes to eliminate the product terms (as in Example 1) and then translate axes to remove the linear terms, if possible. After rotating axes, we may determine the appropriate translation by completing the squares in those variables whose squares (in the transform) have nonzero coefficients.

Example 2. Rotate axes so as to remove the product terms. Then translate axes so as to remove as many linear terms as possible from the transform.

$2x^2 + y^2 + 2z^2 + 2xy - 2yz + 2y + 2z + 1 = 0$

SOLUTION. In matrix for, the equation is

$(x \ y \ z) \begin{pmatrix} 2 & 1 & 0 \\ 1 & 1 & -1 \\ 0 & -1 & 2 \end{pmatrix} \begin{pmatrix} x \\ y \\ z \end{pmatrix} + (0 \ 2 \ 2) \begin{pmatrix} x \\ y \\ z \end{pmatrix} + 1 = 0$

Thus,

$Q = \begin{pmatrix} 2 & 1 & 0 \\ 1 & 1 & -1 \\ 0 & -1 & 2 \end{pmatrix}, \quad S = (0 \ 2 \ 2)$

Since Q is the same as in Example 1, we may use the rotation matrix P computed in that example to eliminate the product terms. Now the transform of the given equation is

$X''(P^tQP)X' + (SP)X' + J = 0$

where

$P^tQP = Q' = \begin{pmatrix} 3 & 0 & 0 \\ 0 & 2 & 0 \\ 0 & 0 & 0 \end{pmatrix}$

and

$$SP = (0 \ 2 \ 2) \begin{pmatrix} \frac{1}{\sqrt{3}} & \frac{1}{\sqrt{2}} & \frac{1}{\sqrt{6}} \\ \frac{1}{\sqrt{3}} & 0 & \frac{-2}{\sqrt{6}} \\ \frac{-1}{\sqrt{3}} & \frac{1}{\sqrt{2}} & \frac{-1}{\sqrt{6}} \end{pmatrix} = (0 \ \sqrt{2} \ -\sqrt{6})$$

Thus, the transform reduces to

$$3x'^2 + 2y'^2 + \sqrt{2}y' - \sqrt{6}z' + 1 = 0$$

Here we may complete the square in x' and y' only. We have

$$3x'^2 + 2\left(y' + \frac{\sqrt{2}}{4}\right)^2 - \sqrt{6}z' + \tfrac{3}{4} = 0$$

Combining the constant term with the linear term, we obtain

$$3x'^2 + 2\left(y' + \frac{\sqrt{2}}{4}\right)^2 - \sqrt{6}\left(z' - \frac{3}{4\sqrt{6}}\right) = 0$$

We now translate axes as follows:

$$\begin{cases} x'' = x' \\ y'' = y' + \dfrac{\sqrt{2}}{4} \\ z'' = z' - \dfrac{3}{4\sqrt{6}} \end{cases}$$

In the x'', y'', z'' system, the equation is

$$3x''^2 + 2y''^2 - \sqrt{6}z'' = 0$$

which is the equation of an elliptic paraboloid.

As with quadratic equations in two variables, the computations are usually simpler if one first translates axes to remove the linear terms (if possible), and then rotates to eliminate the product terms. The linear terms are removable by a translation whenever $|Q| \neq 0$. (If $|Q| = 0$, it may or may not be possible to remove the linear terms. See Problem 2.)

Example 3. Remove the linear terms (if possible) by a translation of axes:

$$2xy + 2xz + 2yz - 4x + 6y - 3 = 0$$

SOLUTION

$$Q = \begin{pmatrix} 0 & 1 & 1 \\ 1 & 0 & 1 \\ 1 & 1 & 0 \end{pmatrix}, \quad |Q| = 2 \neq 0$$

Substituting the translation equations $x = x' + l$, $y = y' + m$, and $z = z' + n$ in the given equation yields

$$2(x' + l)(y' + m) + 2(x' + l)(z' + n) + 2(y' + m)(z' + n)$$
$$- 4(x' + l) + 6(y' + m) - 3 = 0$$

$$2x'y' + 2x'z' + 2y'z' + (2m + 2n - 4)x' + (2l + 2n + 6)y'$$
$$+ (2l + 2m)z' + 2lm + 2ln + 2mn - 4l + 6m - 3 = 0$$

Setting the coefficients of x', y', and z' equal to 0, we have

$$\begin{cases} 2m + 2n - 4 = 0 \\ 2l \phantom{{}+{}} + 2n + 6 = 0 \\ 2l + 2m \phantom{{}+{}} = 0 \end{cases}$$

This system is equivalent to

$$\begin{cases} m + n = 2 \\ l \phantom{{}+{}} + n = -3 \\ l + m \phantom{{}+{}} = 0 \end{cases}$$

The solution is $l = -\frac{5}{2}$, $m = \frac{5}{2}$, and $n = -\frac{1}{2}$. The transform of the equation under this translation is

$$2x'y' + 2x'z' + 2y'z' + \frac{19}{2} = 0$$

The rotation of axes is left as an exercise. (See Problem 3(b).)

EXERCISE 3

1. Verify Equation (2).
2. (a) Show that the transform of Equation (3) under the translation of axes $X = X' + L$, where

$$L = \begin{pmatrix} l \\ m \\ n \end{pmatrix}$$

is $X'{}^t Q X' + (2L^t Q + S)X' + (L^t Q L + SL + J) = 0$.
(Hint: Since $X'{}^t Q L$ is a one-by-one matrix, it is equal to its transpose. In fact, $X'{}^t Q L = L^t Q X'$.)
(b) Show that if $|Q| \neq 0$, it is possible to determine L so that the coefficient of X' is zero; that is, $2L^t Q + S = 0$.

3. Simplify the following equations by appropriate translations and rotations of axes. Identify the quadric surfaces. (See Section 1.)
   (a) $2xy + 2xz + 2y - 5 = 0$
   (b) $2xy + 2xz + 2yz - 4x + 6y - 3 = 0$ (See Example 3.)
   (c) $2x^2 + 2y^2 + 2z^2 - xy + xz - yz - 6 = 0$
   (d) $2x^2 + 2y^2 + 3z^2 + 2xy - 4x + y - 12z + 7 = 0$
   (e) $3z^2 + 4xy + 2xz + 2yz - 4x + 10y - 2z + 8 = 0$
   (f) $3x^2 - y^2 - 3z^2 - 4xz + 2 = 0$

4. ENUMERATION OF THE QUADRIC SURFACES

The quadric surfaces may be enumerated by examining $|Q|$ associated with the general quadratic equation. As before, we consider

$$X'QX = Ax^2 + By^2 + Cz^2 + 2Dxy + 2Exz + 2Fyz \tag{1}$$

Let $X' = P'X$ be a rotation of axes chosen so as to eliminate the product terms. Then the transform of Equation (1) is

$$X''Q'X = \lambda_1 x'^2 + \lambda_2 y'^2 + \lambda_3 z'^2 \tag{3}$$

where

$$Q' = P'QP = \begin{pmatrix} \lambda_1 & 0 & 0 \\ 0 & \lambda_2 & 0 \\ 0 & 0 & \lambda_3 \end{pmatrix}$$

and $\lambda_1, \lambda_2, \lambda_3$ are the eigenvalues of Q. We have $|Q| = |Q'| = \lambda_1 \lambda_2 \lambda_3$. We may enumerate all the quadric surfaces by considering all possible cases for the product $|Q| = \lambda_1 \cdot \lambda_2 \cdot \lambda_3$, bearing in mind that at least one eigenvalue must be nonzero. We assume, then, that the transform of the general quadratic equation may be written (after dropping the primes) in the form

$$\lambda_1 x^2 + \lambda_2 y^2 + \lambda_3 z^2 + a_1 x + a_2 y + a_3 z + a_4 = 0$$

Moreover, by a translation of axes, we may eliminate all those linear terms whose corresponding quadratic terms have a nonzero coefficient. For example, if $\lambda_1 \neq 0$, $\lambda_2 \neq 0$, and $\lambda_3 = 0$, the resulting equation is (after dropping the primes) in the form

$$\lambda_1 x^2 + \lambda_2 y^2 + a_3 z + K = 0$$

Furthermore, if $a_3 \neq 0$, we may translate in such a way that the constant term K is eliminated. The equation then takes the form (again dropping primes)

$$\lambda_1 x^2 + \lambda_2 y^2 + a_3 z = 0$$

We consider the three cases: $|Q| > 0$, $|Q| = 0$, and $|Q| < 0$.
(1) $|Q| > 0$. Q has three nonzero eigenvalues with a positive product. There are two subcases:
  (a) All eigenvalues are positive. The equation is
  $$\lambda_1 x^2 + \lambda_2 y^2 + \lambda_3 z^2 + K = 0$$
  This may be reduced to
  $$\frac{x^2}{a^2} + \frac{y^2}{b^2} + \frac{z^2}{c^2} = 1 \quad \text{(ellipsoid)}$$
  $$\frac{x^2}{a^2} + \frac{y^2}{b^2} + \frac{z^2}{c^2} = 0 \quad \text{(point)}$$
  or
  $$\frac{x^2}{a^2} + \frac{y^2}{b^2} + \frac{z^2}{c^2} = -1 \quad \text{(no graph)}$$
  according as $K < 0$, $K = 0$, or $K > 0$, respectively. (Simply divide through by K, if $K \neq 0$.)
  (b) One eigenvalue is positive, two negative. Letting $\lambda_1 > 0$, $\lambda_2 < 0$, $\lambda_3 < 0$, and reducing as in (a), we have
  $$\frac{x^2}{a^2} - \frac{y^2}{b^2} - \frac{z^2}{c^2} = 1 \quad \text{(hyperboloid of two sheets)}$$
  $$\frac{x^2}{a^2} - \frac{y^2}{b^2} - \frac{z^2}{c^2} = 0 \quad \text{(elliptic cone)}$$
  or
  $$\frac{x^2}{a^2} - \frac{y^2}{b^2} - \frac{z^2}{c^2} = -1 \quad \text{(hyperboloid of one sheet)}$$
  according as $K < 0$, $K = 0$, or $K > 0$, respectively.
(2) $|Q| = 0$. At least one eigenvalue is zero. There are five subcases:
  (a) One eigenvalue is 0 and the other two positive.
  (b) One eigenvalue is 0 and the other two negative.
  (c) One eigenvalue is 0, one positive, and one negative.
  (d) Two eigenvalues are 0, one positive.
  (e) Two eigenvalues are 0, one negative.
We shall discuss (a). (See Problem 1.) Let $\lambda_1 = 0$; $\lambda_2, \lambda_3 > 0$. The quadratic equation is
$$\lambda_2 y^2 + \lambda_3 z^2 + K = 0 \quad \text{or} \quad \lambda_2 y^2 + \lambda_3 z^2 + a_1 x = 0$$
These may be reduced to one of the following:
$$\frac{y^2}{a_2} + \frac{z^2}{b^2} = 1 \quad \text{(elliptic cylinder)}$$

Enumeration of the Quadric Surfaces

$$\frac{y^2}{a^2} + \frac{z^2}{b^2} = 0 \qquad \text{(line: } y = 0, z = 0; \text{ that is, the } x\text{-axis)}$$

$$\frac{y^2}{a^2} + \frac{z^2}{b^2} = -1 \qquad \text{(no graph)}$$

$$\frac{y^2}{a^2} + \frac{z^2}{b^2} - x = 0 \quad \text{or} \quad \frac{y^2}{a^2} + \frac{z^2}{b^2} + x = 0 \qquad \text{(elliptic paraboloid)}$$

(3) $|Q| < 0$. All subcases may be reduced to those of Case 1. (See Problem 2.)

All in all, there are seventeen distinct cases, three of which yield no graph. The fourteen real quadric surfaces are listed below:

(1) elliptic paraboloid
(2) ellipsoid
(3) hyperboloid of one sheet
(4) hyperboloid of two sheets
(5) hyperbolic paraboloid
(6) elliptic cone
(7) parabolic cylinder
(8) elliptic cylinder
(9) hyperbolic cylinder
(10) two intersecting planes
(11) two parallel planes
(12) plane
(13) line
(14) point

EXERCISE 4

1. The following refer to Case 2: $|Q| = 0$.
   (a) Show that the quadric surfaces of subcase (b) are precisely those of subcase (a). (Let $\lambda_1 = 0; \lambda_2, \lambda_3 < 0$.)
   (b) Discuss subcase (c). (Let $\lambda_1 = 0, \lambda_2 > 0, \lambda_3 < 0$.)
   (c) Discuss subcase (d). (Let $\lambda_1 = \lambda_2 = 0; \lambda_3 > 0$.)
   (d) Show that the quadric surfaces of subcase (e) are the same as those of subcase (d). (Let $\lambda_1 = \lambda_2 = 0; \lambda_3 < 0$.)
2. Show that the quadric surfaces of Case 3 are precisely those of Case 1.
3. (a) Write the general quadratic equation in the four variables: x, y, z, and u.
   (b) Write the equation of Part (a) in matrix form $X'QX + SX + R = 0$, where

$$X = \begin{pmatrix} x \\ y \\ z \\ u \end{pmatrix}$$

Q is a 4×4 symmetric matrix, S is a 1×4 matrix, and R is the constant term.

Appendix

We shall give the basic definitions concerning *matrices* and *determinants* and state some theorems without proof.

MATRICES

Let m and n be fixed positive integers. The $m \times n$ (read "m by n") *matrix* A is a rectangular array of real numbers consisting of m rows and n columns:

$$A = \begin{pmatrix} a_{11} & a_{12} & \cdots & a_{1j} & \cdots & a_{1n} \\ a_{21} & a_{22} & \cdots & a_{2j} & \cdots & a_{2n} \\ \vdots & \vdots & & \vdots & & \vdots \\ a_{i1} & a_{i2} & \cdots & a_{ij} & \cdots & a_{in} \\ \vdots & \vdots & & \vdots & & \vdots \\ a_{m1} & a_{m2} & \cdots & a_{mj} & \cdots & a_{mn} \end{pmatrix} = (a_{ij}) \quad (i = 1, 2, \ldots, m; j = 1, 2, \ldots, n)$$

with the jth column indicated above and the ith row indicated on the left.

The number a_{ij} in the ith row and jth column is called the (i, j)th *element* or *entry* of A. A has mn entries (not necessarily distinct). We denote matrices by capital letters $A, B, C, \ldots$ and entries by lower-case letters $a_{ij}, b_{ij}, c_{ij}, \ldots$.

The sequence $(a_{11}, a_{22}, \ldots, a_{ll})$, where l is the smaller of m and n, is called the *main diagonal* of A. If $m = n$, the main diagonal is $(a_{11}, a_{22}, \ldots, a_{nn})$ and A is called a *square matrix*. An $m \times n$ matrix is said to have *size* $m \times n$. If $A = (a_{ij})$ has size $m \times n$ and $B = (b_{ij})$ has size $p \times q$, then $A = B$ iff $m = p$, $n = q$, and $a_{ij} = b_{ij}$ for all ordered pairs (i, j).

Addition of Matrices. Let $A = (a_{ij})$ and $B = (b_{ij})$ be matrices of the same size. We define the *sum* of A and B by

$$A + B = (a_{ij} + b_{ij})$$

Thus, $A + B$ is the $m \times n$ matrix $C = (c_{ij})$, where $c_{ij} = a_{ij} + b_{ij}$ for each (i, j).

The $m \times n$ *zero matrix* is $(0) = (a_{ij})$, where $a_{ij} = 0$ for all (i, j). We write

$$(0) = \begin{pmatrix} 0 & 0 & 0 & \ldots & 0 \\ 0 & 0 & 0 & \ldots & 0 \\ \vdots & \vdots & & \ddots & \vdots \\ 0 & 0 & 0 & \ldots & 0 \end{pmatrix}$$

If A is $m \times n$, then $A + (0) = (0) + A = A$. The *additive inverse* (or *negative*) of A is the matrix $-A = (-a_{ij})$.

Basic Properties of Matrix Addition. For all $m \times n$ matrices A, B, and C, the following laws hold:

(1) $A + B = B + A$ (commutative law)
(2) $(A + B) + C = A + (B + C)$ (associative law)
(3) $A + (0) = (0) + A = A$ (additive identity law)
(4) $A + (-A) = (-A) + A = (0)$ (additive inverse law)

We define *subtraction* by $A - B = A + (-B)$. Thus, $A - B = (a_{ij} - b_{ij})$. The product of A and a number c is the matrix $cA = (ca_{ij})$.

Multiplication of Matrices. If A has size $m \times n$ and B has size $n \times r$ (that is, the number of rows of B equals the number of columns of A), we say B is *conformable with* A, and define the product AB (in that order) to be the $m \times r$ matrix $C = (c_{ij})$, where

$$c_{ij} = a_{i1}b_{1j} + a_{i2}b_{2j} + \ldots + a_{in}b_{nj} \qquad (i = 1, 2, \ldots, m; j = 1, 2, \ldots, r)$$

Thus, c_{ij} is the sum of the products of the elements of the ith row of A with the corresponding elements of the jth column of B. Denoting the ith row of A by

$$R_i = (a_{i1}, a_{i2} \ldots a_{in})$$

and the jth column of B by

$$C_j = \begin{pmatrix} b_{1j} \\ b_{2j} \\ \vdots \\ b_{nj} \end{pmatrix}$$

we write $c_{ij} = R_i \circ C_j = a_{i1}b_{1j} + a_{i2}b_{2j} + \ldots + a_{in}b_{nj}$, so that

$$AB = \begin{pmatrix} R_1 \circ C_1 & R_1 \circ C_2 & \ldots & R_1 \circ C_r \\ R_2 \circ C_1 & R_2 \circ C_2 & \ldots & R_2 \circ C_r \\ \vdots & \vdots & \ddots & \vdots \\ R_m \circ C_1 & R_m \circ C_2 & \ldots & R_m \circ C_r \end{pmatrix}$$

Example 1

$$A = \begin{pmatrix} 3 & -1 & 0 & 2 \\ 1 & -2 & 3 & 0 \end{pmatrix}, \quad B = \begin{pmatrix} 1 & 2 \\ -3 & -1 \\ 4 & 3 \\ 0 & -5 \end{pmatrix}$$

$$AB = \begin{pmatrix} 6 & -3 \\ 19 & 13 \end{pmatrix}, \quad BA = \begin{pmatrix} 5 & -5 & 6 & 2 \\ -10 & 5 & -3 & -6 \\ 15 & -10 & 9 & 8 \\ -5 & 10 & -15 & 0 \end{pmatrix}$$

Note that $AB \neq BA$.

If A, B, and C are $m \times n$, $n \times r$, and $r \times s$ matrices, respectively, then $(AB)C = A(BC)$. That is, matrix multiplication is *associative*. If A is $m \times n$ and B and C are $n \times r$, then $A(B + C) = AB + AC$. If B and C are $m \times n$ and A is $n \times r$, then $(B + C)A = BA + CA$. Matrix multiplication is *distributive* with respect to addition.

Further Properties of Matrices. The *identity matrix* of order n is the $n \times n$ matrix

$$I_n = \begin{pmatrix} 1 & 0 & 0 & \ldots & 0 \\ 0 & 1 & 0 & \ldots & 0 \\ \vdots & \vdots & \vdots & \ddots & \vdots \\ 0 & 0 & 0 & \ldots & 1 \end{pmatrix}$$

(1's along the main diagonal and 0's elsewhere). If A is $m \times n$, then $AI_n = A = I_m A$. An $n \times n$ (square) matrix A is said to be *nonsingular* iff there exists an $n \times n$ matrix B such that $AB = BA = I_n$. The (unique) matrix B is called the *inverse* of A and is written $B = A^{-1}$. Clearly, B is also nonsingular, and $A = B^{-1}$. Thus,

$$AA^{-1} = A^{-1}A = I_n$$

We shall subsequently give a formula for the inverse of a nonsingular matrix.

If $A = (a_{ij})$ is an $m \times n$ matrix, the *transpose* of A is the $n \times m$ matrix $A^t = (b_{ij})$, where $b_{ij} = a_{ji}$. Thus, A^t is that $n \times m$ matrix whose kth column is the kth row of A ($k = 1, 2, \ldots, m$). It follows that A and A^t have the same main diagonal and that $(A^t)^t = A$. For example, if

$$A = \begin{pmatrix} 3 & -1 \\ -4 & 2 \\ 6 & 5 \end{pmatrix}$$

then

$$A^t = \begin{pmatrix} 3 & -4 & 6 \\ -1 & 2 & 5 \end{pmatrix}$$

Theorem 1. *If B is conformable with A, then A^t is conformable with B^t, and*

$$(AB)^t = B^t A^t$$

(The transpose of a product is the product of the transposes in the reverse order.) (See Problem 7(b).)

Definition. A is *symmetric* iff $A = A^t$. For example,

$$\begin{pmatrix} 2 & -1 & 3 \\ -1 & 4 & 5 \\ 3 & 5 & -6 \end{pmatrix}$$

is symmetric.

Theorem 2. *If A is an $n \times n$ symmetric matrix and B is $m \times n$, then*

$$(BAB^t)^t = BAB^t$$

that is, BAB^t is symmetric.

Theorem 3. *If A and B are nonsingular and have the same order, then AB is nonsingular and*

$$(AB)^{-1} = B^{-1}A^{-1}$$

(The inverse of a product is the product of the inverses in the reverse order.) (See Problem 8.)

Definition. An $n \times n$ matrix A is *orthogonal* iff $A^t = A^{-1}$.

Thus, A is orthogonal iff $AA^t = A^t A = I_n$. (See Problem 3(a).) For the cases $n = 2, 3$, an orthogonal matrix has an interesting interpretation in terms of vectors: A 2×2 or 3×3 matrix A is orthogonal iff the rows of A, regarded as vectors, form an *orthonormal set;* that is, each row-vector has length 1, and any two (distinct) row-vectors are orthogonal (perpendicular). A similar statement holds for columns.

Theorem 4. *If A, B are orthogonal matrices of the same size, then AB is orthogonal. (See Problem 3(b).)*

DETERMINANTS

The Determinant of a Square Matrix. If A is $n \times n$, we say A is of *order* n. We shall associate with each square matrix a unique number $|A|$, called the *determinant* of A. We write

$$\begin{vmatrix} a_{11} & a_{12} & \cdots & a_{1n} \\ a_{21} & a_{22} & \cdots & a_{2n} \\ \vdots & \vdots & & \vdots \\ a_{n1} & a_{n2} & \cdots & a_{nn} \end{vmatrix}$$

or simply $|A| = |a_{ij}|$. The definition of $|A|$ is by induction on n. For $n = 1$, $A = (a_{11})$, and $|A| = a_{11}$ by definition. For $n = 2$,

$$A = \begin{pmatrix} a_{11} & a_{12} \\ a_{21} & a_{22} \end{pmatrix}$$

and $|A| = a_{11}a_{22} - a_{12}a_{21}$. Now suppose that $|A|$ has been defined for all matrices of order $1, 2, 3, \ldots, n-1$ for $n \geq 2$. Let A be an nth order matrix. For each $i, j = 1, 2, \ldots, n$, let M_{ij} be the determinant of the matrix obtained from A by deleting the ith row and the jth column. Now M_{ij} is the determinant of a (square) matrix of order $n - 1$, and by our assumption, M_{ij} is defined. M_{ij} is called the *minor* of the element a_{ij}. Now let $A_{ij} = (-1)^{i+j}M_{ij}$, called the *cofactor* of a_{ij}. We define $|A|$ as follows:

$$|A| = a_{11}A_{11} + a_{12}A_{12} + \ldots + a_{1j}A_{1j} + \ldots + a_{1n}A_{1n}$$

where the first factors in the terms of the sum are the elements of the *first row* of A, and the second factors are the corresponding cofactors. This is called the expansion of $|A|$ by minors along the first row.

Example 2. Evaluate the following determinants.

(a) $|-3|$ (1st order)

(b) $\begin{vmatrix} 2 & -4 \\ 3 & -5 \end{vmatrix}$ (2nd order)

(c) $\begin{vmatrix} 2 & 0 & -1 \\ -3 & 4 & 5 \\ 1 & -6 & -2 \end{vmatrix}$ (3rd order)

(d) $\begin{vmatrix} 1 & -2 & 0 & 3 \\ 4 & 0 & 2 & -1 \\ -3 & 4 & 5 & -2 \\ 0 & 3 & 1 & 6 \end{vmatrix}$ (4th order)

SOLUTION

(a) $|-3| = -3$

(b) $\begin{vmatrix} 2 & -4 \\ 3 & -5 \end{vmatrix} = 2 \cdot (-5) - (-4)(3) = 2$

(c) $\begin{vmatrix} 2 & 0 & -1 \\ -3 & 4 & 5 \\ 1 & -6 & -2 \end{vmatrix} = 2 \cdot \begin{vmatrix} 4 & 5 \\ -6 & -2 \end{vmatrix} - 0 \cdot \begin{vmatrix} -3 & 5 \\ 1 & -2 \end{vmatrix}$

$+ (-1) \cdot \begin{vmatrix} -3 & 4 \\ 1 & -6 \end{vmatrix}$

$= 2(-8 - (-30)) + 0 + (-1)(18 - 4) = 30$

(d) $\begin{vmatrix} 1 & -2 & 0 & 3 \\ 4 & 0 & 2 & -1 \\ -3 & 4 & 5 & -2 \\ 0 & 3 & 1 & 6 \end{vmatrix} = 1 \cdot \begin{vmatrix} 0 & 2 & -1 \\ 4 & 5 & -2 \\ 3 & 1 & 6 \end{vmatrix} - (-2) \cdot \begin{vmatrix} 4 & 2 & -1 \\ -3 & 5 & -2 \\ 0 & 1 & 6 \end{vmatrix}$

$+ 0 \cdot \begin{vmatrix} 4 & 0 & -1 \\ -3 & 4 & -2 \\ 0 & 3 & 6 \end{vmatrix} - 3 \cdot \begin{vmatrix} 4 & 0 & 2 \\ -3 & 4 & 5 \\ 0 & 3 & 1 \end{vmatrix}$

(The student should complete the example by evaluating the third-order determinants as in Part (c).)

Theorem 5. $|A| = |A^t|$ (see Problem 6(a))

Theorem 6. Let B be the matrix obtained from A by interchanging two rows R_k and R_l ($k \neq l$), or two columns C_k and C_l ($k \neq l$). Then $|B| = -|A|$. (See Problem 6(b).)

Theorem 7. If A has two equal rows (or two equal columns), then $|A| = 0$.

Theorem 8
$|A| = a_{i1}A_{i1} + a_{i2}A_{i2} + \ldots + a_{in}A_{in}$ (for each $i = 1, 2, \ldots, n$)
$= a_{1j}A_{1j} + a_{2j}A_{2j} + \ldots + a_{nj}A_{nj}$ (for each $j = 1, 2, \ldots, n$)

That is to say, $|A|$ is equal to the expansion by minors along any row (or column).

Let $R_i = (a_{i1} \; a_{i2} \ldots a_{in})$ be the ith row of A. The product of R_i by a number c is the *row-matrix*

$cR_i = (ca_{i1} \; ca_{i2} \ldots ca_{in})$

Theorem 9. Let B be the matrix obtained from A by replacing the ith row R_i by cR_i (or the jth column C_j by cC_j). Then $|B| = c|A|$.

Corollary 1. $|cA| = c^n|A|$. *(See Problem 6(c).)*

Corollary 2. If $k \neq l$ and $R_k = cR_l$ (or $C_k = cC_l$), then $|A| = 0$.

Theorem 10. Let R_k and R_l ($k \neq l$) be two rows of A, and let c be a number. Let B be the matrix obtained from A by replacing R_k by

$$R_k + cR_1 = (a_{k1} + ca_{l1}\ a_{k2} + ca_{l2} \ldots a_{kn} + ca_{ln})$$

that is, c times the lth row is added to the kth row. Then $|B| = |A|$, *(A similar statement holds for columns.)*

Theorem 11 *(The product theorem)*

$$|AB| = |A| \cdot |B|$$

that is, the determinant of a product of two (square) matrices is the product of their determinants. *(See Problem 7(a).)*

Theorem 12. A is nonsingular iff $|A| \neq 0$. Moreover, if A is nonsingular, then

$$A^{-1} = \frac{1}{|A|} \cdot \begin{pmatrix} A_{11} & A_{12} & \ldots & A_{1n} \\ A_{21} & A_{22} & \ldots & A_{2n} \\ \vdots & \vdots & & \vdots \\ A_{n1} & A_{n2} & \ldots & A_{nn} \end{pmatrix}^t$$

Example 3. Find the inverse of the following:

$$A = \begin{pmatrix} 1 & -1 & 2 \\ 0 & 2 & 1 \\ 1 & 0 & -1 \end{pmatrix}$$

SOLUTION. $|A| = -7$; so A is nonsingular.

$$A^{-1} = \frac{1}{-7} \cdot \begin{pmatrix} \begin{vmatrix} 2 & 1 \\ 0 & -1 \end{vmatrix} & -\begin{vmatrix} 0 & 1 \\ 1 & -1 \end{vmatrix} & \begin{vmatrix} 0 & 2 \\ 1 & 0 \end{vmatrix} \\ -\begin{vmatrix} -1 & 2 \\ 0 & -1 \end{vmatrix} & \begin{vmatrix} 1 & 2 \\ 1 & -1 \end{vmatrix} & -\begin{vmatrix} 1 & -1 \\ 1 & 0 \end{vmatrix} \\ \begin{vmatrix} -1 & 2 \\ 2 & 1 \end{vmatrix} & -\begin{vmatrix} 1 & 2 \\ 0 & 1 \end{vmatrix} & \begin{vmatrix} 1 & -1 \\ 0 & 2 \end{vmatrix} \end{pmatrix}^t$$

$$= \frac{1}{-7} \cdot \begin{pmatrix} -2 & 1 & -2 \\ -1 & -3 & -1 \\ -5 & -1 & 2 \end{pmatrix}^t$$

$$= \begin{pmatrix} \frac{2}{7} & \frac{1}{7} & \frac{5}{7} \\ -\frac{1}{7} & \frac{3}{7} & \frac{1}{7} \\ \frac{2}{7} & \frac{1}{7} & -\frac{2}{7} \end{pmatrix}$$

The student should verify that $AA^{-1} = A^{-1}A = I_3$.

EXERCISE

1. Compute AB or BA (or both).

   (a) $A = \begin{pmatrix} -1 & 0 & 3 \\ 0 & -2 & 1 \\ 2 & 0 & -2 \end{pmatrix}$ $B = \begin{pmatrix} 1 & 2 & 3 \\ 0 & -2 & 4 \\ 1 & 3 & -2 \end{pmatrix}$

   (b) $A = \begin{pmatrix} 2 & -1 & 0 \\ 4 & 3 & -2 \\ 0 & 2 & 1 \end{pmatrix}$ $B = \begin{pmatrix} 3 \\ 4 \\ 2 \end{pmatrix}$

2. Solve for a, b, c, and d.

   (a) $\begin{pmatrix} 1 & -2 \\ 3 & -1 \end{pmatrix} \begin{pmatrix} a & b \\ c & d \end{pmatrix} = \begin{pmatrix} 2 & 1 \\ -1 & 2 \end{pmatrix}$

   (b) $\begin{pmatrix} \frac{1}{3} & 1 \\ -2 & \frac{1}{3} \end{pmatrix} \begin{pmatrix} a & b \\ c & d \end{pmatrix} = \begin{pmatrix} 1 & 0 \\ 0 & 1 \end{pmatrix}$

3. Let

   $A = \begin{pmatrix} \frac{-1}{\sqrt{2}} & \frac{1}{\sqrt{2}} & 0 \\ \frac{1}{\sqrt{3}} & \frac{1}{\sqrt{3}} & \frac{1}{\sqrt{3}} \\ \frac{1}{\sqrt{6}} & \frac{1}{\sqrt{6}} & \frac{-2}{\sqrt{6}} \end{pmatrix}$

   $B = \begin{pmatrix} \frac{1}{\sqrt{2}} & 0 & \frac{1}{\sqrt{2}} \\ 0 & -1 & 0 \\ \frac{1}{\sqrt{2}} & 0 & \frac{-1}{\sqrt{2}} \end{pmatrix}$

   (a) Show that A and B are orthogonal matrices.
   (b) Compute AB and show that AB is orthogonal (Theorem 4).

4. Evaluate the following:

   $\begin{vmatrix} \sec\theta & \tan\theta \\ \tan\theta & \sec\theta \end{vmatrix}$

5. Evaluate by expanding by minors along a row. Check your answer by expanding along a column.

   (a) $\begin{vmatrix} 2 & -1 & 3 \\ -1 & 0 & -2 \\ 1 & 4 & 0 \end{vmatrix}$

   (b) $\begin{vmatrix} 3 & 1 & 2 & 3 \\ 4 & -1 & 2 & 4 \\ 1 & -1 & 1 & 1 \\ 4 & -1 & 2 & 5 \end{vmatrix}$

6. Let
$$A = \begin{pmatrix} a_{11} & a_{12} & a_{13} \\ a_{21} & a_{22} & a_{23} \\ a_{31} & a_{32} & a_{33} \end{pmatrix}$$

Verify that
(a) $|A| = |A^t|$ (Theorem 5).
(b) $|B| = -|A|$, where
$$B = \begin{pmatrix} a_{31} & a_{32} & a_{33} \\ a_{21} & a_{22} & a_{23} \\ a_{11} & a_{12} & a_{13} \end{pmatrix}$$

(Theorem 6).
(c) $|cA| = c^3 \cdot |A|$, where c is a number (Corollary 1 of Theorem 9).

7. Let
$$A = \begin{pmatrix} 2 & -1 & 0 \\ 0 & 3 & -2 \\ 1 & 4 & 5 \end{pmatrix} \quad B = \begin{pmatrix} 1 & -1 & 2 \\ 0 & 2 & -3 \\ 1 & 0 & 4 \end{pmatrix}$$

Verify that
(a) $|AB| = |A| \cdot |B|$ (Theorem 11).
(b) $(AB)^t = B^t A^t$ (Theorem 1).

8. Let
$$A = \begin{pmatrix} 2 & 1 \\ -1 & 2 \end{pmatrix} \quad B = \begin{pmatrix} 2 & 3 \\ 4 & 1 \end{pmatrix}$$

Show that A and B are nonsingular and verify that $(AB)^{-1} = B^{-1} A^{-1}$ (Theorem 3).

9. Find the inverse of the following matrix (if it exists):
$$\begin{pmatrix} -1 & 0 & 3 \\ 0 & -2 & 1 \\ 2 & 0 & -2 \end{pmatrix}$$

Answers To Selected Exercises

CHAPTER 1

Exercise 1

1. F
3. T
5. F
7. T
9. T

Exercise 2

1. (a) $A \cap B = \{3, 8\}$
    $A \cup C = \{1, 4, 3, 8, 7, 2, 5, 9\}$
    $B - C = \{2, 5, 9, 10\}$
    $C - B = \emptyset$
   (c) $C \times D = \{(3, 2), (3, 5), (3, 9), (0, 2), (0, 5), (0, 9), (8, 2), (8, 5), (8, 9)\}$
    $D \times C = \{(2, 3), (5, 3), (9, 3), (2, 0), (5, 0), (9, 0), (2, 8), (5, 8), (9, 8)\}$
2. (a) $x = \frac{1}{2}, y = -\frac{5}{2}$
   (c) $x = 2, y = 3, z = 1$

Exercise 3

1. (a) $x > -4$

Answers to Selected Exercises

(c) $x < \dfrac{8}{3(3 - \sqrt{2})}$

2. (a) $2 < x < 4$
   (c) All x
   (e) $-\dfrac{\sqrt{5} - 1}{2} a \leq x \leq \dfrac{\sqrt{5} + 1}{2} a$

3. (a) $x > \sqrt{\tfrac{3}{5}}$ or $-\sqrt{\tfrac{3}{5}} < x < 0$
   (c) $x < 2$ or $x > \tfrac{19}{7}$

4. (a) $x = 7, -\tfrac{1}{3}$
   (c) $x = \pm 1, \pm \sqrt{7}$

Exercise 4

1. (a) $-\sqrt{2}$
   (c) $\sqrt{2}$
   (e) -1

2. (a) $210°$
   (c) $-120°$
   (e) $\dfrac{315°}{2}$

3. (a) $\tfrac{2}{3}\pi$
   (c) $\tfrac{25}{36}\pi$
   (e) $-\tfrac{5}{6}\pi$

4. (a) $\sin \theta = \dfrac{3}{4}$ $\csc \theta = \dfrac{4}{3}$

   $\cos \theta = -\dfrac{\sqrt{7}}{4}$ $\sec \theta = -\dfrac{4}{\sqrt{7}}$

   $\tan \theta = -\dfrac{3}{\sqrt{7}}$ $\cot \theta = -\dfrac{\sqrt{7}}{3}$

   (c) $\sin \theta = \pm \dfrac{1}{\sqrt{5}}$ $\csc \theta = \pm \sqrt{5}$

   $\cos \theta = \dfrac{2}{\sqrt{5}}$ $\sec \theta = \dfrac{\sqrt{5}}{2}$

   $\tan \theta = \pm \dfrac{1}{2}$ $\cot \theta = \pm 2$

   (e) $\sin \theta = -\dfrac{1}{\sqrt{10}}$ $\csc \theta = -\sqrt{10}$

   $\cos \theta = -\dfrac{3}{\sqrt{10}}$ $\sec \theta = -\dfrac{\sqrt{10}}{3}$

   $\tan \theta = 3$ $\cot \theta = \dfrac{1}{3}$

CHAPTER 2

Exercise 1
1. (a) 0
   (c) 12
   (e) $3\pi^2$
   (g) $3(a^2 + 2ah + h^2)$
3. (a) $-4 \leq x \leq 4$. y is not a function of x.
   (c) $x \geq 0$. y is not a function of x.
   (e) All x. y is a function of x.
   (g) All x. y is a function of x.
   (i) $x \geq -2$. y is not a function of x.
4. (a) All x
   (c) $x \geq 0$
   (e) All x
7. (a) $y = \sqrt{\dfrac{x^2 - 1}{2}}$
   (c) $y = x + 1$
   (e) $y = x$

Exercise 2
1. $D_{f+g} = R$, $D_{f-g} = R$, $D_{f \cdot g} = R$, $D_{f/g} = \{x \in R | x \neq 0\}$
3. (a) $f(g)(x) = \sqrt{|x|}$; $\quad D_{f(g)} = R$
       $g(f)(x) = \sqrt{x}$; $\quad D_{g(f)} = \{x \in R | x \neq 0\}$
   (c) $f(g)(x) = \dfrac{x^2 + 2x - 1}{(x+1)^2}$; $\quad D_{f(g)} = \{x \in R | x \neq -1\}$
       $g(f)(x) = \dfrac{1}{x^2 - 1}$; $D_{g(f)} = \{x \in R | x \neq \pm 1\}$
   (e) $f(g)(x) = \dfrac{1}{x^2 + 2}$; $D_{f(g)} = R$
       $g(f)(x) = \dfrac{1 + 2x^2}{x^2}$; $D_{g(f)} = \{x \in R | x \neq 0\}$

Exercise 3
1. $f^{-1}(x) = (x + 2)^2$
3. $f^{-1}(x) = \dfrac{-x + 5}{3}$
5. $f^{-1}(x) = (x + 5)^{1/3}$
7. $f^{-1}(x) = -\sqrt{x - 2}$
9. $f^{-1}(x) = 2 + \sqrt{x - 3}$
11. $f^{-1}(x) = x^2 - 5; x \geq 0$

Exercise 4

1. (a) Intercepts: $(\pm\sqrt{3}, 0), (0, \pm\sqrt{2})$
   Symmetry: about x- and y-axes and origin
   Extent: $-\sqrt{3} \le x \le \sqrt{3}; -\sqrt{2} \le y \le \sqrt{2}$
   (c) Intercepts: $(3, 0)$
   Symmetry: about x-axis
   Extent: $x \ge 3$
2. (a) Union of graphs of $x = \pm 1$ and $y = 0$
   (c) Union of graphs of $x^2 + y^2 = 4$ and $y = x$
3. (a) Intercepts: none
   Symmetry: about y-axis
   Extent: $y > 0$
   Asymptotes: $x = 0, y = 0$
   (c) Intercepts: $(0, -\frac{1}{6})$
   Extent: $x \ne 2, -3$
   Asymptotes: $x = 2, x = -3, y = 0$
4. (a) Extent: $x > 3$
   Symmetry: about x-axis
   Asymptotes: $x = 3, y = 0$
   (c) Intercepts: $(0, \frac{1}{9})$
   Extent: $y > 0$
   Asymptotes: $x = 3, y = 0$
5. (a) Symmetry: about x- and y-axes and origin
6. (a) Intercepts: $(0, 0)$
   Asymptotes: $x = 2, x = 3, y = \sqrt[3]{\frac{1}{2}}$
7. (a) Extent: $y \ne 0$
   Asymptotes: $y = \frac{1}{2}, x = \frac{1}{2}$
   (c) Intercepts: $(1, 0)$
   Symmetry: about x-axis
   Asymptotes: $x = 0$
   Extent: $0 < x \le 1$

Exercise 5

1. (a) $y = 7 + 6(x + 2)^2$
   (c) $\dfrac{x^2}{25} + \dfrac{(y-3)^2}{36} = 1$
   (e) $y = -x$
   (g) $x^2 + \dfrac{y^2}{9} = 1$
   (i) $y = x^2$
2. (a) Simple
   (c) Simple
   (e) Simple
3. (a) Not closed
   (c) Not simple
   (e) Closed

Answers to Selected Exercises

4. (a)

θ	0	$\frac{1}{2}\pi$	π	$\frac{3}{2}\pi$	2π
x	0	$(\frac{1}{2}\pi - 1)r$	πr	$(\frac{3}{2}\pi + 1)r$	$2\pi r$
y	0	r	$2r$	r	0

5. $xy = x^3 + y^3$

7. $x = \dfrac{3a}{4}\cos\varphi + \dfrac{a}{4}\cos 3\varphi$

   $y = \dfrac{3a}{4}\sin\varphi - \dfrac{a}{4}\sin 3\varphi$

CHAPTER 3

Exercise 1

1. (a) $|\overrightarrow{P_1P_2}| = 2, \alpha = \frac{7}{4}\pi, m = -1$
   (c) $|\overrightarrow{P_1P_2}| = 4\sqrt{2}, \alpha = \frac{3}{4}\pi, m = -1$
   (e) $|\overrightarrow{P_1P_2}| = 6, \alpha = 0, m = 0$
3. (a) $(-5, 5\sqrt{3})$
   (c) $\left(-\dfrac{2 + 5\sqrt{2}}{2}, -\dfrac{8 + 5\sqrt{2}}{2}\right)$
4. (a) $(7, -4)$

Exercice 2

1. (a) $(1, 7)$
2. (a) Equivalent
   (c) Equivalent
3. (c) $|\mathbf{v}_1| = \sqrt{13}, |\mathbf{v}_2| = \sqrt{37}, |\mathbf{v}_1 + \mathbf{v}_2| = 2\sqrt{17}, |\mathbf{v}_1 - \mathbf{v}_2| = 4\sqrt{2}$
4. (a) $(-6, 5)$
   (c) $(\frac{8}{3}, 12)$
   (e) $\mathbf{v}_1 = (\frac{1}{7}, -\frac{10}{7}), \mathbf{v}_2 = (\frac{-5}{7}, -\frac{6}{7})$

Exercise 3

1. (a) $315°$
   (c) $180°$
   (e) $\tan\alpha = -\frac{3}{2}$, 2nd quadrant
2. (a) $90°$
   (c) $45°$
   (e) $180°$
3. (a) Perpendicular
   (c) Perpendicular
   (e) Perpendicular
4. (b) $-2\mathbf{i} + 3\mathbf{j}$ and $2\mathbf{i} - 3\mathbf{j}$
12. (a) $a\mathbf{i}$
    (c) $\mathbf{0}$
    (e) $2\mathbf{i} + \mathbf{j}$
    (g) $4\mathbf{i} + 2\mathbf{j}$

Exercise 4

1. (a) $\begin{cases} x = 4 - 9t \\ y = -1 + 3t, 0 \leq t \leq 1 \end{cases}$
   (c) $(\frac{7}{4}, -\frac{1}{4})$
   (e) $-\frac{1}{3}$
2. (a) Noncollinear
   (c) Collinear
   (e) Collinear
   (g) Collinear
3. (a) $-\frac{5}{3}$
4. (a) $(0, \frac{24}{7})$
   (c) $\left(-1 + \frac{35}{\sqrt{58}}, 3 + \frac{15}{\sqrt{58}}\right), \left(-1 - \frac{35}{\sqrt{58}}, 3 - \frac{15}{\sqrt{58}}\right)$
5. (a) $(\frac{3}{5}, \frac{24}{5})$
6. (a) None
   (c) None
   (e) Equilateral
9. $(4, 2)$
14. (c) $(-6, -6), (4, 14), (8, 10)$
15. $(1, 8), (9, 0), (-3, -4)$

CHAPTER 4

Exercise 1

3. (1) (a) $x = -5 + 12t, y = 1 - 4t$
   (b) $y - 1 = \frac{1 - (-3)}{-5 - 7}(x - (-5))$
   $y - 1 = -\frac{1}{3}(x - (-5))$
   $\frac{x}{-2} + \frac{y}{-\frac{2}{3}} = 1$
   $y = -\frac{1}{3}x - \frac{2}{3}$

   (3) (a) $x = -2 + 5t, y = -5 + 9t$
   (b) $y - (-5) = \frac{4 - (-5)}{3 - (-2)}(x - (-2))$
   $y - (-5) = \frac{9}{5}(x - (-2))$
   $\frac{x}{\frac{7}{9}} + \frac{y}{-\frac{7}{5}} = 1$
   $y = \frac{9}{5}x - \frac{7}{5}$

   (5) (a) $x = 3 + 0t, y = 10 - 11t$
   (b) $x = 3$; all other forms undefined
4. (a) $11x - y + 73 = 0$
   (c) $5x + y = 0$
   (e) $x + 6 = 0$
   (g) $4x + y = 0$

Answers to Selected Exercises

5. (a) $5x + 8y - 41 = 0$
   (c) $8x - 5y + 29 = 0$
6. (a) $6x - 2y + 15 = 0$
   (c) $6x - y + 2 = 0$

Exercise 2

1. (a) $x = 5 + 2t, y = 0 + 6t; 3x - y - 15 = 0$
   (c) $x = 0 + t, y = 0 + t; x - y = 0$
2. (a) $\mathbf{i} + \mathbf{j}; -\mathbf{i} + \mathbf{j}$
   (c) $4\mathbf{i} + 3\mathbf{j}; -3\mathbf{i} + 4\mathbf{j}$
   (e) $2\mathbf{i}; \mathbf{j}$
   (g) $\mathbf{i} - 6\mathbf{j}; 6\mathbf{i} + \mathbf{j}$
3. (a) $(1, 3)$
   (c) $(1, 1)$
   (e) No solution

Exercise 3

1. (a) $\frac{1}{6}\pi$
   (c) $\frac{3}{4}\pi$
   (e) $\tan \alpha = 0.25, \alpha \approx 14°$
   (g) 0
2. (a) $\tan \varphi = \frac{19}{9} \approx 2.11, \varphi \approx 64.7°$
   (c) $\tan \varphi = 0.5, \varphi \approx 26.5°$
3. (a) Parallel
   (c) Perpendicular
   (e) Parallel
   (f) $\varphi = 60°$
5. (a) $2x - 5y - 22 = 0$
   (c) $2x - y + 6 = 0$
   (e) $3x + y + 3 = 0$ and $x - 3y - 9 = 0$

Exercise 4

1. (a) $\dfrac{7\sqrt{5}}{10}$
   (c) $\dfrac{49\sqrt{13}}{26}$
   (e) $\frac{7}{3}$
2. (a) $x = 5; y = 1$
   (c) $x + (8 - 5\sqrt{3})y + 5\sqrt{3} - 24 = 0; 11x + (8 + 5\sqrt{3})y - (24 + 5\sqrt{3}) = 0$
   (e) $2y + 3\sqrt{3} - 7 = 0; 2\sqrt{3}x + \sqrt{3} + 1 = 0$
3. (a) $\dfrac{\sqrt{3}}{2}x + \dfrac{y}{2} - 4 = 0; \frac{1}{6}\pi; 4$
   (c) $y - \frac{4}{3} = 0; \frac{1}{2}\pi; \frac{4}{3}$
   (e) $-\dfrac{x}{\sqrt{2}} - \dfrac{y}{\sqrt{2}} - \sqrt{2} = 0; 225°; \sqrt{2}$

(g) $-\dfrac{x}{\sqrt{10}} - \dfrac{3}{\sqrt{10}}y - \dfrac{4}{\sqrt{10}} = 0; \cos\omega = \dfrac{-\sqrt{10}}{10}, \sin\omega = \dfrac{-3\sqrt{10}}{10}; \dfrac{2\sqrt{10}}{5}$

(i) $x - \tfrac{5}{2} = 0; 0°; \tfrac{5}{2}$

(k) $\dfrac{x}{\sqrt{2}} + \dfrac{y}{\sqrt{2}} = 0; \tfrac{1}{4}\pi; 0$

4. (a) $x + a^2y - a = 0; x + y + 1 = 0$ and $9x + 4y - 6 = 0$

(c) $x\cos\omega + y\sin\omega - 5 = 0; \dfrac{x}{\sqrt{5}} - \dfrac{2}{\sqrt{5}}y - 5 = 0$

and $-\dfrac{x}{\sqrt{5}} + \dfrac{2}{\sqrt{5}}y - 5 = 0$

(e) $y + 2 = m(x - 4); y + 2 = -(x - 4)$ and $y + 2 = -\tfrac{1}{2}(x - 4)$

5. (a) $(2h + 3k)x + (k - h)y - (4h + 11k) = 0; 3x - 4y - 1 = 0$

(c) $(2h + k)x + (h + k)y + (h - 2k) = 0$
$7x + 3y + 6 = 0$ and $101x + 81y - 102 = 0$

Exercise 5

1. (a) $x^2 + y^2 = 16; x^2 + y^2 - 16 = 0$

(c) $(x - 4)^2 + (y + \tfrac{1}{2})^2 = 2; 4x^2 + 4y^2 - 32x + 4y + 57 = 0$

(e) $(x - 5)^2 + (y - \tfrac{3}{4})^2 = \tfrac{1}{25}; 400x^2 + 400y^2 - 4000x - 600y + 10{,}209 = 0$

2. (a) $(x - 3)^2 + (y + 4)^2 = 5; (3, -4); \sqrt{5}$

(c) $(x + 5)^2 + (y - \tfrac{2}{3})^2 = -45$; no locus

(e) $(x - 5)^2 + (y - \tfrac{1}{5})^2 = \tfrac{1}{5}; (5, \tfrac{1}{5}); \dfrac{1}{\sqrt{5}}$

3. (a) $x + 2y - 13 = 0$

(c) $y + 4 = 0$

4. (a) $2x - 3y - 13 = 0$ and $2x + 3y - 13 = 0; \dfrac{3\sqrt{13}}{2}$

(c) No solution

(e) $2x + y = 0$ and $x - 2y + 5 = 0; \sqrt{5}$

5. (a) $(x - 3)^2 + (y - 1)^2 = 13$

(c) $(x + 2)^2 + (y + 3)^2 = \tfrac{49}{4}$

6. (e) $(x - 1)^2 + (y - 1)^2 = 13$

Exercise 6

1. (a) $(x - h)^2 + (y - (5 - h))^2 = r^2; (x - 3)^2 + (y - 2)^2 = 9$

and $(x + 3)^2 + (y - 8)^2 = 9$

(c) $\left(x - \left(\dfrac{k + \sqrt{5}|k|}{2}\right)\right)^2 + (y - k)^2 = k^2;$

$(x - (2 + 2\sqrt{5}))^2 + (y - 4)^2 = 16$
$(x - (2 - 2\sqrt{5}))^2 + (y - 4)^2 = 16$
$(x - (2\sqrt{5} - 2))^2 + (y + 4)^2 = 16$
$(x + 2\sqrt{2} + 2)^2 + (y + 4)^2 = 16$

Answers to Selected Exercises

(e) $(x-h)^2 + (y-3(1-h))^2 = r^2; (x-\frac{1}{2})^2 + (y-\frac{3}{2})^2 = \frac{5}{2}$

2. (a) $(x-\frac{19}{14})^2 + (y-\frac{5}{7})^2 = \frac{1105}{196}$
   (c) $19x^2 + 19y^2 - 11x - 43y - 260 = 0$
   (e) No solution

3. (1) (a) $(s+t)x^2 + (s+t)y^2 - (s+8t)x + (3s-4t)y - 4s + 10t = 0$
   (b) $x^2 + y^2 + 2x + 6y - 10 = 0$
   (c) $x + y - 2 = 0$
   (3) (a) $(2s+2t)x^2 + (2s+2t)y^2 + (12s+16t)x - (5s+11t)y = 0$
   (b) $2x^2 + 2y^2 + 13y = 0$
   (c) $2x - 3y = 0$

Exercise 7

5. (1) (b) $x = -5 + 8t, y = 2 - 3t; \frac{10}{39}$
   (3) (b) $x = -4 + 7t, y = 5 - 4t; \frac{1}{2}$
   (5) (b) $x = 1 + t, y = -1 + 3t$; no solution

CHAPTER 5

Exercise 1

1. (a) $(\sqrt{3}, -1)$
   (c) $(1, 0)$
   (e) $(0, 0)$
   (g) $(-\sqrt{2}, \sqrt{2})$

2. (a) $(\sqrt{2}, \frac{1}{4}\pi), (-\sqrt{2}, \frac{5}{4}\pi), (\sqrt{2}, \frac{9}{4}\pi)$
   (c) $(2, \frac{4}{3}\pi), (-2, \frac{1}{3}\pi), (2, \frac{2}{3}\pi)$
   (e) $(2, 0), (-2, \pi), (-2, -\pi)$
   (g) $(\sqrt{2}, \frac{3}{4}\pi), (-\sqrt{2}, -\frac{1}{4}\pi), (\sqrt{2}, \frac{11}{4}\pi)$

Exercise 3

1. (a) $r = \dfrac{a}{\sin\theta + \cos\theta}$
   (c) $r^2 = \dfrac{36}{9\cos^2\theta + 4\sin^2\theta}$
   (e) $r = 12\sec\theta, r = 0$
   (g) $r^2 + a^2 \sec 2\theta = 0$
   (i) $r = \pm \sin 2\theta$
   (k) $r = \pm 2 \csc 2\theta$

2. (a) $x^2 + y^2 = 100$
   (c) $x = a$
   (e) $x^4 + 2x^2y^2 + y^4 - 2xy = 0$
   (g) $3x^2 + 4y^2 - 4x - 4 = 0$

3. (a) $r\cos(\theta - \frac{1}{6}\pi) = 4$
   (c) $r\cos(\theta - \frac{1}{4}\pi) = -2$
   (e) $r\cos(\theta + 150°) = 6$
4. (a) $r^2 - 8r\cos(\theta - \frac{1}{3}\pi) = 0$
   (c) $r = 5$
5. $r\cos(\theta - \omega) = r[\cos\theta\cos\omega + \sin\theta\sin\omega] = p$
   $\cos\omega(r\cos\theta) + \sin\omega(r\sin\theta) = p$
   $(\cos\omega)x + (\sin\omega)y = p$
7. (a) $\sqrt{13}$
   (c) $\sqrt{65}$
   (e) $\sqrt{60}$
8. (a) $x = (1 - \cos\theta)\cos\theta$
       $y = (1 - \cos\theta)\sin\theta$
   (c) $x = 5\cos\theta$
       $y = 5\sin\theta$
   (e) $x = \frac{1}{2}\csc\theta$
       $y = \frac{1}{2}\sec\theta$
9. $t = \frac{x}{2} - 1; y = \left(\frac{x}{2} - 1\right)\left(\frac{x}{2} + 1\right) = \frac{x^2}{4} - 1$

   $r - r\sin\theta = 2; \sqrt{x^2 + y^2} - y = 2; x^2 + y^2 = 4 + 4y + y^2; y = \frac{x^2}{4} - 1$

Exercise 4

1. $(1, 0), (-\frac{1}{2}, \frac{2}{3}\pi), (-\frac{1}{2}, \frac{4}{3}\pi)$
3. $\left(\frac{2 - \sqrt{2}}{2}, \frac{\pi}{4}\right); \left(\frac{2 + \sqrt{2}}{2}, \frac{\pi}{4}\right)$
5. $(\frac{1}{2}, \frac{1}{6}\pi); (\frac{1}{2}, \frac{5}{6}\pi)$
7. $(-2, \frac{1}{6}\pi), (-2, \frac{5}{6}\pi)$
9. $(6, \frac{1}{6}\pi), (6, \frac{5}{6}\pi)$
11. $\left(\pm\frac{1}{\sqrt{2}}, \frac{\pi}{8}\right), (0, 0)$

CHAPTER 6

Exercise 1

1. $x' = x + 3$
   $y' = y - 4$
   (a) $(3, -4)$
   (c) $(7, -6)$
2. $x' = -\frac{1}{2}x + \frac{\sqrt{3}}{2}y$
   $y' = -\frac{\sqrt{3}}{2}x - \frac{1}{2}y$

Answers to Selected Exercises

(a) $(0, 0)$
(c) $(-2 - \sqrt{3}, 1 - 2\sqrt{3})$

3. $(-4, 6)$
4. (a) $4x'^2 + y'^2 = 4$; transl. to $(2, -3)$
   (c) $x'^2 - 2y'^2 = \frac{35}{4}$; transl. to $(\frac{5}{2}, \frac{3}{2})$
   (e) $x'^3 - 3y'^3 = 7$; transl. to $(4, -5)$
5. $\left(\frac{1}{2} - \sqrt{3}, -1 - \frac{\sqrt{3}}{2}\right)$
6. (a) $x'^2 + y'^2 = 5$
   (c) $-3\sqrt{3}x'^2 + 3x'^2y' - \sqrt{3}x'y'^2 + y'^3 - 4x'^2 - 8\sqrt{3}x'y' - 12y'^2 = 0$
   (e) $\dfrac{y'^2}{4} + \dfrac{x'^2}{9} = 1$
8. (a) $x'' = \dfrac{1}{2}x - \dfrac{\sqrt{3}}{2}y - \left(\dfrac{3 + 4\sqrt{3}}{2}\right)$

   $y'' = -\dfrac{\sqrt{3}}{2}x + \dfrac{1}{2}y + \left(\dfrac{4 + 3\sqrt{3}}{2}\right)$

Exercise 2

1. $\begin{pmatrix} x'' \\ y'' \end{pmatrix} = \begin{pmatrix} -\dfrac{1}{\sqrt{2}}x + \dfrac{1}{\sqrt{2}}y + 3 \\ -\dfrac{1}{\sqrt{2}}x - \dfrac{1}{\sqrt{2}}y - 2 \end{pmatrix}$; $(3 - \sqrt{2}, -2 - 2\sqrt{2})$

   $\begin{pmatrix} x'' \\ y'' \end{pmatrix} = \begin{pmatrix} -\dfrac{1}{\sqrt{2}}(x + 3) + \dfrac{1}{\sqrt{2}}(y - 2) \\ -\dfrac{1}{\sqrt{2}}(x + 3) - \dfrac{1}{\sqrt{2}}(y - 2) \end{pmatrix}$; $\left(-\dfrac{7}{\sqrt{2}}, -\dfrac{5}{\sqrt{2}}\right)$

5. (a) $\begin{pmatrix} 1 & 0 & -h \\ 0 & 1 & -k \\ 0 & 0 & 1 \end{pmatrix} \begin{pmatrix} x \\ y \\ 1 \end{pmatrix} = \begin{pmatrix} x - h \\ y - k \\ 1 \end{pmatrix} = \begin{pmatrix} x' \\ y' \\ 1 \end{pmatrix}$ so that $x' = x - h, y' = y - k$

CHAPTER 7

Exercise 1

1. (a) $y^2 - 6y - 4x + 5 = 0$
   (c) $x^2 + 2xy + y^2 - 4x - 12y + 20 = 0$
   (e) $16x^2 - 24xy + 9y^2 + 96y + 72x - 144 = 0$
2. (a) $5x^2 + 9y^2 - 24x - 36 = 0$
   (c) $15x^2 + 16y^2 + 68x - 128y + 316 = 0$
   (e) $x^2 + y^2 + 16xy + 66x - 84y - 27 = 0$

Answers to Selected Exercises

4. (a) Parabola; 1; $x = -3$
   (c) Hyperbola; $\frac{4}{3}$; $y = -\frac{9}{4}$
   (e) Hyperbola; 5; $x = \frac{7}{5}$

5. $e = 1, \left(\frac{p}{2}, 0\right), x' = -\frac{p}{2}; e < 1, (-ea, 0), x' = -(p + ea); e > 1, (ea, 0), x' = ea - p$

Exercise 2

1. (a) $V(0, 0); F(0, -1);$ axis $x = 0;$ directrix $y = 1; \overline{(-2, -1)(2, -1)}$
   (c) $V(0, 0); F(-\frac{2}{3}, 0);$ axis $y = 0;$ directrix $x = \frac{2}{3}; \overline{(-\frac{2}{3}, \frac{4}{3})(-\frac{2}{3}, -\frac{4}{3})}$
2. (a) $V(3, -2); F(4, -2);$ axis $y = -2;$ directrix $x = 2; \overline{(4, 0)(4, -4)}$
   (c) $V(-\frac{1}{2}, -3); F(-1, -3);$ axis $y = -3;$ directrix $x = 0; \overline{(-1, -2)(-1, -4)}$
3. (a) $(y + 3)^2 = -2(x - 1); V(1, -3); F(\frac{1}{2}, -3);$ axis $y = -3;$ directrix $x = \frac{3}{2};$
      chord $\overline{(\frac{1}{2}, -2)(\frac{1}{2}, -4)}$
   (c) $(y - \frac{3}{2})^2 = -4(x + 2); V(-2, \frac{3}{2}); F(-3, \frac{3}{2});$ axis $y = \frac{3}{2};$ directrix $x = -1;$
      chord $\overline{(-3, \frac{7}{2})(-3, -\frac{1}{2})}$
   (e) $(y - \frac{3}{2})^2 = \frac{1}{4};$ parallel lines $y = 1$ and $y = 2$
   (g) $y^2 = 3(x + 2); V(-2, 0); F(-\frac{5}{4}, 0);$ axis $y = 0;$ directrix $x = -\frac{11}{4};$
      chord $\overline{(-\frac{5}{4}, \frac{3}{2})(-\frac{5}{4}, -\frac{3}{2})}$
   (i) $(y + 1)^2 = -8(x - 4); V(4, -1); F(2, -1),$ axis $y = -1;$ directrix $x = 6;$
      chord $\overline{(2, 3)(2, -5)}$
5. (a) $y^2 = 16x$
   (c) $(x + 1)^2 = 8y$
   (e) $(y - 3)^2 = -4(x + 1)$ and $(y - 3)^2 = 4(x + 3)$
   (g) $(y - 2)^2 = 4(x + 1)$
7. (a) $(x - 2)^2 = -8(y - 1)$

Exercise 3

1. (a) $C(0, 0); a = 2, b = 1; F_1(0, \sqrt{3}), F_2(0, -\sqrt{3}); V_1(0, 2), V_2(0, -2); e = \dfrac{\sqrt{3}}{2}$
   (c) $C(0, 0); a = 5, b = 1; F_1(-3, 0), F_2(3, 0); V_1(-5, 0), V_2(5, 0); e = \frac{3}{5}$
   (e) $C(0, 0); a = 2, b = \sqrt{3}; F_1(-1, 0), F_2(1, 0); V_1(-2, 0), V_2(2, 0); e = \frac{1}{2}$
2. (a) $\dfrac{(x + 1)^2}{4} + \dfrac{(y - 1)^2}{1} = 1; C(-1, 1); a = 2, b = 1; F_1(-1 - \sqrt{3}, 1),$
      $F_2(-1 + \sqrt{3}, 1); V_1(-3, 1), V_2(1, 1); e = \dfrac{\sqrt{3}}{2}$
   (c) $\dfrac{(x - 3)^2}{4} + \dfrac{(y - \frac{3}{2})^2}{16} = 1; C(3, \frac{3}{2}); a = 4, b = 2; F_1(3, \frac{3}{2} - 2\sqrt{3}),$
      $F_2(3, \frac{3}{2} + 2\sqrt{3}); V_1(3, -\frac{5}{2}), V_2(3, \frac{11}{2}); e = \dfrac{\sqrt{3}}{2}$
   (e) $2(x - 2)^2 + 3(y - 1)^2 = 0;$ point ellipse; $C(2, 1)$
3. (a) $\dfrac{x^2}{81} + \dfrac{y^2}{5} = 1$

(c) $\dfrac{x^2}{49} + \dfrac{y^2}{16} = 1$

(e) $\dfrac{(x+2)^2}{9} + \dfrac{(y+3)^2}{1} = 1$

7. (a) $\dfrac{x^2}{36} + \dfrac{y^2}{20} = 1$

Exercise 4

1. (a) $\dfrac{x^2}{1} - \dfrac{y^2}{4} = 1$; $V_1(-1, 0), V_2(1, 0); F_1(-\sqrt{5}, 0), F_2(\sqrt{5}, 0); y = \pm 2x$

(c) $\dfrac{y^2}{16} - \dfrac{x^2}{25} = 1$; $V_1(0, -4), V_2(0, 4); F_1(0, -\sqrt{41}), F_2(0, \sqrt{41}); y = \pm \tfrac{4}{5}x$

(e) $\dfrac{x^2}{4} - \dfrac{y^2}{3} = 1$; $V_1(-2, 0), V_2(2, 0); F_1(-\sqrt{7}, 0), F_2(\sqrt{7}, 0); y = \pm \dfrac{\sqrt{3}}{2} x$

2. (a) $\dfrac{(x+1)^2}{9} - \dfrac{(y-2)^2}{4} = 1$; $V_1(-4, 2), V_2(2, 2); F_1(-1 - \sqrt{13}, 2),$
$F_2(-1 + \sqrt{13}, 2); y - 2 = \pm \tfrac{2}{3}(x+1)$

(c) $y - 1 = \pm \tfrac{9}{2}(x + 3)$, degenerate hyperbola

(e) $y + 1 = \pm \dfrac{x-2}{3}$, degenerate hyperbola

3. (a) $\dfrac{x^2}{9} - \dfrac{y^2}{16} = 1$

(c) $\dfrac{x^2}{16} - \dfrac{y^2}{20} = 1$

(e) $\dfrac{y^2}{9} - \dfrac{x^2}{4} = 1$

4. (a) $\dfrac{(x-4)^2}{3} + \dfrac{(y - \tfrac{2}{3})^2}{9} = 1$; ellipse

(c) $\dfrac{(x - \tfrac{2}{3})^2}{1} + \dfrac{(y+3)^2}{4} = 1$; ellipse

(e) $(y - \tfrac{4}{5})^2 = 3(x - 2)$; parabola

5. (a) $\dfrac{x^2}{4} - \dfrac{y^2}{12} = 1$

(c) $2xy = 1$

Exercise 5

1. (a) hyperbola; $x = x' - 7, y = y' + 7; x' = \dfrac{1}{\sqrt{2}}(x'' - y''), y' = \dfrac{1}{\sqrt{2}}(x'' + y'')$;

$\dfrac{y''^2}{108} - \dfrac{x''^2}{108} = 1$

(c) parabola; $x = x' + 2$; $y = y'$; $x' = \dfrac{1}{\sqrt{2}}(x'' - y'')$, $y' = \dfrac{1}{\sqrt{2}}(x'' + y'')$;

$x'' = 0$, degenerate parabola

(e) parabola; $x = x' + 3$, $y = y'$; $x' = \dfrac{1}{\sqrt{5}}(2x'' - y'')$, $y' = \dfrac{1}{\sqrt{5}}(x'' + 2y'')$;

$(x'' + \sqrt{5})^2 = -4\sqrt{5}\,(y'' - 2\sqrt{5})$

Exercise 6

1. $P = \begin{pmatrix} \dfrac{1}{\sqrt{2}} & -\dfrac{1}{\sqrt{2}} \\ \dfrac{1}{\sqrt{2}} & \dfrac{1}{\sqrt{2}} \end{pmatrix}$; $\tfrac{1}{2}(x')^2 + \tfrac{3}{2}(y')^2 = 5$; ellipse

3. $P = \begin{pmatrix} \dfrac{1}{2} & \dfrac{\sqrt{3}}{2} \\ -\dfrac{\sqrt{3}}{2} & \dfrac{1}{2} \end{pmatrix}$; $(y')^2 = 4$; two parallel lines (degenerate parabola)

5. $P = \begin{pmatrix} \dfrac{\sqrt{3}}{2} & -\dfrac{1}{2} \\ \dfrac{1}{2} & \dfrac{\sqrt{3}}{2} \end{pmatrix}$; $4(x')^2 - 4(y')^2 = 36$; hyperbola

CHAPTER 8

Exercise 1

1. (a) Not equivalent
   (c) Equivalent

Exercise 2

1. (b) (2, 6, 5); (0, −4, 1); (0, 4, −1)
   (d) (−3, −15, −6)
2. (a) (−6, 5, 1)
3. (a) Parallel
   (c) Not parallel

Exercise 3

1. (a) $\sqrt{19}$
   (c) 6
3. (a) $|\mathbf{v}_1 \circ \mathbf{v}_2| = |0| < \sqrt{14}\sqrt{21} = |\mathbf{v}_1||\mathbf{v}_2|$
       $|\mathbf{v}_1 + \mathbf{v}_2| = \sqrt{35} < \sqrt{14} + \sqrt{21} = |\mathbf{v}_1| + |\mathbf{v}_2|$
       since $(\sqrt{14} + \sqrt{21})^2 = 35 + 2\sqrt{14 \cdot 21} > (\sqrt{35})^2$

4. (a) $|\overline{P_1P_2}| = \sqrt{19}$
$|\overline{P_2P_3}| = \sqrt{35}$ So $|\overline{P_1P_3}| < |\overline{P_1P_2}| + |\overline{P_2P_3}|$
$|\overline{P_1P_3}| = \sqrt{8}$

Exercise 4
1. (a) 0
   (c) $\frac{1}{2}\pi$
   (e) $\frac{1}{6}\pi$
2. (a) $\cos \alpha = \frac{2}{\sqrt{21}}; \cos \beta = -\frac{1}{\sqrt{21}}; \cos \gamma = \frac{4}{\sqrt{21}}$
   (c) $\cos \alpha = \cos \beta = \cos \gamma = \frac{1}{\sqrt{3}}$
3. (a) $-5\mathbf{i} + 4\mathbf{j} - 6\mathbf{k}, \cos \alpha = -\frac{5}{\sqrt{77}}; \cos \beta = \frac{4}{\sqrt{77}}; \cos \gamma = -\frac{6}{\sqrt{77}}$
   (c) $-\mathbf{i} - 2\mathbf{j} - 3\mathbf{k}, \cos \alpha = -\frac{1}{\sqrt{14}}; \cos \beta = -\frac{2}{\sqrt{14}}; \cos \gamma = -\frac{3}{\sqrt{14}}$
4. (a) Parallel, opposite direction
   (c) Perpendicular
   (e) Parallel, same direction
5. (a) $5\mathbf{i} - 2\mathbf{j} + 6\mathbf{k}$
   (c) 0
   (e) $\frac{1}{2}\mathbf{i} + \frac{7}{3}\mathbf{j} - 7\mathbf{k}$
6. (a) $\gamma = 90°$
   (c) $\left(-\frac{\sqrt{3}}{2}, \frac{\sqrt{2-\sqrt{3}}}{2}, -\frac{\sqrt{\sqrt{3}-1}}{2}\right)$

Exercise 5
1. (a) $-\mathbf{i} + 3\mathbf{j} + 5\mathbf{k}$
   $\mathbf{i} - 3\mathbf{j} - 5\mathbf{k}$
   (c) $-2\mathbf{i} + \frac{1}{3}\mathbf{k}$
   $2\mathbf{i} - \frac{1}{3}\mathbf{k}$
   (e) $\mathbf{a} \times \mathbf{b} = \mathbf{b} \times \mathbf{a} = 0$
2. (a) $\mathbf{b} = \mathbf{i} - \mathbf{j}; \mathbf{c} = \mathbf{i} + \mathbf{j} - 2\mathbf{k}$
   (c) $\mathbf{b} = \mathbf{i} - 2\mathbf{j} + 2\mathbf{k}; \mathbf{c} = 9\mathbf{j} + 9\mathbf{k}$
   (e) $\mathbf{b} = \mathbf{i} - \mathbf{j} - \mathbf{k}; \mathbf{c} = 5\mathbf{i} + 4\mathbf{j} + \mathbf{k}$
3. $\mathbf{a} = \frac{1}{3}\mathbf{b}; \mathbf{a} \times \mathbf{b} = 0$
6. (a) 2
   (c) 3

CHAPTER 9

Exercise 1
1. (a) $x = 2 - 2t$
   $y = 1 + 7t$
   $z = 3 - 5t$
   (b) $x = 3$
   $y = 4$
   $z = 7 - 9t$

Answers to Selected Exercises

(e) $x = 0$
$y = 1 + 4t$
$z = 0$

2. (1) (a) $(\frac{3}{2}, 1, -2)$, (b) $(\frac{5}{3}, \frac{1}{3}, -\frac{4}{3})$, (c) $(\frac{7}{4}, 0, -1)$
   (3) (a) $(1, 0, \frac{3}{2})$, (b) $(1, \frac{1}{3}, \frac{4}{3})$, (c) $(1, \frac{1}{2}, \frac{5}{4})$
3. (a) Collinear
   (c) Collinear
   (e) Collinear
4. (a) Nonintersecting
   (c) Nonintersecting

Exercise 2

1. (a) $A \notin \ell$
   (c) $A \in \ell$

2. (a) $\dfrac{x}{2} = \dfrac{y + 2}{3} = \dfrac{z - 4}{-3}$

   (c) $\dfrac{x - 2}{8} = \dfrac{z - 5}{3}$; $y = -3$

   (e) $\dfrac{x}{-4} = \dfrac{y - 1}{2} = \dfrac{z - 2}{3}$

3. (a) $\dfrac{1}{\sqrt{14}} : \dfrac{-2}{\sqrt{14}} : \dfrac{-3}{\sqrt{14}}$

   (c) $\dfrac{1}{\sqrt{11}} : \dfrac{-3}{\sqrt{11}} : \dfrac{1}{\sqrt{11}}$

   (e) $\dfrac{2}{\sqrt{13}} : \dfrac{-3}{\sqrt{13}} : 0$

4. (a) $\frac{1}{2}\pi$; $\ell_1 \perp \ell_2$

   (c) $\cos^{-1} \dfrac{3}{2\sqrt{21}}$; neither

5. (a) 1. (i) $x = -2 + 2s$
              $y = -2 - s$
              $z = 4 + 3s$
         (ii) $x = -2$
              $y = -2 + 6r$
              $z = 4 + 2r$
       2. $\sqrt{40}$

   (c) 1. (i) $x = 2 - 2t$
              $y = 5 + 7t$
              $z = -7 + t$
         (ii) $x = 2 - 32s$
              $y = 5 + s$
              $z = -7 + 37s$

       2. $\sqrt{\dfrac{133}{2}}$

Answers to Selected Exercises

6. (a) Let $v_1 = -2i + j + 5k$; $v_2 = -\frac{3}{5}i + \frac{3}{10}j + \frac{3}{2}k$

   $v_2 = \frac{10}{3}v_1$ so $\ell_1 \| \ell_2$. $d = \frac{\sqrt{46}}{2}$.

7. $x = x_0 + 3t$
   $y = y_0 - 10t$ for any (x_0, y_0, z_0)
   $z = z_0 - 18t$

9. $(1, -2, 1)$

Exercise 3

1. (a) $x = 2 + 2s - t$
       $y = -1 + s + 2t$ $x - 7y + 5z - 24 = 0$
       $z = 3 + s + 3t$

   (c) $x = 2 + 4s + 9t$
       $y = 1 - s + 20t$ $9x + 20y - 8z - 70 = 0$
       $z = -4 + 2s - 8t$

   (e) $x = -1 + 2s + t$
       $y = -2 + s + t$ $x + y - z + 7 = 0$
       $z = 2 + 3s + 2t$

2. (a) $x = x$
       $y = y$
       $z = \dfrac{-2 - 2x + 3y}{4}$

   (c) $x = x$
       $y = 3x$
       $z = z$

3. (a) $5x + 3y + 2z - 13 = 0$
   (c) Points are collinear.

4. (a) $\ell_1 : \begin{cases} x = 1 - 3t \\ y = -1 + 5t \\ z = 1 + 11t \end{cases}$ $2(1 - 3t) - (-1 + 5t) + (1 + 11t) = 4$

Exercise 4

1. (a) $x = 4t$
       $y = 1 + 8t$
       $z = \frac{5}{4} + 5t$

   (c) $x = -6 + 7t$
       $y = 1 - 2t$
       $z = 5 - 5t$

2. (b) (1) $\dfrac{x}{-4} + \dfrac{y}{-4} + \dfrac{z}{-4} = 1$

   (3) $\dfrac{x}{-6} + \dfrac{y}{3} + \dfrac{z}{-4} = 1$

   (5) $\dfrac{y}{5} - \dfrac{z}{2} = 1$

3. (a) $\pi_1 \perp \pi_2$

(c) $\cos^{-1}\dfrac{1}{\sqrt{3}}$

(e) $\cos^{-1}\dfrac{2}{\sqrt{5}}$

4. (a) $2x + 3y - 7z - 22 = 0$
   (c) $4x + 5y + 2z - 19 = 0$
   (e) $9x - 3y + 11z - 18 = 0$

5. (a) $x = 2$
       $y = 2t$ $\quad \sqrt{13}$
       $z = 3t$
   (c) $x = 4 + 4v$
       $y = 1 + 3v$ $\quad$ Lines intersect at $(4, 1, -3)$.
       $z = -3 + v$

Exercise 5

1. (a) $\dfrac{2}{\sqrt{14}}$

   (c) $\dfrac{7}{\sqrt{13}}$

   (e) $\dfrac{2}{\sqrt{11}}$

2. (a) $\dfrac{6}{\sqrt{70}}x - \dfrac{3}{\sqrt{70}}y + \dfrac{5}{\sqrt{70}}z - \dfrac{30}{\sqrt{70}} = 0; d = \dfrac{30}{\sqrt{70}};$
       $\mathbf{n} = \dfrac{6}{\sqrt{70}}\mathbf{i} - \dfrac{3}{\sqrt{70}}\mathbf{j} + \dfrac{5}{\sqrt{70}}\mathbf{k}$

   (c) $-\dfrac{3}{\sqrt{14}}x + \dfrac{2}{\sqrt{14}}y - \dfrac{1}{\sqrt{14}}z = 0; d = 0;$ none

   (e) $\dfrac{2}{\sqrt{29}}x + \dfrac{4}{\sqrt{29}}y - \dfrac{3}{\sqrt{29}}z - \dfrac{12}{\sqrt{29}} = 0; d = \dfrac{12}{\sqrt{29}};$
       $\mathbf{n} = \dfrac{2}{\sqrt{29}}\mathbf{i} + \dfrac{4}{\sqrt{29}}\mathbf{j} - \dfrac{3}{\sqrt{29}}\mathbf{k}$

3. (a) $d = \tfrac{14}{9}$

5. (a) $\dfrac{x}{5} + \dfrac{y}{2} + \dfrac{z}{c} = 1; \dfrac{x}{5} + \dfrac{y}{2} + \dfrac{2z}{5} = 1$

6. (a) $9x - 19y + 6z - 18 = 0$
   (c) $5x - 7y + 2z - 10 = 0$

7. (a) $2x + y - z - 1 = 0 \quad$ and $\quad 4x - 3y + 5z - 11 = 0$

Exercise 6

3. (a) $x' = \sqrt{10}(x - 2) - \dfrac{6}{\sqrt{10}}(y + 3)$

       $y' = \dfrac{\sqrt{14}}{\sqrt{10}}(y + 3)$

(b) $P_1 = (2, -3, 1)$

$$P_2 = \left(2 + \frac{1}{\sqrt{10}} - \frac{12}{\sqrt{10}\sqrt{14}}, -3 - \frac{2\sqrt{10}}{\sqrt{14}}, 1 - \frac{3}{\sqrt{10}} - \frac{4}{\sqrt{10}\sqrt{14}}\right)$$

$$P_3 = \left(2 - \frac{3}{\sqrt{10}} + \frac{24}{\sqrt{10}\sqrt{14}}, -3 + \frac{4\sqrt{10}}{\sqrt{14}}, 1 + \frac{3}{\sqrt{10}} + \frac{8}{\sqrt{10}\sqrt{14}}\right)$$

$|P_1P_2| = |P'_1P'_2| = \sqrt{5}$
$|P_1P_3| = |P'_1P'_3| = 5$
$|P_2P_3| = |P'_2P'_3| = \sqrt{52}$

Exercise 7
1. (a) $(x-5)^2 + (y+2)^2 + (z-1)^2 = 36$;
$x^2 + y^2 + z^2 - 10x + 4y - 2z - 6 = 0$
   (c) $(x+4)^2 + (y+2)^2 + (z+3)^2 = 0$;
$x^2 + y^2 + z^2 + 8x + 4y + 6z + 29 = 0$
2. (a) $(x-1)^2 + (y+3)^2 + (z-1)^2 = 2$; center $(1, -3, 1)$; radius $= \sqrt{2}$
   (c) $(x-1)^2 + (y-2)^2 + (z-3)^2 = -4$
3. (a) $y - 2z - 2 = 0$
   (c) $x + 2y + z = 6$ and $11x - 13y - 2z - 42 = 0$
5. (a) Not tangent
6. (a) $(-2, 3, 2)$

Exercise 8
1. (a) $3 \cdot 1 - 2(-2) - (12) + 4 = -1 < 0$ while $3(-2) - 2(-3) - 1 + 4 = 3$
2. (a) $-6(0) + (2) - 2(1) + 4 = 4 > 0$ while $-6(4) + 0 - 2(-3) + 4 = -14 < 0$

CHAPTER 10

Exercise 1
1. (a) No symmetry; traces $-3y + 6z = 12$, $2x + 6z = 12$, $2x - 3y = 12$
   (c) Symmetric about all coordinate planes;
traces $z = \pm 3$, $x^2 + z^2 = 9$, $x = \pm 3$
   (e) Symmetric about y-z plane and x-z plane; traces $y^2 = z$, $x^2 = z$, $(0, 0, 0)$
2. (a) Symmetric about x-z plane and x-y plane; traces $y = 0$, $x = 0$, $y^2 = x$
   (c) Symmetric about y-z plane and x-z plane; traces $y^2 = 9z$, $x^2 = 4z$, $(0, 0, 0)$
   (e) Symmetric about all coordinate planes;
traces $\frac{y^2}{9} - \frac{z^2}{16} = 1$, $\frac{x^2}{4} - \frac{z^2}{16} = 1$, $\frac{x^2}{4} + \frac{y^2}{9} = 1$
3. (a) $3x^2 + 3y^2 + 3z^2 - 62x - 10y - 14z + 273 = 0$
   (c) $x^2 + z^2 + 10y = 25$
   (e) $(2\sqrt{3} - \sqrt{14})x + (\sqrt{14} - \sqrt{3})y + (3\sqrt{3} - \sqrt{14})z = 4\sqrt{3} - \sqrt{14}$ or
$(2\sqrt{3} + \sqrt{14})x - (\sqrt{14} + \sqrt{3})y + (3\sqrt{3} + \sqrt{14})z = 4\sqrt{3} + \sqrt{14}$
5. (a) $x^2 + y^2 + 2z^2 - 2xy - 18x - 18y - 83 = 2\sqrt{2}|x + y + 9|$
7. $25x^2 + 25y^2 + 9z^2 = 225$

Answers to Selected Exercises

Exercise 2
1. (a) $(x - z)^2 + (y + 2z)^2 = 9$
   (c) $2x + 2z + y = 8$
   (e) $\dfrac{x^2}{4} - \dfrac{y^2}{9} = 1$
3. $g\left(x - \dfrac{ay}{b}, z - \dfrac{cy}{b}\right) = 0$

Exercise 3
1. (a) $x^2 + y^2 = 16; x = 4 \sin \dfrac{z}{2}; y = 4 \cos \dfrac{z}{2}$
   (c) $x = z \cos z; y = z \sin z;$ the curve lies on the cone $x^2 + y^2 - z^2 = 0$
2. (a) $2x^2 + 2y^2 = 9; z = \pm \dfrac{9}{\sqrt{2}}$
   (c) $x^2 + y^2 + (x + y)^2 = 16; x^2 + (x + z)^2 + z^2 = 16;$
       $(y + z)^2 + y^2 + z^2 = 16$
   (e) $\dfrac{x^2}{16} + \dfrac{y^2}{9} + \dfrac{(x + y)^2}{4} = 1; \dfrac{x^2}{16} + \dfrac{(z - x)^2}{9} + \dfrac{z^2}{4} = 1; \dfrac{(z - y)^2}{16} + \dfrac{y^2}{9} + \dfrac{z^2}{4} = 1$
   (g) $x^2 + y^2 = 16; z^2 + y^2 = 16; z = x$
3. (a) $\left(0, \dfrac{12}{13}, \dfrac{18}{13}\right); \sqrt{\dfrac{170}{13}}$
   (c) $(-1, 2, 4); 4$

Exercise 4
1. (a) (i) $x^2 + y^2 + z^2 = 4$
       (ii) $x^2 + y^2 + z^2 = 4$
   (c) $y^2 + z^2 = \sin^2 x$
   (e) (i) $9y^2 + 9z^2 + 4x^2 = 36$
       (ii) $4x^2 + 4y^2 + 9z^2 = 36$
2. (a) (i) $x^2 + z^2 + (y - 2)^2 = 1$
       (ii) $(x^2 + y^2 - 2)^2 + z^2 = 1$
   (c) (i) $4x^2 + (y - 4)^2 + 4z^2 = 4$
       (ii) $(x^2 + y^2 - 4)^2 + z^2 = 4$
   (e) (i) $4x^2 + 4z^2 - y^2 + 4 = 0$
       (ii) $x^2 + y^2 - 4z^2 = 4$
3. (a) y-axis; $z = \sqrt{y}$ or $x = \sqrt{y}$
   (c) z-axis; $y^2 + z^2 = 25$
   (e) y-axis; $\dfrac{y^2}{4} - \dfrac{z^2}{1} = 1$
   (g) y-axis; $z = \dfrac{1}{y}$

Answers to Selected Exercises

Exercise 5

1. (a) $(2, \frac{2}{3}\pi, 7)$
   (c) $\left(3, \tan^{-1}\frac{\sqrt{5}}{2}, -2\right)\left(\tan^{-1}\frac{\sqrt{5}}{2} \approx 48°10'\right)$
2. (a) $(4, \frac{2}{3}\pi, \frac{1}{3}\pi)$
   (c) $(9, 143°8', 109°28')$
3. (a) $\left(-5\frac{\sqrt{3}}{2}, \frac{5}{2}, -6\right)$
   (c) $(-4, 2\sqrt{5}, 0)$
4. (a) $\left(\frac{7\sqrt{6}}{4}, \frac{7\sqrt{2}}{4}, \frac{7\sqrt{2}}{2}\right)$
   (c) $(1, \sqrt{8}, 3\sqrt{8})$
5. (a) $r = 3; \rho^2 \sin^2 \varphi = 9$
   (c) $r^2 = z^2; \varphi = \frac{1}{4}\pi, \varphi = \frac{3}{4}\pi$ or $\rho = 0$
   (e) $r^2 = z; \rho \sin^2 \varphi = \cos \varphi$
6. (a) $r^2(9\cos^2\theta + 4\sin^2\theta) = 36z; \rho \sin^2\varphi(9\cos^2\theta + 4\sin^2\theta) = 36 \cos\varphi$
   (c) $(r^2 + z^2)^{3/2} = z; \rho^2 = \cos \varphi$
   (e) $z = \cos\theta \sin\theta; \rho \cos\varphi = \sin\theta \cos\theta$
7. (a) $(x^2 + y^2)^{3/2} = 2xy$
   (c) $(x^2 + y^2)z^2 + x^2 = (x^2 + y^2)^2$
   (e) $(x^2 + y^2)z = 3x^2y - y^3$
8. (a) $x^2 + y^2 + (z - \frac{1}{2})^2 = \frac{1}{4}$
   (c) $xz = (y + 3)\sqrt{x^2 + y^2}$
   (e) $xy = z$
9. $\begin{cases} \rho = \sqrt{r^2 + z^2} \\ \theta = \theta \\ \varphi = \cos^{-1}\dfrac{z}{\sqrt{r^2 + z^2}} \end{cases}$; $\begin{cases} r = \rho \sin \varphi \\ \theta = \theta \\ z = \rho \cos \varphi \end{cases}$
10. (a) $x = \cot\varphi \cos\theta, y = \cot\varphi \sin\theta, z = \cot^2\varphi$
    (c) $x = 3\sin\varphi \cos\theta, y = 3\sin\varphi \sin\theta, z = 3\cos\varphi$
    (e) $x = \tan^{1/3}\varphi \cos\theta, y = \tan^{1/3}\varphi \sin\theta, z = \cot^{2/3}\varphi$

CHAPTER 11

Exercise 2

1. (d) $\left(\dfrac{1 - 2\sqrt{3}}{\sqrt{6}}, \dfrac{1 + 2\sqrt{3}}{\sqrt{6}}, \dfrac{4}{\sqrt{6}}\right)$
3. (a) $(0, \sqrt{3}, 2)$

Answers to Selected Exercises

Exercise 3

3. (a) $X = \begin{pmatrix} 0 & -\frac{1}{\sqrt{2}} & \frac{1}{\sqrt{2}} \\ \frac{1}{\sqrt{2}} & \frac{1}{2} & \frac{1}{2} \\ -\frac{1}{\sqrt{2}} & \frac{1}{2} & \frac{1}{2} \end{pmatrix} X'; X' = X'' + \begin{pmatrix} \frac{20 + \sqrt{2}}{4\sqrt{2}} \\ \frac{1}{2\sqrt{2}} \\ -\frac{1}{2\sqrt{2}} \end{pmatrix};$

$-(y'')^2 + (z'')^2 + x'' = 0$, hyperbolic paraboloid

(c) $X = \begin{pmatrix} \frac{1}{\sqrt{3}} & \frac{2}{\sqrt{6}} & 0 \\ -\frac{1}{\sqrt{3}} & \frac{1}{\sqrt{6}} & \frac{1}{\sqrt{2}} \\ \frac{1}{\sqrt{3}} & -\frac{1}{\sqrt{6}} & \frac{1}{\sqrt{2}} \end{pmatrix} X'; 3(x')^2 + \frac{3}{2}(y')^2 + \frac{3}{2}(z')^2 - 6 = 0$, ellipsoid

(e) $X = X' + \begin{pmatrix} -\frac{5}{2} \\ \frac{1}{2} \\ 1 \end{pmatrix}; X' = \begin{pmatrix} \frac{1}{\sqrt{3}} & \frac{1}{\sqrt{6}} & \frac{1}{\sqrt{2}} \\ \frac{1}{\sqrt{3}} & \frac{1}{\sqrt{6}} & -\frac{1}{\sqrt{2}} \\ -\frac{1}{\sqrt{3}} & \frac{2}{\sqrt{6}} & 0 \end{pmatrix} X'';$

$(x'')^2 + 4(y'')^2 - 2(z'')^2 + 14 = 0$, hyperboloid of two sheets

APPENDIX

1. (a) $AB = \begin{pmatrix} 2 & 7 & -9 \\ 1 & 7 & -10 \\ 0 & -2 & 10 \end{pmatrix}, BA = \begin{pmatrix} 5 & -4 & -1 \\ 8 & 4 & -10 \\ -5 & -6 & 10 \end{pmatrix}$

2. (a) $a = -\frac{4}{5}, b = \frac{3}{5}, c = \frac{-7}{5}, d = -\frac{1}{5}$

3. $A^t = \begin{pmatrix} -\frac{1}{\sqrt{2}} & \frac{1}{\sqrt{3}} & \frac{1}{\sqrt{6}} \\ \frac{1}{\sqrt{2}} & \frac{1}{\sqrt{3}} & \frac{1}{\sqrt{6}} \\ 0 & \frac{1}{\sqrt{3}} & -\frac{2}{\sqrt{6}} \end{pmatrix} \quad AA^t = I_3$

$B^t = \begin{pmatrix} \frac{1}{\sqrt{2}} & 0 & \frac{1}{\sqrt{2}} \\ 0 & -1 & 0 \\ \frac{1}{\sqrt{2}} & 0 & -\frac{1}{\sqrt{2}} \end{pmatrix} \quad BB^t = I_3$

Answers to Selected Exercises

5. (a) 6
7. (a) $|AB| = 336$
$|A| = 48$
$|B| = 7$

9. $\begin{pmatrix} \frac{1}{2} & 0 & \frac{3}{4} \\ \frac{1}{4} & -\frac{1}{2} & \frac{1}{8} \\ \frac{1}{2} & 0 & \frac{1}{4} \end{pmatrix}$

Index

Abscissa, 17
Absolute value, 7
Addition, of matrices, 283
  of ordinates, 23
Addition formulas, trigonometric, 12
Addition law for inequalities, 6
Angle, 10
  between two lines, 73 ff.
  between two planes, 208
  between two vectors, 50, 181
  bisector, 79
  central, 10
  degree measure, 10
  radian measure, 10
  related, 11
Antecedent, 2
Arrow(s), 40 ff.
  components of, 40, 170
  direction of, 41, 42, 186
  equivalent, 44, 171, 177, 186
  horizontal, 43
  initial point of, 40, 170
  slope of, 43
  terminal point of, 40, 170
  vertical, 43
Asymptote, 30
  horizontal, 31
  vertical, 31
Axis of revolution, 246
Axis of symmetry, parabola, 137

Cartesian product, 5
Cauchy's Inequality, 54, 177, 179
Characteristic equation, 161, 273
Characteristic value, 160, 272
Characteristic vector, 160, 272
Chord, 90
Circle, 10, 85, 111
  arc of, 10
  chord of, 90
  diameter of, 90
  equation of, 85, 86
  exterior of, 100
  interior of, 99

Codomain, 17
Cofactor, 286
Collinear points, 58, 61, 196
Conclusion, 2
Cone, 234, 280, 281
Conformable matrices, 283
Congruence, 128, 221
Conic sections, 130 ff.
   polar form of, 132, 135
Conjunction, 1
Connective, 1
Consequent, 2
Converse, 2
Convex set, 96, 137, 225
Coordinate axes, 17, 167
Coordinate planes, 168
Coordinates, 17, 167 ff.
   cylindrical, 248
   polar, 101 ff.
   spherical, 250
Cosines, law of, 14
Cross product (vector product), 189 ff.
Curve, 34, 241
   closed, 38
   parametric equations for, 35, 241
   plane, 34
   simple, 37
Cycloid, 36
Cylinder, 237 ff.
   right, 239
Cylindrical coordinates, 248
Cylindrical helix, 241

Degenerate conic, 140
Degenerate ellipse, 146
Degenerate hyperbola, 150
Degenerate parabola, 140
Determinant, 286 ff.
   product theorem for, 288
Diameter, 90, 223
Direction angles, 184
Direction cosines, 184
Direction numbers, 200
Directrix, 131, 136, 237
Discriminant, 154, 155
Disjunction, 1
Distance, 41, 177, 195
   from a point to a line, 78

   from a point to a plane, 214
Domain, 16
Dot product, 51, 52, 160, 179
Double-angle formulas, 12

Eccentricity, 131, 144, 147
Eigenvalue, 160, 272
Eigenvector, 160, 272
Element, 2
   of cylinder, 237
Eliminant, 35, 230, 243
Eliminating parameter, 35, 230
Ellipse, 130, 131, 133, 136, 142 ff., 166
   center of, 142, 145
   degenerate, 146
   eccentricity of, 144
   foci of, 142, 145
   interior of, 143
   major axis of, 143, 144
   minor axis of, 143, 144
   right focal chord of, 147
   standard form of, 145
   vertices of, 143, 145
Ellipsoid, 233, 255, 257, 280, 281
Elliptic cylinder, 280, 281
Elliptic hyperboloid, 257
   of one sheet, 257, 280, 281
   of two sheets, 258, 280, 281
Elliptic paraboloid, 256, 281
Empty set, 3
Equivalence, 2
   of arrows, 44, 171, 177, 186
   of polar coordinates, 104

Family, 2
   of circles, 90, 93
   of lines, 82
   of planes, 216
Focus, 131, 136, 138
   of ellipse, 142, 145
   of hyperbola, 147, 150
   of parabola, 136, 139
Function, 16 ff.
   algebra of, 22
   decreasing, 27, 137, 142
   implicit, 20
   increasing, 27, 137, 147
   inverse, 24

Index

Function (*continued*)
  one-to-one, 24
  onto, 17
  polynomial, 30
  rational, 30
  trigonometric, 10

Graph, 17
  asymptotes of, 30
  extent of, 28
  intercepts of, 29
  symmetry of, 28

Half plane, 96, 97
Hyperbola, 130, 131, 134, 136, 147 ff., 166
  asymptotes of, 148, 152
  center of, 147
  conjugate axis of, 148
  eccentricity of, 147
  foci of, 147, 150
  interior of, 148
  right focal chord of, 152
  standard form of, 150
  transverse axis of, 148, 150
  vertices of, 148
Hyperbolic cylinder, 281
Hyperbolic paraboloid, 259, 281
Hyperboloid, 235, 255, 257
  of one sheet, 257, 280, 281
  of two sheets, 235, 258, 280, 281
Hypothesis, 2

Implication, 2
Inequality, 8
  linear, 8
  quadratic, 9
Inner product, *see* Dot product
Integers, 6
Intercepts, 29, 256
Interior, 137
  of circle, 99
  of ellipse, 143
  of hyperbola, 148
  of parabola, 100, 137
Interval, 8
  closed, 8
  infinite, 8
  open, 8

Irrational numbers, 6
Isometry, 127, 220, 261

Line(s), 65, 111, 195
  angle of inclination of, 72, 73
  general form of, 66
  intercept form of, 65
  normal form of, 79, 80, 81
  parallel, 74, 201
  parametric form of, 68, 195
  perpendicular, 75, 201
  point-slope form of, 65
  skew, 201
  slope of, 65
  slope-intercept form of, 66
  symmetric equations of, 199
  two-point form of, 65
Line of centers, 93
Line segment(s), 55 ff., 194 ff.
  angle between, 57
  endpoints of, 56, 194
  length of, 57, 195
  midpoint of, 56, 195
  parallel, 57, 195
  parametric equations of, 56, 195
  perpendicular, 57
  slope of, 57
Linear combination, 183, 204
Linear equation, 68, 69, 206

Mapping, 17
  linear, 263
Matrix, 282 ff.
  characteristic equation of, 161, 273
  column, 159, 272
  eigenvalues of, 160, 272
  eigenvectors of, 160, 272
  identity, 124, 284
  inverse, 124, 285
  multiplication of, 283
  nonsingular, 284
  orthogonal, 124, 157 ff., 261, 264, 285
  rotation, 123, 157, 261, 264
  symmetric, 157, 158, 162, 166, 271, 285
  transpose, 157, 285

Negation, 2

Octants, 169
Ordered pair, 4
Ordered triple, 4
Ordinate, 17
Origin, 17, 168
Orthonormal vectors, 183, 286

Parabola, 130, 131, 133, 136, 166
  axis of, 137, 139
  degenerate, 140
  directrix of, 36, 138, 139
  focus of, 136, 138, 139
  right focal chord of, 137
  standard form of, 138, 139
  vertex of, 137, 139
Parabolic cylinder, 281
Paraboloid, 255, 281
Parameter, 34, 90
Parametric equations, 34, 112, 195, 205, 230, 241, 252
Perpendicular bisector, 75
Plane(s), 17, 204 ff.
  angle between, 208
  equation of, 205
  intercept form of, 212
  normal form of, 215
  normal vector, 205, 215
  parallel, 209
  perpendicular, 209
Point, 17, 167
  abscissa of, 17
  ordinate of, 17
Polar coordinates, 101 ff.
Polar curve, 105
Polar equation, 105
Polynomial, 30
Projecting cylinder, 241
Projection, 55, 182

Quadratic equation, general, 152 ff., 255, 270
  matrix form of, 157, 270
Quadratic form, 153, 270
  discriminant of, 154, 155
  symmetric matrix of, 157, 270 ff.
  transform of, 153, 159, 165, 271
Quadrant, 17
Quadrilateral, 58

Radical axis, 93

Radius vector, 102
Range of a function, 16
Rational numbers, 6
Ray, 63, 194
Real-number system, 5
Reduction formulas, 12
Reflection, 127
Relation, 17
  domain of, 19
Right focal chord, 137, 147, 152
Rotation of axes, 119 ff., 153, 265

Scalar product, 41, 173
Segment, *see* Line segment
Set(s), 2
  complement of, 3
  convex, 96, 137, 225
  difference of, 3
  disjoint, 4
  equality of, 4
  inclusion of, 3
  intersection of, 3
  union of, 3
  void, 3
Sines, law of, 14
Sphere, 222, 223, 225
Spherical coordinates, 250
Subset, 3, 4
Surface, 229
  of revolution, 246
Symmetry, 25, 28, 106, 230, 256

Tangent line, 86, 225
Tangent plane, 223
Traces, 211, 230, 256
Transform, 118, 121, 153, 159
Transitive law, 6
Translation of axes, 117, 123, 267
Triangle, 13, 58 ff.
Triangle inequality, 54, 177
Trichotomy law, 5
Trigonometric identities, 12

Vector(s), 44 ff., 171 ff.
  angle between, 50, 51
  column, 159, 272
  cross product of, 189
  difference of, 47, 174
  direction of, 49, 184
  dot product of, 51, 52, 179

Vector(s) (*continued*)
  length of, 46, 177, 184
  orthogonal (perpendicular), 51, 160, 181
  parallel, 51, 175, 176, 185
  scalar product of, 46, 173
  slope of, 50
  sum of, 46, 172
  unit, 162, 182

Vector product (cross product), 189 ff.